Microbiology

Microbiology

by Jennifer C. Stearns, PhD,
Michael G. Surette, PhD,
and Julienne C. Kaiser, MSc

Microbiology For Dummies®

Published by: **John Wiley & Sons, Inc.**, 111 River Street, Hoboken, NJ 07030-5774, www.wiley.com

Published simultaneously in Canada

For general information on our other products and services, please contact our Customer Care Department within the U.S. at 877-762-2974, outside the U.S. at 317-572-3993, or fax 317-572-4002. For technical support, please visit https://hub.wiley.com/community/support/dummies.

Wiley publishes in a variety of print and electronic formats and by print-on-demand. Some material included with standard print versions of this book may not be included in e-books or in print-on-demand. If this book refers to media such as a CD or DVD that is not included in the version you purchased, you may download this material at http://booksupport.wiley.com. For more information about Wiley products, visit www.wiley.com.

Library of Congress Control Number: 2019931894

ISBN: 978-1-119-54442-5; ISBN: 978-1-119-54476-0 (ebk); ISBN: 978-1-119-54441-8 (ebk)

Manufactured in the United States of America

SKY10063975_010524

Contents at a Glance

Table of Contents

PART 6: NEW FRONTIERS IN MICROBIOLOGY 293

PART 7: THE PART OF TENS.............319

Introduction

The world around us is full of tiny invisible living things that affect us every day. Diving into the study of that world is what this book is all about, and we're happy that you'd like to come along. Microbiology as a whole can feel overwhelming, but when you break it down into parts it can be straightforward and even interesting.

Whether you're taking a microbiology course for credit or studying microbiology on your own time, we've written this book with you, the beginner, in mind. This book walks you through the tricky concepts in microbiology while covering the forms, functions, and impacts of microbes in nature and on our lives.

About This Book

Microbiology For Dummies is an overview of the material covered in a typical first-year microbiology course. Some courses cover more medical, molecular, or environmental microbiology than others, so we've included them all here.

In this book, you find clear explanations of

- The characteristics that microorganisms share
- The things that make microbes different from one another and the rest of life on earth
- The processes important to microbial life
- The diversity of microbial life
- How microbes affect us

If you're a visual learner, you'll appreciate the many illustrations. And if you like to organize material into categories, you'll find the lists and tables useful. With this book, you'll be able to explain what makes microorganisms unique and identify where and how they live. You'll also have the skills to delve into specialized areas of microbiology that this book covers in an introductory way.

This book is a reference, which means you don't have to memorize it — unlike your microbiology course, there is no test at the end. Use it as a reference, dipping into whichever chapter or section has the information you need. Finally, sidebars and sections marked with the Technical Stuff icons are skippable. They offer a more in-depth discussion of a topic, extra detail, or interesting cases that are related to the main material of the chapter.

Foolish Assumptions

We don't assume that you have any background knowledge in microbiology except what may be covered in an introductory biology course. In fact, many of the concepts learned in a biology course are also presented here, so we don't expect you to know much of that, either. We assume that you are new to microbiology or other science courses where an introduction to microbiology is beneficial, and we've written this will book in a way that will provide you with the background you need.

The science of microbiology involves knowing a bit of biochemistry, cell biology, molecular biology, and environmental science, so we explain those concepts as needed, but you may like to peruse guides on those topics for a fuller understanding.

Other than that we only assume that you transcend the idea of microorganisms as "bad" and consider them as important members of our world, especially because they outnumber us about 200 million trillion to one!

Icons Used in This Book

Icons appear in the left margin to draw your attention to things that occur on a regular basis. Here's what each icon means:

TIP

The Tip icon marks material that's useful for thinking about a concept in another way or helping you to remember something.

REMEMBER

The Remember icon highlights concepts that are important to keep in mind. Often these concepts come up more than once in the book.

The Warning icon points out places where it can be easy to get confused. We usually know this because there is confusion in the general public about the concept or, worse, in the scientific community. Sometimes the Warning icon points to areas of debate in microbiology so that you don't have to feel confused if other sources disagree with our explanation.

Nonessential but helpful and interesting information is marked by the Technical Stuff icon. You can skip these bits of text if you don't want to get into the details just yet.

Beyond the Book

In addition to the material in the print or e-book that you're reading right now, this book also has some useful digital content, available on the web.

Some facts in microbiology are handy to have at your fingertips, either to study for an exam or to refresh your memory on the spot. To get the free Cheat Sheet, simply go to `www.dummies.com` and search for "Microbiology For Dummies Cheat Sheet" by using the Search box for tips on identifying microbes, remembering the basic differences between them, and figuring out the naming system used in microbiology.

Ever wonder what all the fuss is about fecal transplants or if the anti-vaccine campaigns are telling you the truth? You can find articles on these topics and more at `www.dummies.com/extras/microbiology`.

Where to Go from Here

We'd like to think that you won't skip anything, but if you're taking a microbiology course right now, then you probably don't need an introduction to the topic and can skip Part 1. Even though each chapter can be read on its own, the material in Part 2 is essential to any student of microbiology and will likely be very useful when covering more advanced topics.

There are many kinds of microbiology, perspectives from which will shape how introductory microbiology is taught. For a human health perspective, focus on chapters in Part 5. For an ecology perspective, you'll likely find chapters in Part 3 useful. If you'd like a reference for specific microorganisms, see Part 4.

No matter where you start or where you end, we hope that you'll come away with an appreciation for microbes and a road map for learning microbiology.

1

Getting Started with Microbiology

IN THIS PART . . .

Get a big-picture view of microbiology, including how microorganisms impact our lives in ways that we can and can't see.

Get acquainted with the history of microbiology from before people knew that microbes existed to our current use of sophisticated techniques to study microorganisms.

Gain an understating of the vastness of microbial lifestyles and how microbes are everywhere living in communities.

Understand microbial diversity and all the different ways these tiny organisms have figured out to get energy from their environments.

IN THIS CHAPTER

» **Seeing the importance of microbiology**

» **Getting to know microorganisms**

» **Listing the tools used to study microbes**

Chapter 1
Microbiology and You

When considering the imperceptibly small, it's sometimes easy to lose sight of the big picture. In this chapter, we put the science of microbiology into perspective for you as it relates to human lives, as well as how it fits in with the other sciences. The goal is to give you an idea of the kinds of thinking you'll use throughout the rest of the book. Don't worry, we explain all that pesky biochemistry and molecular biology as it comes up in each chapter.

Why Microbiology?

The question of why to study microbiology is a good one — the impacts of microorganisms on your life may not be immediately obvious. But the truth is, microorganisms not only have a *huge* impact but are literally everywhere, covering all the surfaces of your body and in every natural and urban habitat. In nature, microorganisms contribute to biogeochemical cycling, as well as turnover of material in soil and aquatic habitats. Some are important plant *symbionts* (organisms that live in intimate contact with their host, with mutual benefit for both organisms) whereas others are important *pathogens* (organisms that cause disease) of both plants and animals.

Although not all microorganisms are bad, the treatment and prevention of the diseases caused by bacteria, viruses, protozoa, and fungi have only been possible because of microbiology. Antibiotics were discovered through microbiology, as were vaccines and other therapeutics.

Other applications of microorganisms include industries like mining, pharmaceuticals, food and beverages, and genetics. Microorganisms are important model organisms for studying principles of genetics and biochemistry.

Many professions require you to learn some microbiology. You may already know this because you're in a micro class as part of the training for one of them. These professions include but are not limited to

- Nursing
- Medicine
- Clinical laboratory work
- Pharmaceuticals
- Brewing and winemaking
- Environmental engineering

Introducing the Microorganisms

So, what are microorganisms exactly? Microorganisms are actually a diverse group of organisms. The fact that they're micro isn't even true of all microorganisms — some of them form multicellular structures that are easily seen with the naked eye.

There are three main kinds of microorganisms, based on evolutionary lines (see Figure 1-1):

- **Bacteria** are a large group of unicellular organisms that scientists loosely group as Gram-negative and Gram-positive, but in reality there are many different kinds.
- **Archaea** are another group of unicellular organisms that evolved along with bacteria several billion years ago. Many are *extremophiles,* meaning that they thrive in very hot or very acidic conditions. Archaea are more closely related to eukaryotes than to bacteria.
- **Eukaryotic microorganisms** are a structurally diverse group that includes protists, algae, and fungi. They all have a nucleus and membrane-bound organelles, as well as other key differences from bacteria and archaea. All the rest of the multicellular organisms on earth, including humans, have eukaryotic cells as well.

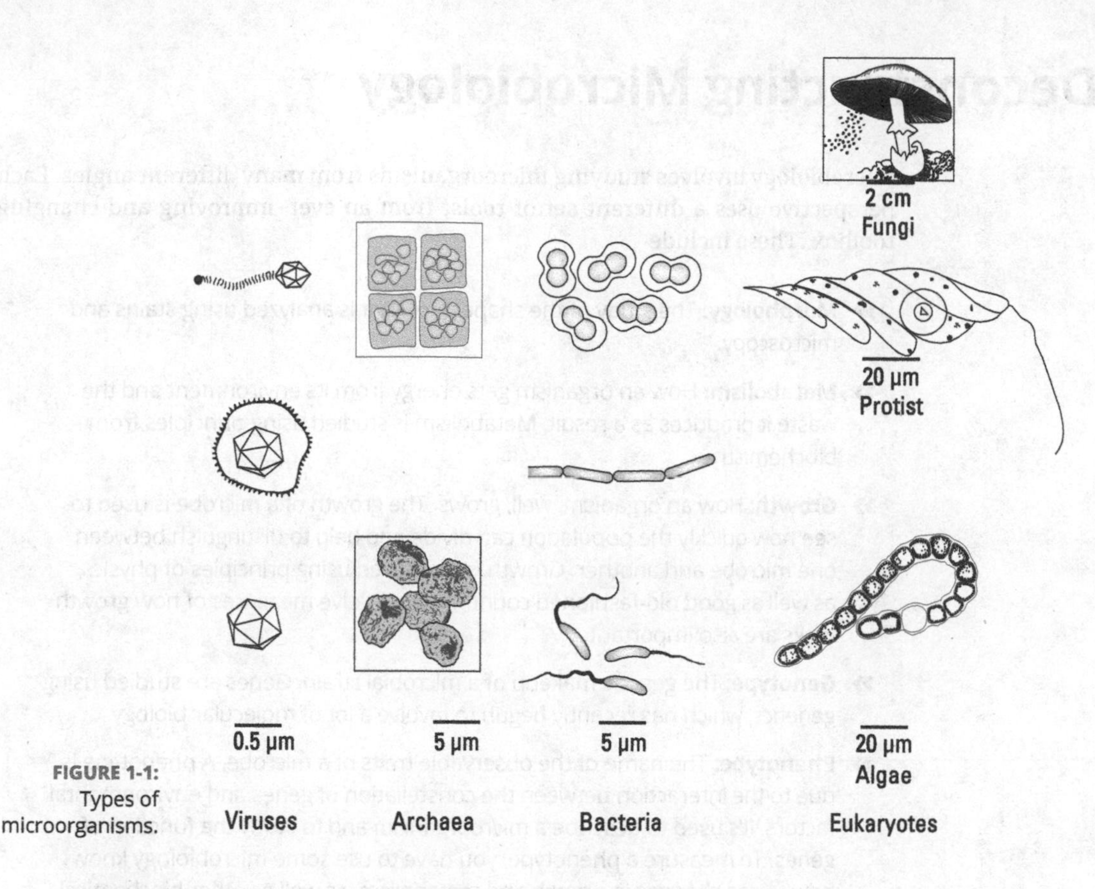

FIGURE 1-1: Types of microorganisms.

REMEMBER

Along with the many eukaryotic microorganisms, the Eukaryotes include all multicellular life on earth, like plants, animals, and humans.

» **Viruses** are smaller than bacteria and are not technically alive on their own — they must infect a host cell to survive. Viruses are made up of some genetic material surrounded by a viral coat, but they lack all the machinery necessary to make proteins and catalyze reactions. This group also includes subviral particles and prions, which are the simplest of life forms, made of naked ribonucleic acid (RNA) or simply protein.

TIP

The bacteria and archaea are often talked about together under the heading of "prokaryotes" because they lack a nucleus. They do share a few characteristics and aren't easily distinguished from one another at first, but they are distinct groups.

Deconstructing Microbiology

Microbiology involves studying microorganisms from many different angles. Each perspective uses a different set of tools, from an ever-improving and changing toolbox. These include

- **Morphology:** The study of the shape of cells. It is analyzed using stains and microscopy.
- **Metabolism:** How an organism gets energy from its environment and the waste it produces as a result. Metabolism is studied using principles from biochemistry.
- **Growth:** How an organism, well, grows. The growth of a microbe is used to see how quickly the population can divide and help to distinguish between one microbe and another. Growth is measured using principles of physics, as well as good old-fashioned counting. Qualitative measures of how growth looks are also important.
- **Genotype:** The genetic makeup of a microbial strain. Genes are studied using genetics, which has recently begun to involve a lot of molecular biology.
- **Phenotype:** The name of the observable traits of a microbe. A phenotype is due to the interaction between the constellation of genes and environmental factors. It's used to describe a microorganism and to study the function of genes. To measure a phenotype, you have to use some microbiology know-how to see changes in growth and metabolism, as well as other biochemical processes for communication and defense.
- **Phylogeny:** The history of the evolution of microorganisms. Phylogeny is important not only because it helps us identify newly discovered microbes but also because it allows us to see how closely related different microbes are to one another. The study of a group's phylogeny involves genetics and molecular biology, as well as evolutionary biology.

When you put all the pieces back together again, you have the science of microbiology. Microbiologists are some of the most creative scientists out there — they have many tools at their disposal that they can use in a variety of ways. The trick is to think up sneaky ways to study microbes, which is why the field is always evolving.

The term *microbiology* is often used to mean the study of mainly bacteria and archaea because the study of other microbes are specialties of their own. For example, the study of viruses is *virology*, the study of fungi is *mycology*, and the study of algae is *phycology*.

IN THIS CHAPTER

» Remembering a time before microbiology

» Discovering microorganisms step-by-step

» Looking forward

Chapter 2 Microbiology: The Young Science

Compared with other more ancient fields of science, microbiology is a relative baby. Physics began in ancient times, mathematics even earlier, but the knowledge of tiny living things, their biology, and their impact on human lives has only been around since the late 19th century. Until about the 1880s, people still believed that life could form out of thin air and that sickness was caused by sins or bad odors.

As with other fields in science, there are two aspects to microbiology research: basic and applied. *Basic microbiology* is about discovering the fundamental rules governing the microbial world and studying all the variety of microbial life and microbial systems. *Applied microbiology* is more about solving a problem and involves using microbes and their genes or proteins for practical purposes such as in industry and medicine.

In this chapter, we introduce the key concepts and experiments that gave rise to the discovery of microbes and their importance in disease. This chapter also highlights the many different areas of study within microbiology and some advances and challenges in the prevention and treatment of infectious diseases.

Before Microbiology: Misconceptions and Superstitions

Medical practices in ancient times were all heavily tinged with supernatural beliefs. Ancient Egypt was ahead of its time in terms of medicine, with physicians performing surgery and treating a wide variety of conditions. Medicine in India was also quite advanced. Ancient Greek physicians were concerned with balancing the body's *humors* (the four distinct body fluids that they believed were responsible for health when in balance, or disease when out of balance), and medicine in medieval Europe was based on this tradition. None, however, had knowledge of the microbial causes of disease.

Opinions about why diseases afflicted people differed between cultures and parts of society, and the treatments differed as well. Diseases were thought to be caused by

- Bad smells, treated by removing or masking the offending odor
- An imbalance in the humors of the body, treated with bleeding, sweating, and vomiting
- Sins of the soul, treated with prayer and rituals

Although the concept of contagion was known, it wasn't attributed to tiny living creatures but to bad odors or spirits, such as the devil. So, simple measures, such as removing sources of infection or washing hands or surgical equipment, were simply not done.

Discovering Microorganisms

Before microorganisms were discovered, life was not known to arise uniquely from living cells; instead, it was thought to spring spontaneously from mud and lakes or anywhere with sufficient nutrients in a process called *spontaneous generation*. This concept was so compelling that it persisted until late into the 19th century.

Robert Hooke, a 17th-century English scientist, was the first to use a lens to observe the smallest unit of tissues he called "cells." Soon after, the Dutch amateur biologist Anton van Leeuwenhoek observed what he called "animacules" with the use of his homemade microscopes.

When microorganisms were known to exist, most scientists believed that such simple life forms could surely arise through spontaneous generation. So, when they heated a container, placed a *nutrient broth* (a mixture of nutrients that supported growth of microorganisms in these early experiments) in the container and then sealed it, and no microorganisms appeared, they believed it had to be due to the absence of either air or the vital force (whatever that was!) necessary to make life.

Debunking the myth of spontaneous generation

The concept of spontaneous generation was finally put to rest by the French chemist Louis Pasteur in an inspired set of experiments involving a goose-necked flask (see Figure 2-1). When he boiled broth in a flask with a straight neck and left it exposed to air, organisms grew. When he did this with his goose-necked flask, nothing grew. The S-shape of this second flask trapped dust particles from the air, preventing them from reaching the broth. By showing that he could allow air to get into the flask but not the particles in the air, Pasteur proved that it was the organisms in the dust that were growing in the broth. This is the principle behind the Petri dish used to grow bacteria on solid growth medium (made by adding a gelling material to the broth), which allows air but not small particles to reach the surface of the growth medium.

REMEMBER

The idea that invisible microorganisms are the cause of disease is called *germ theory*. This was another of the important contributions of Pasteur to microbiology. It emerged not only from his experiments disproving spontaneous generation but also from his search for the infectious organism (typhoid) that caused the deaths of three of his daughters.

Around the same time that Pasteur was doing his experiments, a doctor named Robert Koch was working on finding the causes of some very nasty animal diseases (first anthrax, and then tuberculosis). He devised a strict set of guidelines — named Koch's postulates — that are still used to this day to definitively prove that a microorganism causes a particular disease. Koch's four postulates are

- The organism causing the disease can be found in sick individuals but not in healthy ones.
- The organism can be isolated and grown in pure culture.
- The organism must cause the disease when it is introduced into a healthy animal.
- The organism must be recovered from the infected animal and shown to be the same as the organism that was introduced.

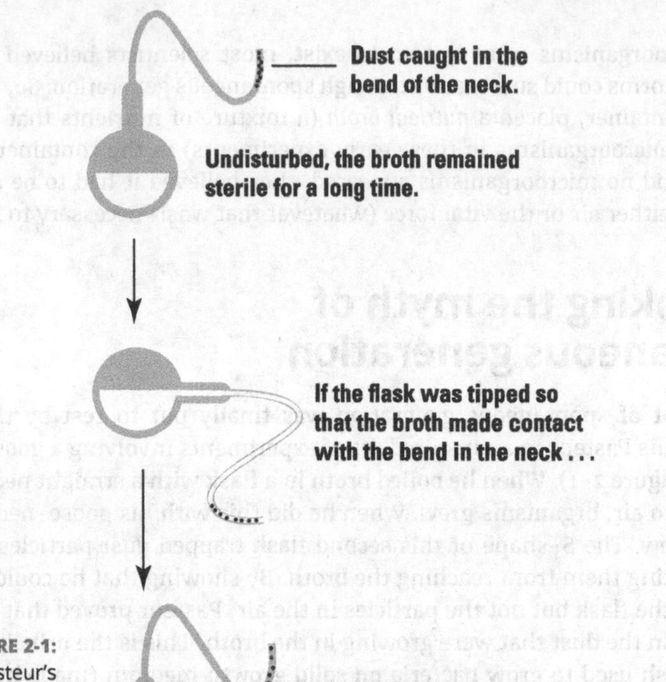
Dust caught in the bend of the neck.
Undisturbed, the broth remained sterile for a long time.
If the flask was tipped so that the broth made contact with the bend in the neck . . .
. . . the broth became contaminated immediately.

FIGURE 2-1: Pasteur's experiments that disproved the theory of spontaneous generation.

Improving medicine, from surgery to antibiotics and more

Once scientists knew that microbes caused disease, it was only a matter of time before medical practices improved dramatically. Surgery used to be as dangerous as not doing anything at all, but once *aseptic* (sterile) technique was introduced, recovery rates improved dramatically. Hand washing and quarantine of infected patients reduced the spread of disease and made hospitals into a place to get treatment instead of a place to die.

Vaccination was discovered before germ theory, but it wasn't fully understood until the time of Pasteur. In the late 18th century, milkmaids who contracted the nonlethal cowpox sickness from the cows they were milking were spared in deadly smallpox outbreaks that ravaged England periodically. The physician Edward Jenner used pus from cowpox scabs to vaccinate people against smallpox. Years later, Pasteur realized that the reason this worked was that the cowpox virus was similar enough to the smallpox virus to kickstart an immune response that would provide a person with long-term protection, or *immunity*.

Antibiotics were discovered completely by accident in the 1920s, when a solid culture in a Petri dish (called a *plate*) of bacteria was left to sit around longer than usual. As will happen with any food source left sitting around, it became moldy, growing a patch of fuzzy fungus. The colonies in the area around the fungal colony were smaller in size and seemed to be growing poorly compared to the bacteria on the rest of the plate, as shown in Figure 2-2.

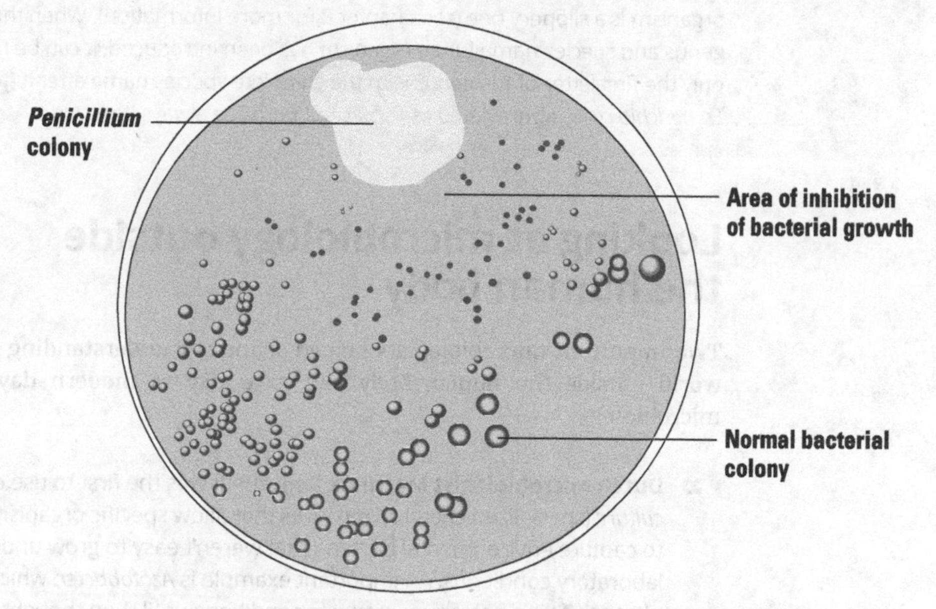

FIGURE 2-2: Antibacterial property of the fungus *Penicillium.*

The compound found to be responsible for this antibacterial action was named penicillin. The first antibiotic, penicillin was later used to treat people suffering from a variety of bacterial infections and to prevent bacterial infection in burn victims, among many other applications.

After bacteria were discovered, the field of molecular biology made great strides in understanding the genetic code, how DNA is regulated, and how RNA is translated into proteins. Until this point, research was focused mainly on plant and animal cells, which are much more complex than bacterial cells. When researchers switched to studying these processes in bacteria, many of the secrets of genes and enzymes started to reveal themselves.

HOW MICROORGANISMS ARE NAMED

Microorganisms are named using the Linnaeus system developed in the 18th century. It uses two-part Latin names for all living things. The first part, which is capitalized, is a genus name given to closely related organisms; the second part is a species name, which is not capitalized, given to define a specific organism. This is more challenging than you may think, even for plants and animals, and the concept of a species of microorganism is a slippery one (see Chapter 8 for more information). When the complete genus and species name for an organism has been introduced, it can be referred to by only the first letter of the genus with the complete species name after it (for example, *Escherichia coli* is abbreviated as *E. coli*), but both are always italicized.

Looking at microbiology outside the human body

Two important microbiologists helped shape our understanding of the microbial world outside the human body and gave rise to modern-day environmental microbiology:

- **Dutch microbiologist Martinus Beijerinck** was the first to use *enrichment culture* (specialized chemical mixtures that allow specific organisms to grow) to capture environmental bacteria that weren't easy to grow under normal laboratory conditions. An important example is *Azotobacter,* which is a nitrogen-fixing bacterium grown in conditions until then thought to be insufficient for life because they contained only nitrogen gas (N_2) as the sole nitrogen source.
- **Russian microbiologist Sergei Winogradsky** described sulfur-oxidizing bacteria called *Beggiatoa* from a hot spring. It was this discovery that convinced the field of microbiology that some microbes get energy from inorganic compounds like hydrogen sulfide (H_2S), a microbial lifestyle called *chemolithotrophy.*

The Future of Microbiology

Today is perhaps the best time in history to be a microbiologist! The development of new experimental techniques and ability to sequence organisms without actually culturing them in the laboratory first has revealed diversity and complexity in the microbial world not previously known. Most microorganisms can't be grown in the lab, so they were previously unknown before the development of DNA sequencing techniques. Exploiting this microbial biodiversity for drug discovery and biotechnology applications is an exciting area of research. With the widespread availability of antibiotics and vaccines in the last half of the 20th century, infectious diseases were thought to be under control. The emergence of antibiotic resistance and the rapid evolution of bacterial and viral pathogens have made medical microbiology an urgent and exciting field of science.

Exciting frontiers

It's an exciting time for microbiology because the tools available to study microbes have improved a lot recently. *Molecular biology* (the study of nucleic acids such as DNA and RNA) has improved so much that microbiologists are now using molecular tools in many branches of the field (see the nearby sidebar). These tools include DNA and RNA sequencing and manipulation, which have allowed microbiologists to understand the function of enzymes and the evolution of microorganisms, and have allowed them to manipulate *microbial genomes* (the genetic material of organisms).

Complete sequencing of microbial genomes is an exciting frontier because it opens the door to knowledge about the varied metabolic diversity in the microbial world. Only a fraction of the many microorganisms on earth have had their entire genomes sequenced, but from those that have, science has learned a lot about microbial genes and evolution.

One interesting example of this is the recent sequencing of the complete genome of the strain of *Yersinia pestis* responsible for the Black Death plague in England that decimated the human population in the 1300s. DNA collected from excavated remains was carefully sequenced to reassemble all the bacterium's genes and showed how this strain is related to the strains of *Y. pestis* still around today.

A rapidly increasing field in microbiology is the study of all the microorganisms and their genes and products from a specific environment, called *microbiome research*. This exciting new frontier of microbiology is possible because of advances in sequencing technology and has opened our eyes to the unseen diversity of microbial life on earth. Recent surveys of oceans, for instance, have revealed many times more species of bacteria and archaea than we expected, with untold new metabolic pathways.

A popular focus of microbiome research is on the microbes that inhabit the human body. The collection of microbes that naturally live in and on the body are present in everyone and potentially play a huge role in human health and disease. Microbiologists think this is the case because these microbes are present in numbers greater than ten times those of the cells of the human body. They account for up to 2½ pounds of the adult body weight and express around 100 times more genes than we do. Research into the microbes of the human body and all their genes is called *human microbiome research* and has found links between it and everything from weight gain to cancer to depression.

Remaining challenges

Microbiology is still a young science, so there are many frontiers yet to explore. For instance, it appears that scientists have described only the tip of the iceberg for the variety of microbial life on earth. In particular, the variety of viruses that infect humans are not all known. Plus, the many varieties of viruses on earth are hard to even estimate.

The study of cures for viral diseases is still a major challenge, with viruses like HIV and influenza remaining a significant challenge. Viruses like polio and measles that have been essentially eradicated in developed countries still kill and disfigure children around the world in developing countries. As recently as 2014, India was declared polio free, only after more than $2 billion was spent mounting a massive vaccination campaign. However, infectious diseases like pneumonia are still the number-one cause of childhood death around the world because vaccines are hard to deliver in developing countries.

Research is ongoing into protection from diseases like malaria and tuberculosis. Vaccination has not proven effective for these diseases that hide from the immune system. Other strategies against malaria include infecting mosquitos (the insects that infect humans with the disease) with bacteria that kill the malaria parasite, but research is ongoing.

Vaccines are effective for prevention of infectious disease, but it's antibiotics that are used to effectively treat active infections. After the golden age of antibiotic discovery came a long period of reliance on antibiotics by modern medicine. They were so effective in treating most infections that we became complacent about their use. We're now entering the antibiotic resistance phase where most, if not all, of the antibiotics we now use are becoming useless against the rise of antibiotic-resistant pathogens. This has become such a serious problem that in the spring of 2014, the World Health Organization declared antibiotic resistance a global health crisis.

THE FIELDS OF MICROBIOLOGY

Since the 19th century, there has been an explosion of great microbiological research, leading to many different branches of microbiology, all of which are both basic and applied in nature. Here's a list of the different fields of microbiology that have developed since the discovery of microorganisms:

- **Aquatic, soil, and agricultural microbiology** study the microorganisms associated with aquatic (including wastewater treatment systems), soil, and agricultural environments, respectively.
- **Bacteriology** is the identification and characterization of bacterial species.
- **Immunology** is the study of the body's response to infection by microorganisms. Included within this field is the area of vaccine research, which aims to develop more and better ways of immunizing people from microorganisms that cause life-threatening infections.
- **Industrial microbiology** applies the large-scale use of microorganisms to make things like antibiotics or alcohol.
- **Medical microbiology** is the study of pathogenic microorganisms that cause infectious disease in humans and animals and ways to prevent and treat infections.
- **Microbial biochemistry** aims to understand the enzymes and chemical reactions inside microbial cells.
- **Microbial biotechnology** is genetically engineering microorganisms to produce a foreign gene or pathway so that it may either make a product for human use (for example, human insulin) or perform a function that we need (for example, degradation of environmental contaminants).
- **Microbial ecology** is the study of microbial diversity in nature, as well as microbial populations and microbial communities and their effects on their environments. This includes nutrient cycling and *biogeochemistry* (biological, chemical, and physical processes that control the composition of the natural environment).
- **Microbial genetics** is the study of the genomes of microorganisms, including how the genetic code varies between microbes and how genes are passed on.
- **Microbial systematics** is the study of how microorganisms diversified through time. It includes the naming and organizing of microbial groups with respect to one another.

(continued)

(continued)

- **Mycology** is the study of fungi, both in terms of their natural habitats and genetics, and in terms of their ability to cause disease in humans, other animals, and plants.
- **Parasitology** is the study of parasites of animals and humans. These are all eukaryotic (not bacterial or archaeal) and include protists and worms.
- **Virology** is the study of viruses and simple nonviral entities, such as viroids (RNA molecules that behave like infectious agents) and prions (proteins that behave like infectious agents).

IN THIS CHAPTER

» Uncovering microbial life

» Seeing the variety of microbial metabolism

» Connecting with microorganisms

Chapter 3 Microbes: They're Everywhere and They Can Do Everything

We tend to think of microorganisms as the causes of diseases (like polio, the plague, and pneumonia) or inconveniences (food spoilage, the common cold, and garden plant diseases), but the truth is that they play a much larger role in our lives. A balanced microbial community is important for the health of an ecosystem, our health, and the health of our pets and gardens. It's convenient to think of the microbial world only as it pertains to our daily lives, but in reality microorganisms far outweigh all other life on earth in terms of genetic variety and the sheer number of cells (some 2.5×10^{30} by recent calculations).

Based on the best estimates of biologists, life appeared on earth almost 4 billion years ago. Multicellular life appeared 2.5 billion years later, but in the meantime, single-celled organisms ruled the earth. Early *prokaryotes* (bacteria and archaea) lived without oxygen — the earth's atmosphere was *anoxic* (without oxygen) and then slowly changed to one where oxygen levels were sufficient to support life dependent on oxygen. The early earth also had a much harsher climate than the planet does today; evidence of microbes that can tolerate extreme conditions are still with us.

Evidence of the existence of these different organisms can be observed if you look at the evolutionary tree of life on earth today (shown in Figure 3-1). You can see that many distant branches of microbes exist. The more distant branches represent the amount of difference in the genetic material between organisms. Looking at the distant branches for microbes shows that the genetic diversity of microbial life is vast compared with that of animals, where the branches are closer together.

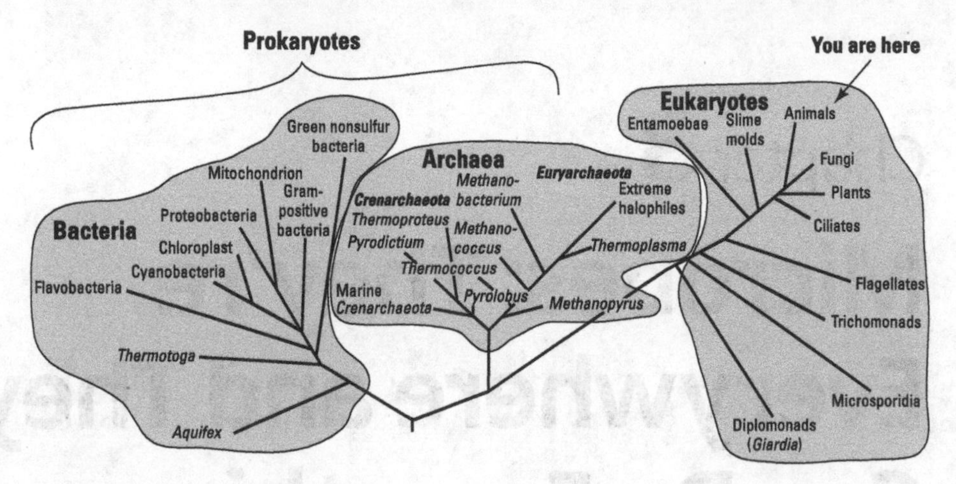

FIGURE 3-1: Genetic relationships among life forms on earth today.

TECHNICAL STUFF

A microbial *population* is a group of cells that are genetically similar to each other (sometimes called a *species*). These populations live together in groups with other microorganism in microbial *communities.* These communities interact extensively with each other and their direct environment, called a *habitat*, consuming nutrients and excreting waste. The environment is the conditions outside the cell and is often discussed in contrast to the conditions inside the bacterial cell. Microbial communities live within the wider context of an *ecosystem*, which can include things like lakes, oceans, or forests. Microorganisms have a profound effect on ecosystems, acting to cycle many of the important elements within it.

In this chapter, we cover two types of diversity among organisms — habitat diversity and metabolic diversity — and then discuss how the presence of microorganisms affects higher life forms like plants and humans.

TECHNICAL STUFF

The term *diversity* describes the variety of possible genes, metabolites, habitats, and so on. *Biodiversity* refers to the variety of live organisms.

Habitat Diversity

The habitat is an important concept in biology and microbiology in particular because microorganisms are greatly affected by where they live. Microbial habitats — including soils, rivers, lakes, oceans, on the surface of living and dead things, inside other organisms, on man-made structures, and everything in between — provide nutrients and protect cells from harsh conditions. The more places we look for microbes, the more microbes we find.

Every environment is *stratified* (arranged in layers) in terms of the degree of temperature, oxygen, nutrients, and sunlight present. These stratifications make up different *niches* to which a specific microorganism, or group of microorganisms, is uniquely suited. Over the billions of years that microorganisms have been on earth, they've evolved to fit perfectly almost every niche. They really are everywhere.

Many habitats on earth have extreme temperatures, pH, salinity, and/or acidity and, because of this, they're inhospitable to most animals and plants. Instead of being devoid of life, these environments are rich in microbial life. Microbes live at the depths of the ocean and in the highest clouds. They thrive at extremely high temperatures, near hydrothermal vents, and at extremely low temperatures inside polar sea ice. Microorganisms can be found at extremes of pH, salt, and dryness, but they're also very abundant and ubiquitous at all conditions in between. In addition, some microorganisms are resistant to substances toxic to most life — the microorganisms often use these substances for energy and, in the process, detoxify them.

Everywhere you look, you find microbes, even in places that you may not expect: surgical suites, NASA *clean rooms* (specially designed rooms with special air handling and disinfection so that they're free of any microbes), the human brain, and subterranean caves. For every system that has been designed to keep out microbes, there is a microbe that has circumvented it. This is because microorganisms are the masters of adaptation and because there are so many different microbes adapted to so many different conditions. What this means is that despite our best efforts, microbes can sometime be very difficult to get rid of.

The number of microbial cells is not the same in each habitat. Some locations host a large number of bacterial cells, called a *biomass*, whereas others have a large number of different bacterial species or groups. For instance, the colons of animals are home to maybe the largest biomass of bacterial cells anywhere, with estimates of 10^{10} cells per gram. However, the number of different bacterial groups are thought to be in the range of 100 or 200. In contrast the most diverse habitat, containing the largest number of bacterial species, is probably soil, with somewhere around half a million species in a gram of soil. Environments with mixed nutrient sources often have high microbial diversity because many different microbes can grow without one taking over completely. So, the total number of microbes is lower, but the number of different species is higher.

In some cases, microorganisms have made their own habitats and grow in large multispecies communities that are obvious to the naked eye. An obvious example of this are *microbial mats*, which can include many different species of bacteria and archaea. Microbial mats contain a wealth of lifestyle and metabolic diversity and are discussed in detail in Chapter 11. Another, slightly different type of microbial community, called a *consortium*, involves a more intimate relationship between a small number of different microbial species. Two examples of this are lichen and the green sulfur bacterial consortium of marine and freshwater environments (see Figure 3-2).

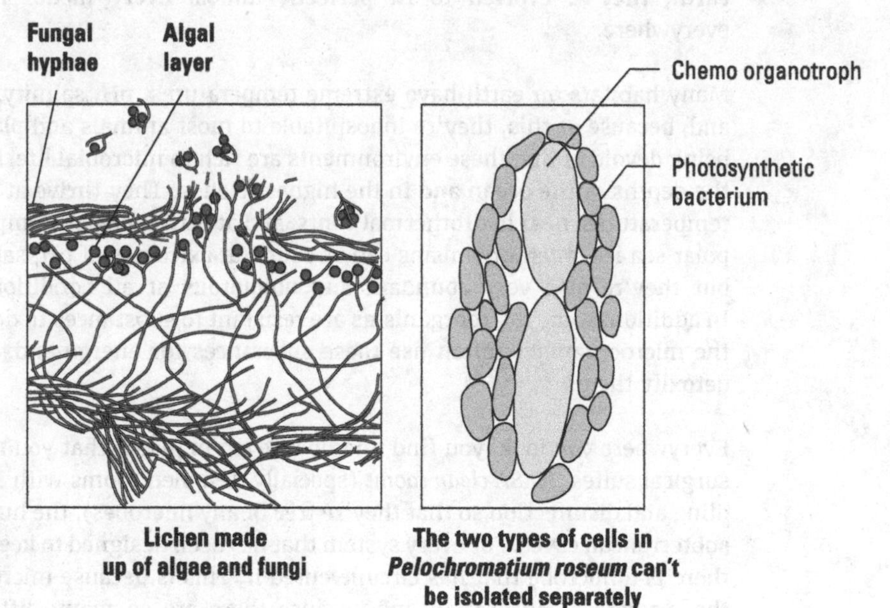

FIGURE 3-2: Microbial consortia such as lichen and *Pelochromatium roseum.*

Metabolic Diversity

Not only are microorganisms extremely widespread, but within the microbial world there is also an impressive number of different metabolic pathways. We know this because of the compounds that they consume and produce, as well as from the study of microbial genes found in nature. Recently, scientists have been able to sequence the full genomes of many microorganisms, giving us access to the sequences of all the genes present. This offers a glimpse into the metabolic potential of a microbe because knowing the genes present can suggest which enzymes the microbe can make and use for its metabolism.

Four broad categories of metabolic diversity include: the main energy-gathering strategy used, strategies for obtaining carbon, essential enzymes for growth, and products not essential for survival called *secondary metabolites.*

Getting energy

There are three sources of energy in nature:

- Organic chemicals (those containing carbon–carbon bonds)
- Inorganic chemicals (those without carbon–carbon bonds)
- Light

Chemoorganotrophy is the type of metabolism where energy comes from organic chemicals, whereas *chemolithotrophy* is the type of metabolism where energy comes from inorganic chemicals. *Phototrophy* involves turning light energy into metabolic energy in a process called *photosynthesis,* and it comes in two main forms:

- **Oxygenic photosynthesis** generates oxygen and is used by the cyanobacteria (a type of bacteria; see Chapter 12) and algae (a eukaryote), as well as all living plants.
- **Anoxygenic photosynthesis** does not make oxygen and is used by the purple and green bacteria (types of bacteria that live in anaerobic aquatic environments; see Chapter 12).

Capturing carbon

All living cells need a lot of carbon, which is part of all proteins, nucleic acids, and cellular structures. Organisms that use organic carbon are called *heterotrophs;* chemorganotrophs fall into this category. Organisms that use carbon dioxide (CO_2) for their carbon needs are called *autotrophs;* most chemolithotrophs and phototrophs are also autotrophs, which makes them *primary producers* in nature because they make organic carbon out of inorganic CO_2 that is then available for themselves, chemoorganotrophs, and eventually all higher life forms. Some organisms can switch between heterotophy when organic carbon is available and autotrophy when food sources run out; these organisms are called *mixotrophs.*

Making enzymes

Few compounds in nature are not degraded by microorganisms. The variety of compounds produced by them is great and not completely known. Their metabolic processes are essential for environmental nutrient cycling, and they are the primary producers that support all other life on earth.

Microbes are specialists at degrading compounds, from the simplest to the most complex and everything in between. They're the only ones able to degrade *resistant plant material* (fiber) made from *cellulose* (building blocks used by plants to make their tough cell walls) and *lignin* (building blocks used by plants for rigid structure, as in wood and straw). The microbes in the *rumen* (part of a cow's or related animal's stomach) of herbivores and the guts of termites are responsible for digesting these tough plant fibers. Fungi and bacteria are the masters of producing special enzymes to degrade complex food sources (hydrolytic enzymes) including all forms of plant and animal tissues, some plastics, and even metals.

Secondary metabolism

Microbial products that are not produced as part of central metabolism and are not essential for everyday activities are called *secondary products*. Many of these products are bioactive compounds useful in interacting with other organisms. Antibiotics are an example of a secondary product used to interact with other microbes. Some plant pathogens produce substances that mimic plant hormones so that they can manipulate plant growth. Other microbes make molecules that are useful in communicating with other microorganisms, insects, and plants.

Knowledge of the metabolism of microorganisms can be used in a variety of ways. One way is to try to isolate them in culture. This isn't always easy — there are many gaps in our knowledge of the metabolic diversity of most microorganisms. It's relatively easy to re-create the temperature and oxygen conditions, but in order to select for the organism you want and select against all the other organisms, you have to know one specific condition that is needed just for your organism of choice. Here are some other ways that the knowledge of microbial metabolism has been useful in the advancements of science:

- **Microbial enzymes are used in molecular biology research.** Bacterial enzymes such as Taq DNA polymerase (used for reproducing sequences of DNA) and restriction enzymes (used to manipulate pieces of DNA in a cut-and-paste fashion) have become invaluable research tools.
- **Microbes are used to express animal proteins or enzymes such as insulin.** When scientists discover that a disease can be cured or treated with a certain protein or enzyme, it becomes very useful and efficient for them to be able to mass-produce the molecule in microbes.

- **Microbial systems are used as part of microscopic machines in synthetic biology.** To conduct further research, scientists make use of what we know to push the envelope of engineering and genetics. Scientists use microbial processes to their fullest potential to create new things within organisms.
- **Industrial processes have taken advantage of the diversity of microbes in the food, pulp and paper, mining, and pharmaceutical industries (to name but a few).** Because some microorganisms are tolerant of extreme conditions, the enzymes they produce are useful in industrial settings where conditions can be harsh.

Because scientists don't know all the metabolic diversity in the microbial world, they haven't been able to isolate in culture a vast number of environmental microbes. This has resulted in huge gaps in knowledge about all the microbial groups that exist. The term *microbial dark matter* has been coined to describe the vast number of microbial lineages for which scientists know very little (and in most cases, almost nothing). Like the dark matter of the universe that makes up the majority of matter, microbial dark matter is enormous and likely outweighs the known biodiversity of the earth by several orders of magnitude.

The Intersection of Microbes and Everyone Else

Microorganisms that live on and inside other organisms have often adapted to interact with their host organism. There are different kinds of relationships between the host organism and microorganisms that live on or in the host (see Figure 3-3):

- **Benign:** Organisms that live in or on us and are neither harmful nor recognized by our bodies. They're present, living happily, and our bodies essentially ignore them most of the time.
- **Friendly:** Organisms that live in or on us and are not harmful. Our bodies recognize that the organisms are present and also recognize that these bacteria are not harmful, so our bodies don't react to them.
- **Mutually beneficial:** Organisms that live in or on us and provide us with a benefit, like making vitamins that we can't make ourselves. We, in turn, provide the organisms with benefits, such as nutrients and a home.
- **Antagonistic:** Organisms that cause us harm, as in the case of viruses where our cells are taken over to produce more virus and are eventually killed.

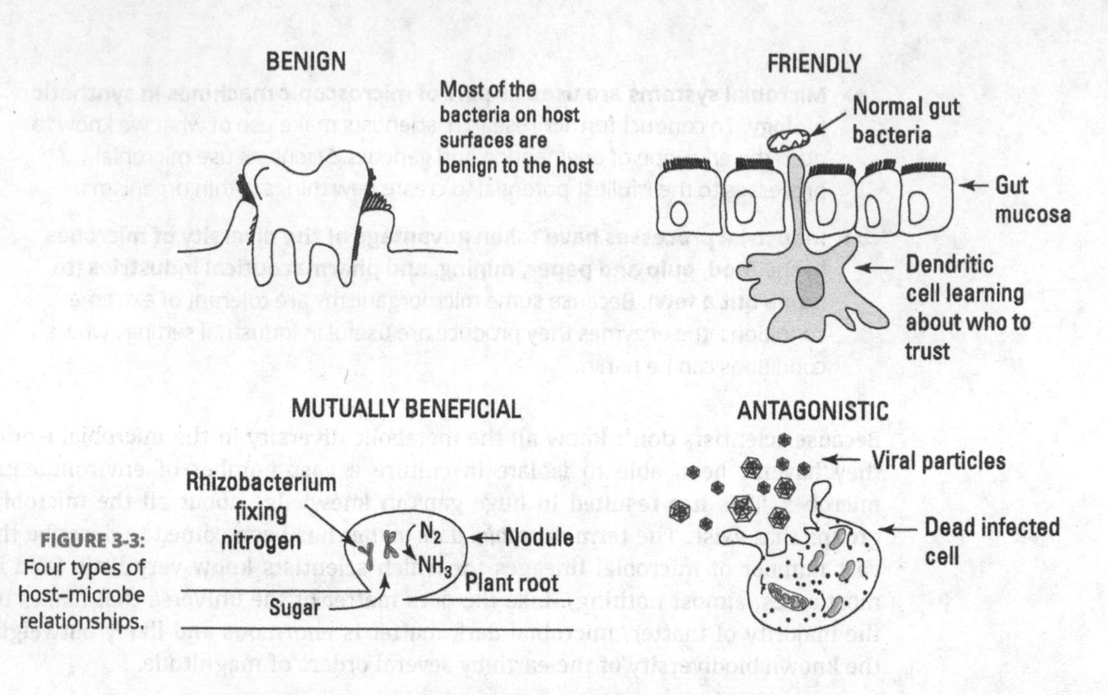

FIGURE 3-3: Four types of host-microbe relationships.

Research is still being done to determine to what extent the human body's immune system recognizes and reacts to the friendly, nonharmful bacteria that live in and on us.

For every living organism on the planet, there is a microbial *pathogen* (a disease-causing organism) perfectly suited to plague it. Some microbes are nasty pathogens that infect their host, subvert their defenses, and wreak havoc in the form of severe illness and sometimes death. Other microbes produce toxins that cause nasty effects on their target. These include microbes that infect animals, plants, insects, invertebrates, and so on. There are even microbes that infect other microbes.

On the other hand, there are accidental infections that result when a person with a compromised immune system comes into contact with a microorganism that is common in the environment but can survive in the warm, moist habitat of the human body. These include yeasts, fungi, and bacteria such as species of *Bacillus*, the pseudomonads, *Acinetobacter*, and *Clostridium*.

Pathogenic microorganisms have even been found to have shaped their hosts' evolution. A chilling example of this was recently discovered by looking for the effects of the plague of the Middle Ages on the evolution of people in Western Europe. Researchers found that genes for immunity were different in people whose ancestors had been exposed to the bacterial pathogen than in those whose ancestors weren't. They also found that genes related to autoimmune disorders were affected, suggesting that a pathogen can have a profound impact on its host's evolution.

2

Balancing the Dynamics of Microbial Life

IN THIS PART . . .

See the different shapes of microbial cells, and get a basic understanding of how cells function.

Get to the bottom of metabolism, including how microbes make energy from their food and store this energy for later.

Understand microbial genetics, from the importance of DNA to how genes can change and the ways in which genes are turned on and off.

Find out about microbial growth, how picky microbes can be, and the lengths scientists will go to in order get the conditions in the lab just right.

Find out about the four different types of "clean" when it comes to microbes and how the presence of unwanted microbes is controlled.

IN THIS CHAPTER

» Getting the basics of cell structure

» Understanding how cells move things in and out

» Finding out about cell division and locomotion

Chapter 4
Understanding Cell Structure and Function

Microbiologists know a lot about microbial cells, considering that they're too tiny to see with the naked eye. Research on bacterial, archaeal, and eukaryotic cells (see Chapter 1) has given us a glimpse into how these cells are put together and how they differ from one another.

In this chapter, we give a bird's-eye view of the structure of microbial cells. Then we go into some of the most important structures in detail. We discuss major differences between microbes — for instance, what differs between eukaryotic and prokaryotic microorganisms, as well as things that all cells have in common.

Seeing the Shapes of Cells

We know that prokaryotic cells come in many different shapes and sizes because we can look at them under a microscope. A description of the shape of a cell is called the *cell morphology*. The most common cell morphologies are *cocci* (spherical) and *bacilli* (rods). *Coccibacillus* are a mix of both, while *vibrio* are shaped like a comma, *spirilla* are shaped like a *helix* (a spiral, sort of like a stretched-out Slinky), and *spirochetes* are twisted like a screw. Figure 4-1 shows these common cell morphologies.

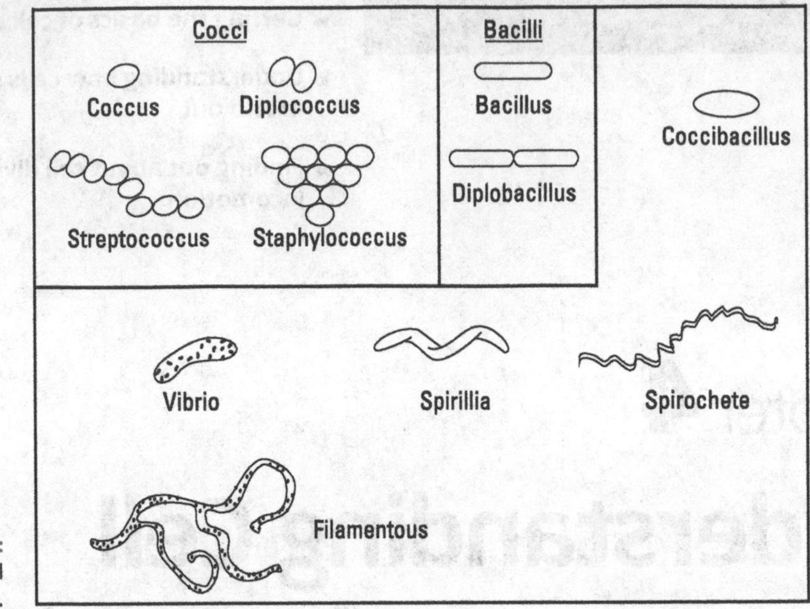

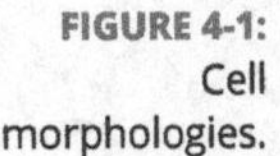

FIGURE 4-1: Cell morphologies.

Although prokaryotes are unicellular organisms, their cells can be arranged in a few different ways, like chains or clusters, depending on how the cells divide:

- Cocci bacteria that divide along a single plane form small chains of two cells called *diplococci* or long chains of multiple cells called *streptococci*.
- Cocci bacteria can also divide along multiple planes to form *tetrads* (two planes), cubelike *sarcinae* (three planes), or grapelike clusters called *staphylococci* (multiple planes).
- Similarly to the cocci, rod-shaped bacteria can divide to form double-celled *diplobacilli* or longer chains called *streptobacilli*.

The shape of a cell is encoded in its genes. Although we generally know how cell shape is controlled, the reason behind the many different shapes remains a mystery.

TIP

You may notice that some of the morphologies are also the names of bacteria — for example, *Streptococcus pneumoniae*, *Staphylococcus aureus*, *Bacillus anthracis*, and *Vibrio cholerae*. That's because morphologies are sometimes characteristic of bacterial genera.

WARNING

Morphology is a descriptive characteristic — it doesn't give you enough information to know exactly what type of bacteria you're looking at or its function.

Life on a Minute Scale: Considering the Size of Prokaryotes

One distinguishing feature of prokaryotic cells is their microscopic size. The average size of a prokaryotic cell ranges from 1 μm to 10 μm (see Figure 4-2). Eukaryotic cells, such as those in the human body, are much larger in comparison and range from 10 μm to 200 μm. Viruses are smaller than prokaryotic cells at 10 nm (0.01 μm) to 100 nm followed by subcellular molecules like proteins or chromosomes that are less than 10 nm in size. The smallest size that is visible to the naked eye is 100 μm, so scientists rely on microscopes to see prokaryotic cells.

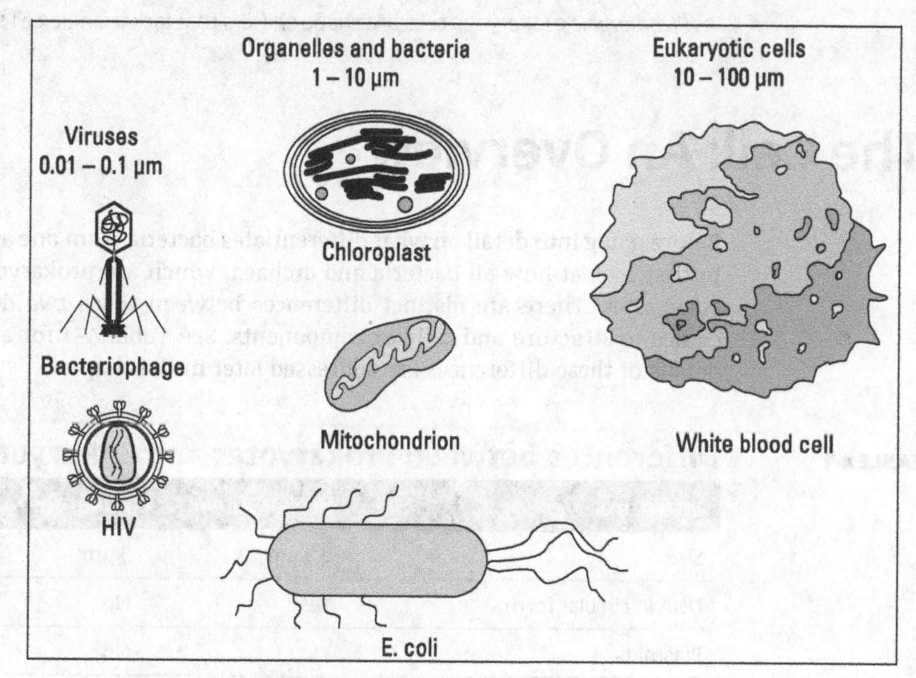

FIGURE 4-2: Comparison of cell sizes.

The size of a cell is actually quite important for its function. Being small comes with its perks! Cells take up nutrients across their surface, and those nutrients then have to diffuse through the cell in order to help the cell grow. Maximizing the surface area and minimizing the total cell volume reduce the distance that nutrients have to travel, decreasing the time it takes to get the nutrients to where they need to go in the cell.

REMEMBER

The higher the *surface-to-volume ratio* is, the better off the cell is. Prokaryotic cells have found the Goldilocks size, where they maximize the surface-to-volume ratio without running the risk of being so small that they crowd their contents and impair biological processes.

ODDS AND ENDS OF SHAPES AND SIZES

Here are some interesting facts about shapes and sizes of cells:

- The smallest bacteria recorded are *nanobacteria.* They range in size from 0.05 μm to 0.2 μm. This size exceeds the lower limit at which scientists had once thought cells could exist!
- The largest bacterium found, *Thiomargarita namibiensis,* measures 750 μm, which is almost 1 millimeter.
- Unusual shapes of bacteria include star shaped, flat square shaped, and pear shaped.
- *Pleomorphic* prokaryotic cells are shape shifters that lack a single cell shape.

The Cell: An Overview

Before going into detail on what differentiates bacteria from one another, we want to first look at how all bacteria and archaea, which are prokaryotes, differ from eukaryotes. There are distinct differences between these two domains when it comes to structure and cellular components. See Table 4-1 for an overview. The details of these differences are addressed later in the chapter.

TABLE 4-1 Differences between Prokaryotes and Eukaryotes

Structure	Prokaryote	Eukaryote
Size	0.5 μm	5 μm
DNA in circular form	Yes	No
Plasmids	Yes	No
Ribosomes	70S	80S
Cell wall	Yes	No
Cell membrane	Yes	Yes
Membrane bound organelles	No	Yes
Growth above 70C	Yes	No
Endospores	Yes	No

Scaling the Outer Membrane and Cell Walls

The barrier between the inside of a cell and its surroundings is the cell wall or the membrane. Because they act as the barrier between the outside world and the cell, these structures need to be strong and function reliably. Cell walls also have to be versatile enough and distinct enough from those in other bacteria to give each bacterium its own identity. Several components make up this unique barrier for the different types of bacteria, and many of these differences are used to help scientists tell bacterial groups apart from one another. By looking at the outer membrane, cell wall, and Gram-stain characteristic (Gram-positive vs. Gram-negative) of bacterial cells, we can tease apart the many bacterial groups. Later in this section, we explain what makes archaea so unique.

Examining the outer membrane

The plasma membrane borders the cell and acts as a barrier between the inside of the cell and the outside environment. The membrane serves many important functions in prokaryotic cells, including the following:

- Providing sites for respiration and/or photosynthesis
- Transporting nutrients
- Maintaining *energy gradients* (the difference in the amount of energy between the inside of the cell and the outside of the cell)
- Keeping large molecules out

The plasma membrane is made of a *phospholipid bilayer*. Individual phospholipids are composed of a charged phosphate "head" group and an uncharged lipid "tail" group. Because phospholipids contain charged and uncharged components, they're *amphipathic* (they can interact with both lipid and aqueous solutions).

TIP

Think of lipid and aqueous solutions as oil and water — they can't really mix and tend to separate. The tails of phospholipids are *hydrophobic* (repellant of water), so they repel water such that in a solution of water they arrange tail-to-tail to limit their contact with it. This tail forms the inner portion of the membrane. In contrast, the head groups are *hydrophilic* (attracted to water). They face the outer portion of the membrane and directly contact the environment or the cytoplasm, both of which are aqueous (see Figure 4-3). The layer nearest to the extracellular environment is referred to as the *outer leaflet*, and the layer nearest to the cytoplasm is referred to as the *inner leaflet*. The hydrophobic nature of the lipids is what makes the phospholipid bilayer impermeable to large, water-soluble molecules.

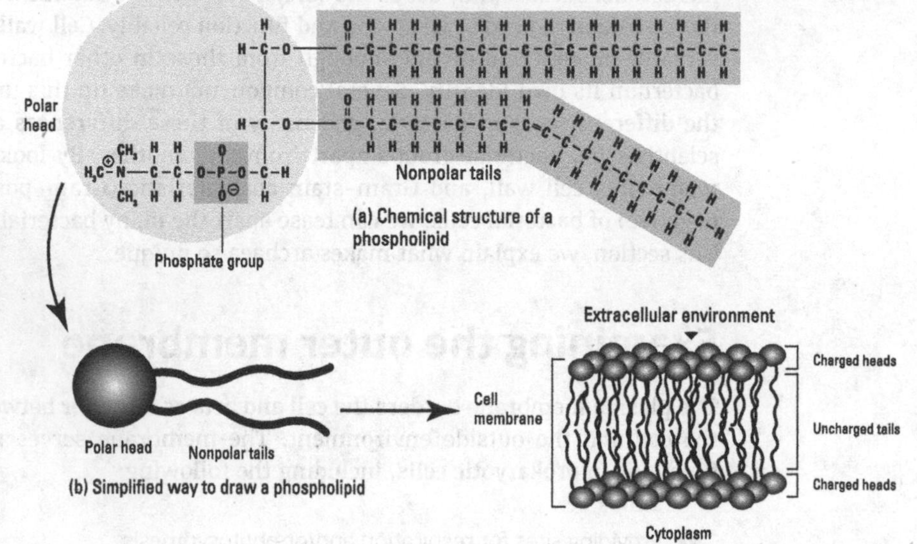

FIGURE 4-3: The structure of the phospholipid bilayer.

Individual phospholipids contain a phosphate group attached to a glycerol molecule that is attached to two fatty acid tails. Sometimes additional molecules are attached to the phosphate group of phospholipids.

Not all membranes are the same

The phospholipid bilayer is common to eukaryota, archaea, and bacteria, but there are several differences that set their membranes apart:

- **Eukaryotic** membranes contain *sterols* (steroid-containing lipids), such as cholesterol, that strengthen the membrane.
- **Bacterial** membranes contain sterol-like molecules called *hopanoids,* which help strengthen the membrane.
- The chemical composition of **archaeal** phospholipids differs from bacterial and eukaryotic phospholipids in four main ways, as shown in Table 4-2. These differences are a major distinction between the domains.

TABLE 4-2 Differences in Phospholipid Structure of Archaea

Structure	Bacteria/Eukaryotic	Archaea
Glyercol chirality	D-glycerol	L-glycerol
Hydrophobic tails	C16 or C18 hydrocarbon chains	Repeating units of 5-carbon isoprene
Glycerol linkage	Ester bond	Ether bond
Branching of side chains	No	Yes

The fat facts

The plasma membrane isn't all just fat. It's almost equally composed of protein. Membrane proteins serve many important biological functions, including motility, adhering to surfaces, sensing and secreting signals, and transporting nutrients.

A membrane protein must contain a hydrophobic region that interacts with the hydrophobic region of phospholipids. There are two classes of membrane proteins:

- **Integral proteins** are insoluble in aqueous environments and are tucked right in the membrane.
- **Peripheral proteins** are soluble in aqueous environments and sit close to the membrane. They contact the membrane itself temporarily or interact with integral proteins.

Integral proteins float within the phospholipid bilayer and are able to move laterally, like a buoy in a lake that moves back and forth with the movement of the water. Proteins are mobile in the membrane because the membrane's consistency is similar to that of cooking oil. The membrane is studded with many proteins that freely move throughout it.

Exploring the cell wall

Cells build up stores of molecules required for growth inside the cell in higher amounts than are found outside the cell. Water naturally wants to flow into the cell to balance the number of molecules inside and outside, but if the cell were to allow this, it would burst like an overfilled balloon. The cell wall allows the cell to withstand this *osmotic pressure*.

In bacteria, the cell wall is made of *peptidoglycan*, a structure not found in either eukaryotes or archaea. This structure forms a meshlike sac around the cell and provides it with rigidity. Peptidoglycan is made up of polysaccharides linked by peptide bridges. Polysaccharides are long sugar chains of alternating N-acetylglucosamine (NAG) and N-acetylmuramic acid (NAM) units, which are firmly attached together by chemical bonds.

As shown in Figure 4-4, polysaccharides arrange like cables that encircle the cell and are connected by peptide bridges made of the four amino acids L-alanine, D-alanine, D-glutamic acid, and either lysine or diaminopimelic acid (DAP). Peptide bridges are covalently linked to NAM sugars. Each individual unit of peptidoglycan is, therefore, a NAG-NAM-tetrapeptide.

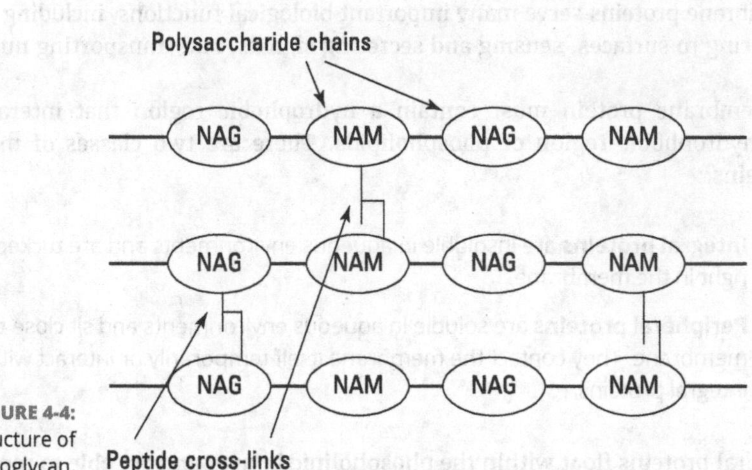

FIGURE 4-4: The structure of peptidoglycan.

The peptide bridge can be made up of different amino acids and creates diversity in the peptidoglycan structure between bacteria.

A CHINK IN THE ARMOR

The necessity of the cell wall for cell survival is a weakness our bodies take advantage of to fight infectious bacteria. Our bodies produce *lysozyme*, an enzyme that is able to split the bond between the NAG-NAM sugars. This weakens the cell wall and allows water to flow in, causing the cell to burst. The antibiotic penicillin also targets several steps in peptidoglycan synthesis that ultimately blocks formation of the cell wall and causes cell death.

Differences at the surface

Bacteria are classified as either Gram-positive or Gram-negative based on the results of the Gram-stain (see Chapter 7). Structural differences in peptidoglycan are the basis for this differentiation. Gram-positive bacteria retain the crystal violet stain in the Gram-stain procedure because they have a thick, multilayer sheet of peptidoglycan. Gram-negative bacteria, on the other hand, do not retain the crystal violet stain in the Gram-stain procedure because they have a single, thin layer of peptidoglycan.

In Figure 4-5, you can see an image of the two types of cell wall structure.

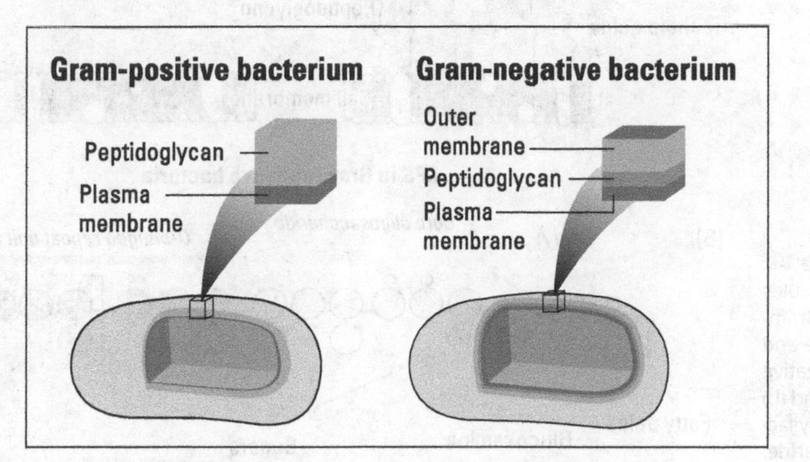

FIGURE 4-5: Structure of Gram-positive and Gram-negative cell walls.

TIP

Some bacteria, like *Mycoplasma*, lack cell walls and are only enclosed in a membrane. *Mycoplasma* species normally live inside other cells where they don't face as much osmotic pressure. Their membranes are often tougher than other bacterial membranes because they contain cholesterol. Most bacterial groups that we know of, however, have a cell wall.

The differences between Gram-positive and Gram-negative bacteria, however, don't stop there. Gram-positive cells have several key features that set them apart from Gram-negative bacteria (see Figure 4-6):

- *Teichoic acid* is a molecule unique to Gram-positive peptidoglycan and runs perpendicular to the layers of polysaccharide. Teichoic acid is directly attached to the peptidoglycan and is occasionally attached to the phospholipids to form lipoteichoic acid.
- The space between the plasma membrane and the peptidoglycan layer is called the *perisplasmic space* and is very thin, almost unnoticeable, in Gram-positive bacteria.

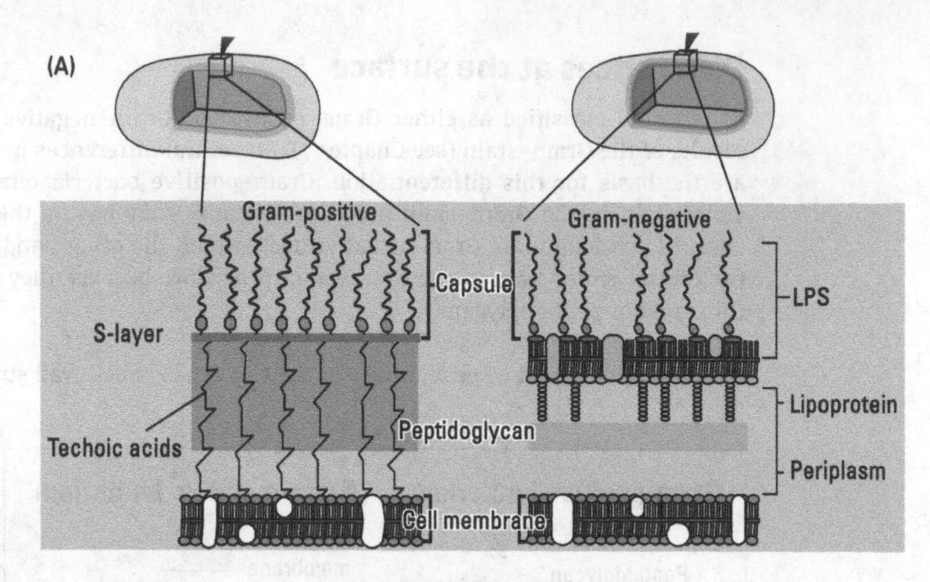

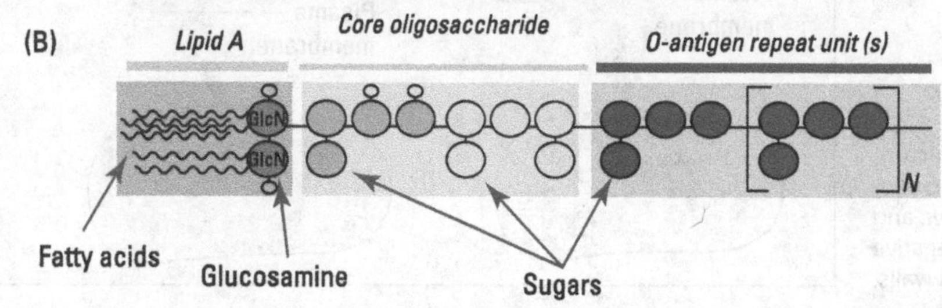

FIGURE 4-6: (a) The molecules within Gram-positive and Gram-negative cell walls and (b) lipopolysaccharide.

- The peptidoglycan of Gram-positive bacteria is often decorated with sugars and proteins that help the bacteria attach to surfaces and interact with the environment.

Gram-negative cells have an additional membrane on the outer surface of the peptidoglycan called the *outer membrane*, but it's not your typical membrane. The inner leaflet contains proteins that anchor it to the cell wall. These tethers are called *lipoproteins*. The outer leaflet is made up of a different kind of phospholipid, *lipopolysaccharide* (LPS). LPS is a large molecule with three main building blocks (see Figure 4-6):

- **Lipid A:** The fatty acid chains in lipid A are attached to a pair of *glucosamine molecules* (a glucose molecule with an amine group) and not a glycerol molecule, as in a typical phospholipid. Each glucosamine is attached to a phosphate head group. Lipid A is an *endotoxin* that activates the immune system, causing many of the severe symptoms of Gram-negative bacterial infections.

- **Core oligosaccharide:** A chain of about five sugars makes up the core oligosaccharide that is attached to one of the glucosamine molecules in lipid A.
- **O polysaccharides:** Chains of sugars that extend further than the core region are called O polysaccharides (or O antigen). They're made up of as many as 200 sugars. The type of sugars in the O polysaccharide vary across different bacteria.

Encasing archaea

Archaea, like other single-cell organisms, have tough cell walls that protect them from the environment, but they differ in that they lack peptidoglycan. The cell wall of archaea is most commonly made of surface-layer proteins, called the S-layer. The S-layer assembles from proteins or glycoproteins that form the outermost layer of the cell.

Some archaeal cell walls, like those of certain *methanogens* (methane producers), contain *pseudomeurin* (*pseudo* = false, *meurin* = wall). The chemical composition of pseudomeurin and peptidoglycan differ in the sugars and the amino acids they contain. Other archaeal cell walls have a layer of polysaccharides and provide protection either alone or with an S-layer as well.

Other Important Cell Structures

Many cellular structures are important for cellular function, and these differ between eukaryotic and prokaryotic cells. Here is a list of structures important to prokaryotic cells.

- **Nucleoid:** The nucleoid is the region in the cell where the tangled mass of genetic material, called the *chromosome,* is found. Unlike the nucleus of eukaryotic cells, the nucleoid does not have a membrane. The chromosome of prokaryotes is often circular but can be linear.
- **Plasmids:** Plasmids also contain genetic material but in smaller circular forms that are independent from the chromosome and replicate on their own.
- **Cytoplasm:** The cytoplasm is the sea in which the rest of the structures swim. It's not just water, though — it's gelatinous and contains filaments that span from one end to the other, providing structure to the cell.
- **Ribosomes:** Ribosomes are involved in making protein and are plentiful in growing cells. They're smaller and less dense in prokaryotes than they are in eukaryotes. Ribosomes are made of subunits — 30S and 50S — that are named for their size and shape. (Well, actually they're named for how far they move through a solution when you spin them very fast, but that's a function of

size and shape.) Many antibiotics target bacterial ribosomes as a way of slowing down their growth.

- **Glycocalyx:** The glycocalyx, an extracellular polymer that surrounds the outside of the cell, is called the *capsule* when firmly attached; when loosely attached, it's called a *slime layer*. The capsule protects bacteria from being *phagocytosed* (taken up) by immune cells. A biofilm is formed when one or several bacterial populations use a glycocalyx called an *extracellular polymeric substance* (EPS) to attach to a surface. Biofilm formation not only helps microbes to form a niche for themselves but also helps them survive environmental conditions such as drought or stress from antibiotics.
- **Inclusion:** An inclusion is the concentration of a substance inside of the prokaryote cell. Many types of inclusions exist, such as those for nutrient storage (such as polysaccharides, lipids, or sulfur granules) or those that have a function such as carboxysomes (for carbon dioxide fixation), vacuoles (filled with gas for buoyancy), or magnetosomes (which contain magnetic compounds for direction).
- **Endospores:** Endospores are formed as a survival mechanism by some bacteria. When nutrients are depleted or conditions start to become unfavorable, an endospore can begin forming inside the growing cell (called the *vegetative cell*). Endospores have a tough spore coat, contain very little water, and take with them only the bare essentials to start growing again when conditions improve. After it's released from the vegetative cell, the endospore doesn't actively metabolize or grow; instead, it lies waiting, sometimes for thousands (or millions!) of years until it can germinate near a suitable food source.

Eukaryotic cells differ from prokaryotic cells in several ways. Although they contain some of the same structures as prokaryotic cells, for the most part eukaryotic cells are larger and more complex. One major difference is that eukaryotic cells contain compartmented structures called *organelles*, membrane-bound structures that usually have a special function. Here are some important organelles:

- **Nucleus:** The nucleus is the largest organelle in the cell and contains the chromosomes. DNA in eukaryotes is associated with DNA-binding proteins called *histones*.
- **Mitochondria:** Mitochondria are the powerhouse of the cell and produce adenosine triphosphate (ATP) through respiration. They have their own DNA and bear an evolutionary relationship to bacteria.
- **Chloroplasts:** Chloroplasts are organelles within some eukaryotic cells (such as algae and plants) that are responsible for the photosynthetic reaction.

Other important organelles include the endoplasmic reticulum (where protein is made), the Golgi body (where proteins are processed), and lysosomes (containing digestive enzymes).

Divining Cell Division

Cell division is a process that is necessary for microbial growth. It starts with a single cell that stretches in size until it separates into two separate cells, in a process called *binary fission.* Each new cell is equipped with the right amount of proteins, nutrients, and importantly, the chromosome, to function as an independent cell.

Several important events take place leading up to cell division (see Figure 4-7):

1. In preparation for cell division, a cell makes more peptidoglycan and membranes to support cell lengthening. In general, the cell increases in volume.
2. The genetic material of prokaryotic cells is typically stored in a single circular chromosome. The cell makes a copy of this chromosome, one for each cell after division.
3. A structure called the *Z ring* is made from the protein FtsZ and assembles around the center of the cell.
4. The chromosomes and other cellular contents are distributed to opposite sides of the cell.
5. The Z ring constricts inward in the middle of the cell, and the *septum* (dividing wall) between the two new cells is formed.
6. The cell wall is pinched off to complete division and gives rise to two new daughter cells.

The time it takes for one cell to divide into two is called the *generation time.* If *E. coli* is given a rich supply of nutrients, its generation time is 20 minutes. This means that if a single cell of *E. coli* is grown for an hour, it will become eight cells (three generations). This increase is called *exponential growth* and is one of the phases of growth discussed in more detail in Chapter 7.

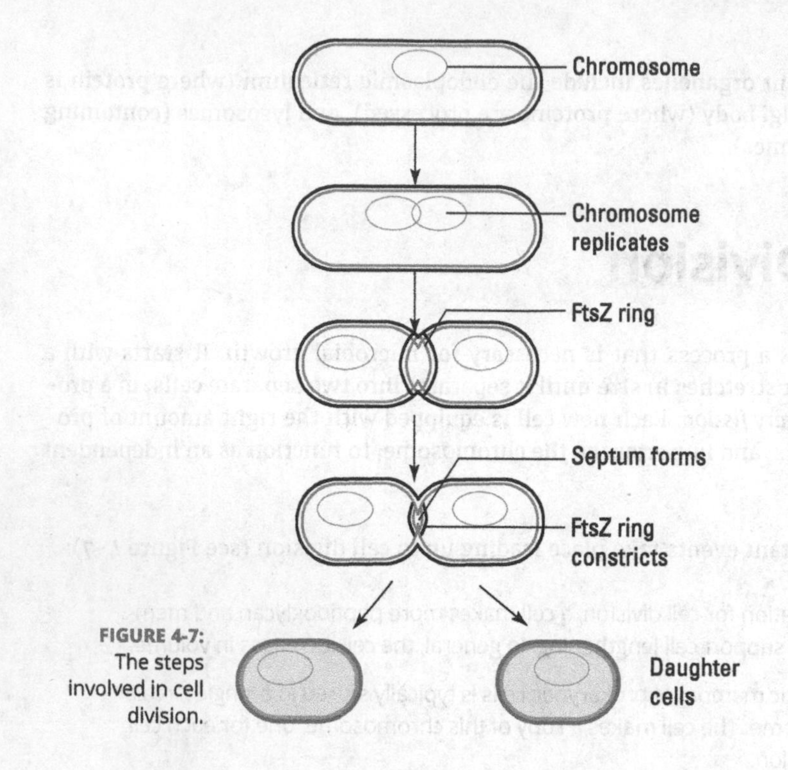

FIGURE 4-7: The steps involved in cell division.

Tackling Transport Systems

The plasma membrane separates the contents of a cell from its external surroundings. It's impermeable to large, water-soluble molecules because it's made of tightly packed lipids that are hydrophobic. Although this physical barrier ensures that important molecules, like those needed for energy, stay inside the cell it also creates a challenge to cells when they want to bring things into the cell (such as sugars, amino acids, or ions). To make things more difficult, nutrients are often present in very low concentrations and hard to scavenge.

Cells are able to take up nutrients from the environment thanks to membrane-spanning transporter proteins. Different types of transporters are required to transport the wide range of nutrients that cells need, but this often comes at an energy cost because the cell must build up internal stores at higher concentrations than those found in the environment. Nevertheless, it's worth the price because these nutrients are vital for survival.

Coasting with the current: Passive transport

Passive transport is the movement of molecules from an area of a high concentration to an area of a lower concentration until equilibrium is reached (see Figure 4-8). Imagine that you're kayaking in a river and you stop paddling. Your kayak will travel in the direction of the current and slow to a stop once you reach a calm area. You don't have to use your paddle. Similarly, passive transport doesn't require energy from the cell.

Passive transport is called *diffusion* and there are two kinds:

- **Passive diffusion** is passage of some small, uncharged molecules, like CO_2, O_2, and H_2O, through the membrane.
- **Facilitated diffusion** involves a carrier protein that allows passive transport of larger molecules, such as glycerol, across the membrane.

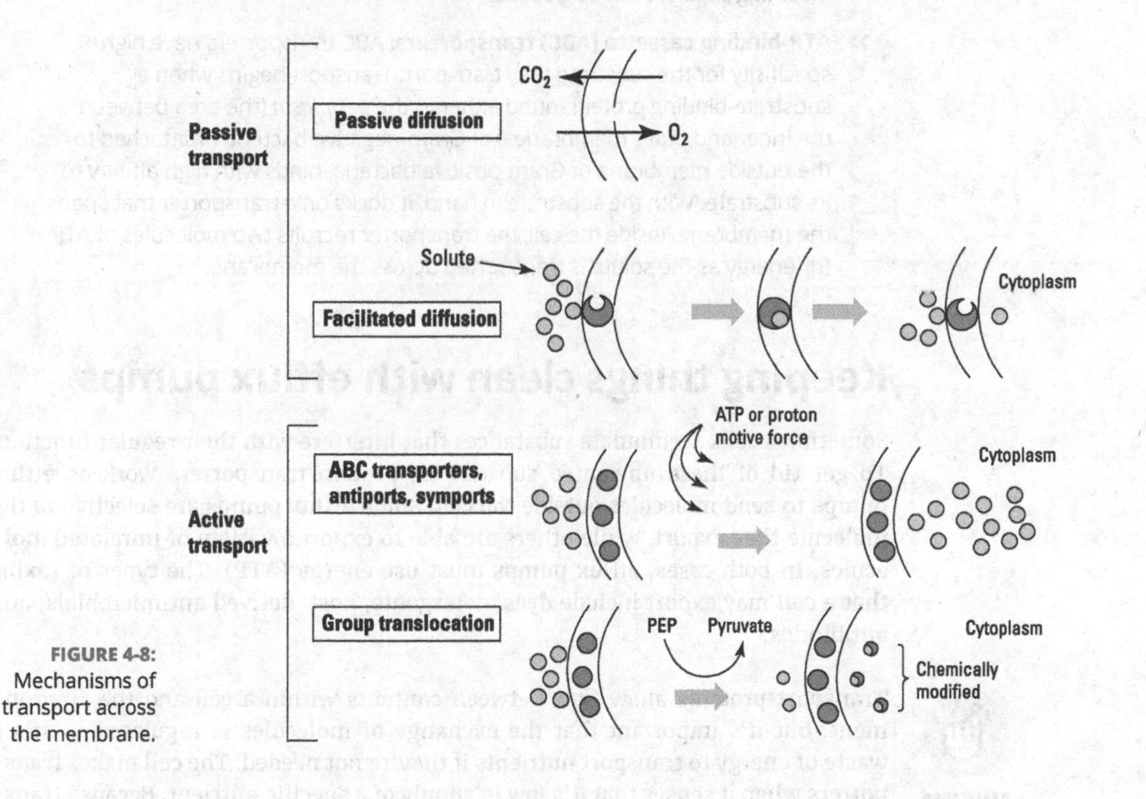

FIGURE 4-8: Mechanisms of transport across the membrane.

Upstream paddle: Active transport

Active transport is the movement of molecules against a concentration gradient and requires energy from the cell (refer to Figure 4-8). In this case, you need to do a lot of paddling in your kayak if you want to go upstream! These types of transporters are usually specific for the molecule or ion they need to transport. This specificity helps the cell to maintain optimal levels of each molecule or ion. Here are some active transport systems:

- **Antiports andsymports:** Antiports and symports use energy from the proton motive force of the cell to drive transport molecules into the cell. As the names suggest, *antiports* transport a molecule and a proton in the opposite direction, whereas *symports* transport a molecule and a proton in the same direction.
- **Group transport:** Group transport involves enzymes that chemically modify a substrate as it is transported into the cell. Energy in the form of phosphoenolpyruvate (PEP) provides a phosphate group that is transferred to incoming sugars, such as glucose.
- **ATP-binding cassette (ABC) transporters:** ABC transporters have high specificity for the substrate they transport. Transport begins when a substrate-binding protein found either in the *periplasm* (the area between the inner and outer membranes) of Gram-negative bacteria or attached to the outside membrane of Gram-positive bacteria, binds with high affinity to its substrate. With the substrate in hand, it docks on a transporter that spans the membrane. Inside the cell, the transporter recruits two molecules of ATP for energy as the solute is transported across the membrane.

Keeping things clean with efflux pumps

Sometimes cells accumulate substances that interfere with their regular function. To get rid of these unwanted substances, protein transporters work as efflux pumps to send molecules outside the cell. Some efflux pumps are selective to the molecule they export, while others are able to export a variety of unrelated molecules. In both cases, efflux pumps must use energy (ATP). The types of toxins that a cell may export include dyes, detergents, host-derived antimicrobials, and antibiotics.

REMEMBER

Transport proteins allow flow between contents within a cell and the environment, but it's important that the exchange of molecules is regulated — it's a waste of energy to transport nutrients if they're not needed. The cell makes transporters when it senses that it's low in supply of a specific nutrient. Because transporters are selective for the substrate they transport, they help build that supply back up, and when their work is done, they're no longer made.

Getting Around with Locomotion

Some microbial cells are stationary, but most of them have a means of getting around, called *locomotion*. Flagella are important for movement. They propel the cell forward or backward. Both prokaryotes and eukaryotes have them, but they're more complex in eukaryotes. In prokaryotic cells, flagella spin around and propel the cells very quickly; in eukaryotic cells, they move in a wave motion and propel the cells more slowly. The different kinds of flagellar arrangement on bacterial cells are shown in Figure 4-9.

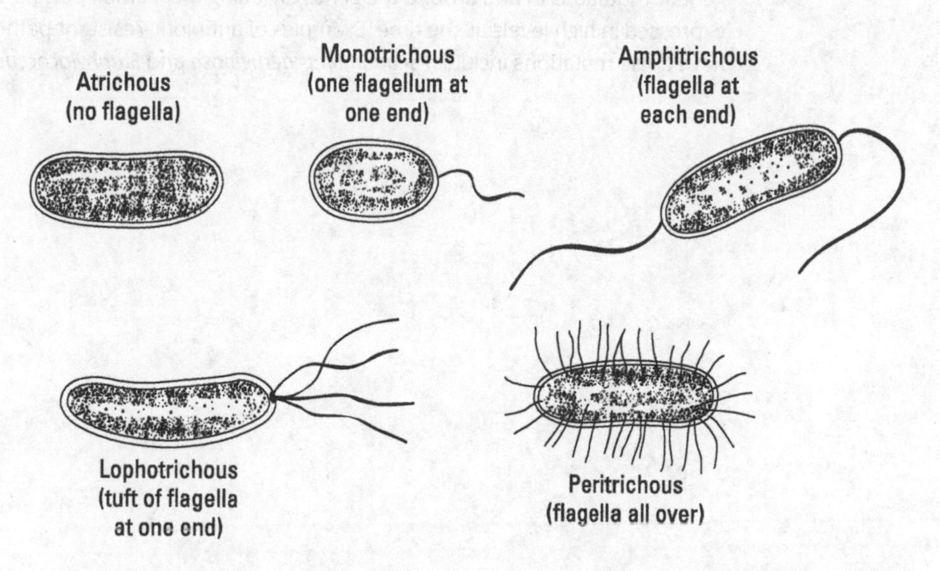

FIGURE 4-9: Flagellum structure and placement on bacterial cells.

Bacteria without flagella can also move around by a type of motion called *gliding*. In some bacteria, gliding is done by secreting slime that sticks to a surface and on which the cell can slide. Another method used is *pili*, which are thin, hairlike projections on the surface of bacterial cells, to twitch across a surface. Still others glide by some unknown method.

Microbes often move toward or away from something based on the chemical gradient nearby in a process called *taxes*. Both chemical gradients (oxygen or sugar) and photogradients (light) can induce taxes, which is also considered a type of microbial communication.

BAILING THEMSELVES OUT

Efflux pumps of human pathogens are of major concern in a clinical setting because they're able to pump antibiotics out of the cell, making the pathogen resistant to the drug. These pumps are referred to as multi-drug resistance (MDR) efflux pumps because they're typically able to export more than one type of antibiotic. This leaves the doctor with few options for treating an infection. Normally, MDR efflux pumps are expressed at low levels and can't pump out antibiotics fast enough to prevent them from doing damage. However, many pathogens that are antibiotic resistant contain genetic mutations in and around the genes encoding MDR efflux pumps so that they're expressed in high levels all the time. Examples of antibiotic-resistant pathogens that have these mutations include *Pseudomonas aeruginosa* and *Staphylococcus aureus.*

IN THIS CHAPTER

» Following the flow of energy

» Breaking down organic compounds for food

» Seeing how microbes synthesize what they need to survive

Chapter 5 Making Sense of Metabolism

Metabolism, in essence, is all the steps involved in consuming something and using the energy from it to make the building blocks you need to survive and thrive. For some microbes, this means living on things we might expect, like sugars; for other microbes, it means living on things we wouldn't normally consider food, like petroleum.

Microorganisms are great at breaking things down. In fact, the vastness of metabolic diversity in the microbial world isn't even completely known. Microbes are very frugal and, in the end, nothing is ever wasted. One microbe's trash is another microbe's treasure!

In this chapter, we cover the common thread of all life — energy! We explain how tiny microbes turn the wheels of metabolism by extracting energy from fuel sources and making the things they need to survive.

Converting with Enzymes

If you imagine the microbial cell as a small factory, then the enzymes are the robots doing all the work. Enzymes are special proteins that are very good at converting things from one form to another. They do this by kicking off the chemical reactions needed for the conversion. The kinds of enzymes a microbe makes determine what type of metabolism the microbe will use to harness energy and grow.

Chemical reactions are at the heart of all cellular processes. Although you can't see them happening, chemical reactions power all living things and many forces of nature. In science, these chemical reactions are written as equations with substances going into the reaction (the *substrates*) on one side of an arrow and the results coming out of the reaction (the *products*) on the other side of the arrow, like this:

Substrate 1 + Substrate 2 → Product

Not all chemical reactions have the same number of substrates and products as the example above, but they always have substrates going in and products coming out.

Chemical reactions are always accompanied by a change in energy. The substrates and products contain different amounts of energy. The difference between the amount of energy contained within the substrates and the amount of energy contained within the products determines whether energy needs to be put in or is released during the reaction. Reactions that consume energy are called *endergonic*; reactions that release energy are called *exergonic*. The energy released from an exergonic reaction can either be stored for later use or be used right away to power an endergonic reaction.

Biologically important chemical reactions are needed for every process in the cell, but by themselves, the substrates in the cell would react so slowly that life would grind to a halt. Catalysts keep chemical reactions moving along quickly.

A *catalyst* is a protein (or sometimes an RNA) molecule that speeds up the rate of a chemical reaction without interfering in the reaction in any other way.

In the cell, these catalysts are called *enzymes* and they work by bringing substrates into close proximity with one another and by bending them into the right shape so they can react with one another and form products. Enzymes are usually much bigger than the substrates and the products of the reaction they catalyze. Enzymes have a pocket specifically for the substrates and the products to fit inside of, called an *active site*. Lysozyme (shown in Figure 5-1) is much bigger than its substrate, peptidoglycan, which fits into its active site and is *cleaved* (split) into two sugars.

Enzymes are usually specific to one reaction or one group of similar reactions. Another interesting thing about enzymes and the reactions they catalyze is that most are reversible. That means that the products made in a reaction can recombine to re-form the substrates with the help of the same enzyme, as long as the energy change is reversed as well. If the reaction produces or requires a lot of energy, there is usually a different enzyme for each direction of the reaction.

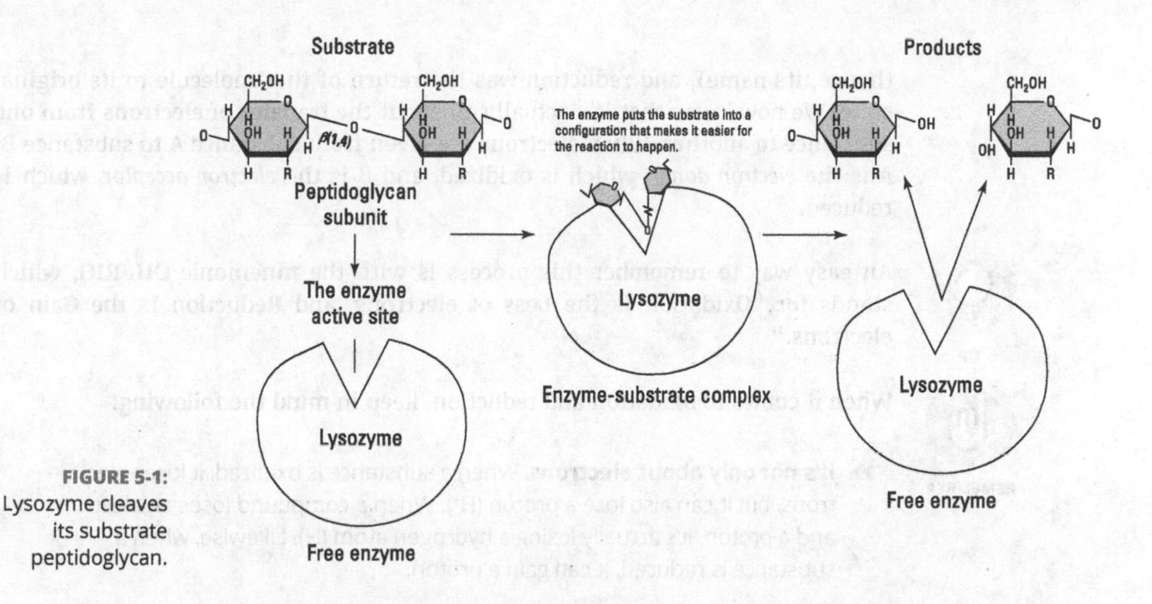

FIGURE 5-1: Lysozyme cleaves its substrate peptidoglycan.

Coenzymes are small molecules that associate with an enzyme and help in its function but aren't a substrate. NAD^+/NADH is an example of a coenzyme that can associate with many different enzymes and helps with electron transfer (see the "Donating and accepting electrons" section, later in this chapter). *Prosthetic groups*, on the other hand, are small molecules that bind permanently and are essential to the enzyme's activity.

In Charge of Energy: Oxidation and Reduction

Energy can't be created or destroyed, so it has to be passed around. Within the cell, energy is reused and recycled very efficiently. The same is true outside the cell, where energy is stored in everything, including leaves and rocks on the ground. The trick is getting energy out from where it's stored.

Another way of thinking about energy is to think about electrons, which are the negatively charged part of atoms. Electrons are essentially carriers of energy that are able to be conveniently passed from one molecule to another. So, for microbes, passing electrons around between molecules — called *oxidation* and *reduction* — is important for gathering energy.

Most of the energy generated for living comes from oxidation and reduction reactions. Oxidation used to be thought of as a molecule combining with oxygen (hence, its name), and reduction was the return of that molecule to its original state. We now know that it's actually all about the transfer of electrons from one substance to another. When electrons are given from substance A to substance B, A is the *electron donor,* which is oxidized, and B is the *electron acceptor,* which is reduced.

TIP

An easy way to remember this process is with the mnemonic **OIL RIG,** which stands for "**Oxidation Is the Loss of electrons, and Reduction Is the Gain of electrons.**"

REMEMBER

When it comes to oxidation and reduction, keep in mind the following:

- **It's not only about electrons.** When a substance is oxidized, it loses electrons, but it can also lose a proton (H^+). When a compound loses an electron and a proton, it's actually losing a hydrogen atom (H). Likewise, when a substance is reduced, it can gain a proton.
- **There's some give and take.** Most substances are flexible and can act as an electron donor in one situation and an electron acceptor in another situation. But often each substance has a preference for one or the other.
- **Donors and acceptors come in pairs.** If a substance is oxidized, the electrons lost don't just hang around in solution but are transferred to an electron acceptor. This is called a *redox reaction,* because all instances of oxidation or reduction happen in pairs.

Redox reactions happen continuously, as long as electron donors and electron acceptors are available, and the energy obtained can power other reactions in the microbial cell. Whether the energy is to be stored for the short term or the long term determines what kind of molecule will be used to transfer this energy.

Donating and accepting electrons

Substances have different tendencies to donate or accept electrons. When a really good donor meets a great acceptor, the chemical reaction releases a lot of energy. Oxygen (O_2) is the best electron acceptor and is used in many *aerobic reactions* (reactions with oxygen). Hydrogen gas (H_2) is a good electron donor. When O_2 and H_2 are combined, along with a catalyst, water (H_2O) is formed. This example of a redox reaction can be written like this:

$$H_2 + ½\, O_2 \rightarrow H_2O$$

REMEMBER

A redox reaction is one in which all instances of oxidation and reduction happen in pairs.

Notice that the reaction has to balance — the total number of atoms of hydrogen and oxygen on one side of the reaction are the same as the number in water on the other side. Balancing the redox reaction is crucial to all biochemical reactions in the cell and can create interesting challenges for microorganisms that live in *anaerobic environments* (environments without oxygen).

Oxygen and hydrogen are at either end of the spectrum of electron acceptors and donors, but there are many substances in between than can readily accept electrons in one situation and donate them in another.

Cells need a lot of primary electron donors and final electron acceptors on hand for the number of chemical reactions going on all the time. In reality, there aren't always unlimited amounts of electron donors and electron acceptors around. And this is where *electron carriers* come in. These convenient little molecules go about accepting electrons and protons (H^+), which they then donate to another reaction. Figure 5-2 shows NAD^+/NADH, which is an electron carrier that is reduced (to NADH) in one reaction after which it is oxidized (to NAD^+) in another reaction. Electron carriers like this one help increase the productivity of the cell by linking incompatible redox donors and acceptors; because they're recycled over and over, the cell only needs a small amount of each one.

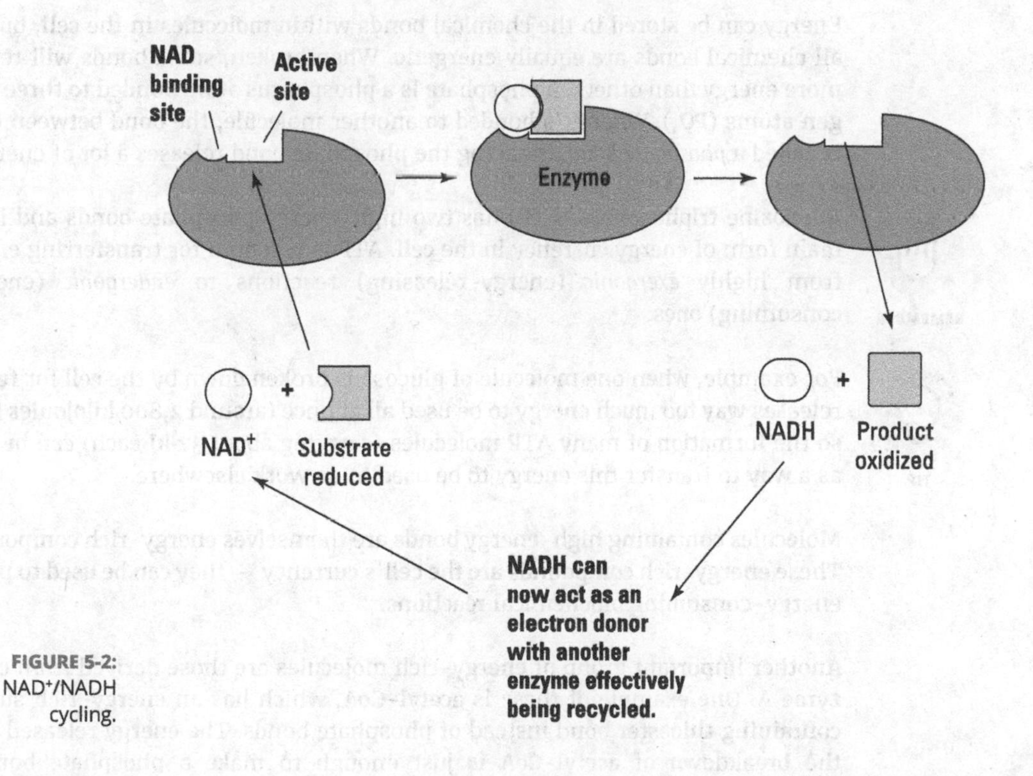

FIGURE 5-2: NAD^+/NADH cycling.

METABOLIC DIVERSITY: THE EXAMPLE OF SYNTROPHS

When it comes to the different kinds of redox reactions, the possibilities seem endless — and this enables a great amount of metabolic diversity in the microbial world. One example of an ingenious microbial lifestyle is *syntrophy,* where two species of bacteria cooperate to break down a substance. The key to syntrophy is the transfer of hydrogen gas (H_2) between the two species, where one member makes H_2 and the other consumes it. Because H_2 is such a good electron donor, it's always in huge demand, especially in anaerobic environments. The syntrophic bacteria *Syntrophomonas* takes advantage of this fact by partnering up with an H_2-consuming bacteria (of which there are many). *Syntrophomonas* lives by fatty-acid oxidation, which releases H_2, but the reaction is energetically unfavorable. It's only when an H_2-consuming bacteria creates a demand for H_2 that fatty-acid oxidation can go ahead. And because of this mutually beneficial arrangement, *Syntrophomonas* is able to survive.

Bargaining with energy-rich compounds

Energy can be stored in the chemical bonds within molecules in the cell, but not all chemical bonds are equally energetic. When broken, some bonds will release more energy than others. A phosphate is a phosphorus atom bonded to three oxygen atoms (PO_3). When it's bonded to another molecule, the bond between them is called a *phosphate bond.* Breaking the phosphate bond releases a lot of energy.

Adenosine triphosphate (ATP) has two high-energy phosphate bonds and is the main form of energy currency in the cell. ATP is essential for transferring energy from highly *exergonic* (energy-releasing) reactions to *endergonic* (energy-consuming) ones.

For example, when one molecule of glucose is broken down by the cell for fuel, it releases way too much energy to be used all at once (around 2,800 kilojoules [kJ]), so the formation of many ATP molecules (carrying about 32 kJ each) can be used as a way to transfer this energy to be used to do work elsewhere.

Molecules containing high-energy bonds are themselves energy-rich compounds. These energy-rich compounds are the cell's currency — they can be used to power energy-consuming biochemical reactions.

Another important group of energy-rich molecules are those derived from coenzyme A. One example of these is acetyl-CoA, which has an energy-rich sulfur-containing thioester bond instead of phosphate bonds. The energy released from the breakdown of acetyl-CoA is just enough to make a phosphate bond in ATP. Although used for all types of metabolism, these molecules are essential for microbes that rely on a type of anaerobic metabolism called *fermentation*, where food is broken down in the absence of oxygen.

Both ATP and acetyl-CoA are short-term storage molecules — they're usually used to power other reactions relatively quickly. For longer-term energy storage, microbes put their energy reserves elsewhere.

Storing energy for later

Long-term storage of energy within the cell is done by producing insoluble molecules called *polymers*. Prokaryotes (bacteria and archaea) make things like glycogen, poly-β-hydroxybutyrate, or polyhydroxyalkanoates, whereas eukaryotes make starch and fats (see Figure 5-3). The goal is to produce energy reserves that the cell can store; then, during leaner times, these polymers can be broken down to yield ATP.

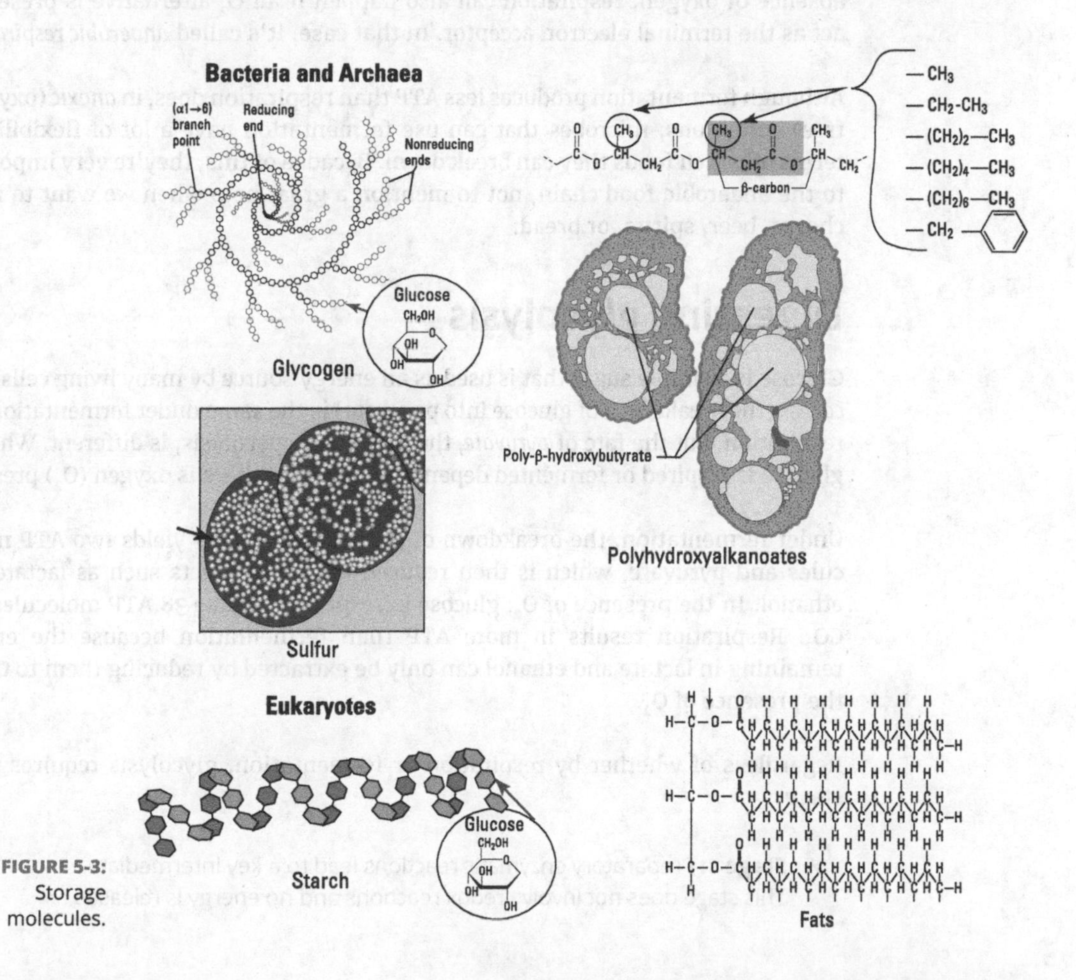

FIGURE 5-3: Storage molecules.

Breaking Down Catabolism

Catabolism (the breakdown of compounds for energy conservation) can happen in different ways for different types of microorganisms. In *chemoorganotrophs* (organisms that derive their energy from organic compounds), there are two forms of catabolic metabolism: fermentation and respiration.

Fermentation is a form of anaerobic catabolism (in the absence of O_2) where the organic substrate acts as both the electron donor and the electron acceptor. Here, ATP is made from energy-rich reaction intermediates through a process called *substrate-level phosphorylation.*

Respiration can occur in the presence or absence of oxygen. During aerobic respiration, O_2 acts as the terminal electron acceptor and ATP is formed at the expense of the *proton motive force* by a process called *oxidative phosphorylation.* In the absence of oxygen, respiration can also happen if an O_2 alternative is present to act as the terminal electron acceptor. In that case, it's called *anaerobic respiration.*

Although fermentation produces less ATP than respiration does, in *anoxic* (oxygen-free) conditions, microbes that can use fermentation have a lot of flexibility in terms of which foods they can break down. Because of this, they're very important to the anaerobic food chain, not to mention a great help when we want to make cheese, beer, spirits, or bread.

Digesting glycolysis

Glucose is a simple sugar that is used as an energy source by many living cells. *Glycolysis* (the breakdown of glucose into pyruvate) is the same under fermentation and respiration, but the fate of *pyruvate,* the product of glycolysis, is different. Whether glucose is respired or fermented depends on whether there is oxygen (O_2) present.

Under fermentation, the breakdown of glucose for energy yields two ATP molecules and pyruvate, which is then reduced to end products such as lactate and ethanol. In the presence of O_2, glucose is respired to make 38 ATP molecules and CO_2. Respiration results in more ATP than fermentation because the energy remaining in lactate and ethanol can only be extracted by reducing them to CO_2 in the presence of O_2.

Regardless of whether by respiration or fermentation, glycolysis requires three stages:

- **Stage 1:** Preparatory enzymatic reactions lead to a key intermediate. This stage does not involve redox reactions and no energy is released.

» **Stage 2:** This is when the redox reactions occur, producing ATP and pyruvate. At this point, the breakdown of glucose is done, but because the redox reactions aren't balanced, another stage is required.

» **Stage 3:** The redox reactions needed to balance the reactions in Stage 2 happen in Stage 3. In fermentation, this is the reduction of pyruvate to its fermentation products that are then excreted as waste and in respiration this is the reduction of pyruvate to CO_2.

Stepping along with respiration and electron carriers

The breakdown of compounds by respiration releases much more energy than does the breakdown of the same compounds by fermentation. This is because the complete reduction of the products of fermentation isn't possible without oxygen or oxygen substitutes to act as terminal electron acceptors.

The star of this phenomenon is the *electron transport chain*, which involves several electron acceptors positioned within a membrane in order of reducing power so that the weakest electron acceptors are at one end of the chain and the strongest electron acceptors are at the other end. It's the specific orientation of electron carriers in the membrane that creates the proton motive force and links ATP synthesis with it.

TECHNICAL STUFF

The electron transport chain reduces organic compounds to CO_2 and conserves some of the energy as electrons are transferred from the carbon substrate (glucose), through several redox reactions to the terminal electron acceptor (O_2).

Some membrane-bound carriers, like the quinones, are nonprotein molecules, but most are oxidation-reduction enzymes and some of these have prosthetic groups that participate in redox reactions. Prosthetic groups are small molecules that are permanently bound to an enzyme and are important to its activity. Before we can talk about the proton motive force and how ATP is made during respiration, we want to list some of the many different electron carriers that take part in the electron transport chain:

» **NADH dehydrogenases:** These are proteins that accept an electron (e^-) and a proton (H^+) from NADH, oxidizing it to NAD^+ and passing them onto a flavoprotein.

» **Flavoproteins:** These are made up of a protein bound to a prosthetic group called *flavin*, which comes from the vitamin riboflavin. The flavin group accepts two e^- and two H^+ but only donates two e^- when oxidized.

- **Cytochromes:** These proteins contain a heme prosthetic group with an iron atom in its center that gains or loses a single e⁻. There are different classes of cytochromes based on the type of heme they contain and labeled with a different letter (for example, cytochrome a). When the same class of cytochrome is slightly different in two organisms, each gets a number attached to the name (for example, cytochromes a_1 and a_2).
- **Iron-sulfur proteins:** These bind iron but without a heme group. Instead, they have clusters of sulfur and iron atoms arranged in the center of the protein. They only accept e⁻ and have a range of reducing power depending on the number of iron and sulfur atoms present.
- **Quinones:** These hydrophobic molecules (not proteins) are free to move around the membrane. They accept two e⁻ and two H^+ and usually act as a link between iron-sulfur proteins and cytochromes.

In Figure 5-4, you can see examples of some of these compounds and how they're physically sitting within the membrane in the right order for electrons to flow from the most electronegative to the most electropositive.

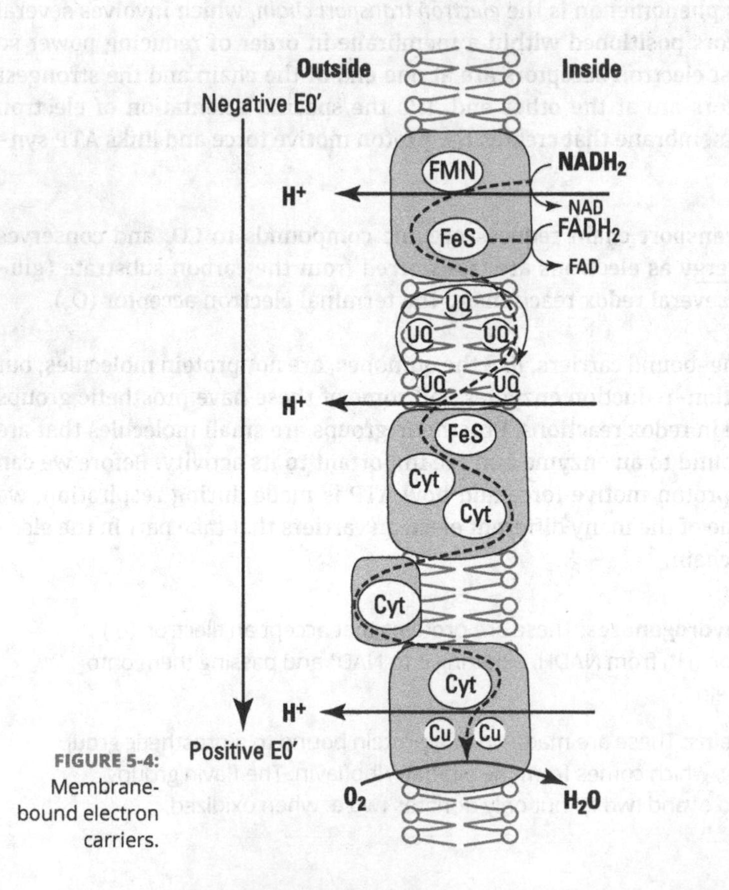

FIGURE 5-4: Membrane-bound electron carriers.

Moving with the proton motive force

The proton motive force occurs when the cell membrane becomes energized due to electron transport reactions by the electron carriers embedded in it. Basically, this causes the cell to act like a tiny battery. Its energy can either be used right away to do work, like power flagella, or be stored for later in ATP. ATP synthesis is linked to the proton motive force through *oxidative phosphorylation*, where a phosphate group is added to ADP.

Although several steps are involved in creating an energized cell membrane, there's one simple concept behind this phenomenon: the separation of positive protons (H^+) on the outside of the membrane and negative hydroxide ions (OH^-) on the inside of the membrane. The fact that they're charged makes it impossible for these ions to cross the membrane on their own. Trapping the ions on either side of the membrane creates two things, which together make the *proton motive force:* a pH and a charge difference. A difference in charge on the inside and the outside of a cell is called an *electrochemical potential* and is a huge source of energy.

TIP

Imagine you're standing on the roof of a tall building, holding an orange. If you let the orange drop over the side of the building, by the time it reaches the ground it will have gained so much speed that it will hit the ground with great force and smash. Because of the large difference in height from where it was dropped and where it landed, there's a great amount of energy. It's sort of the same thing with electrochemical potential, where the great difference in charge creates a lot of potential energy.

How this happens is that during electron transport, H^+ are pushed to the outside of the membrane. The H^+ come from both NADH and the dissociation of H_2O into H^+ and OH^-. As you'll see in a minute, it's a bit more complicated, but the overall result is accumulation of a positive charge outside the cell and a negative charge within.

In the example in Figure 5-4, the proton motive force is created by a series of complexes within the cell membrane. These complexes are made up of the e^- carriers mentioned earlier, the exact combination and number of which differ between organisms. The proton motive force has two possible beginnings:

- **Complex I:** One way the proton motive force begins is with the donation of H^+ from NADH to flavin mononucleotide (FMN) to make $FMNH_2$. $4H^+$ move to the outside of the cell when $FMNH_2$ donates $2e^-$ to the Fe/S proteins in complex I.
- **Complex II:** The other way the proton motive force begins is through complex II, where FADH feeds e^- and H^+ from the oxidation of *succinate,* a product of the citric acid cycle, to the quinones. Complex II is less efficient than complex I.

Once electrons enter the cycle through Complex I or II they move on to Complex III and eventually IV:

- **Complex III:** Quinones are reduced in the *Q-cycle* (a series of oxidation and reduction reactions of the coenzyme Q that result in the release of additional H+ to the outside of the membrane). Then electrons are passed one at a time from the Q-cycle to complex III, which contains the heme-containing proteins (specifically, cytochrome bc_1) and an FeS protein.
- **Complex IV:** Cytochrome bc_1 transfers e^- to cytochrome c, which then passes them to cytochromes a and a_3 in complex IV. This is the end of the line where O_2 is reduced to H_2O.

At almost every step, H+ are pumped to the outer surface of the membrane, increasing the strength of the proton motive force.

REMEMBER

Electron transport chains differ among organisms, but they always have three things in common:

- Electron carriers arranged in order of increasing reducing power
- Alternating $H^+ + e^-$ and e^--only carriers
- Generation of a proton motive force

The production of adenosine triphosphate (ATP) from aerobic respiration is called *oxidative phosphorylation*, and it's carried out by a complex of proteins called *ATP synthase*. This complex is made up of two subunits, F_1 and F_0, each of which is actually a rotary motor. Protons (H^+), driven back through the membrane from the outside by the proton motive force, go through F_0 and place pressure (or torque) on F_1. A molecule of adenosine diphosphate (ADP, with only two phosphate groups) along with a free phosphate group (P_i) bind to F_1, and when the torque is released, energy is free to power the formation of ATP through the bonding of ADP and P_i. Like many enzymes, ATP synthase is reversible so that it can contribute to the proton motive force instead of weakening it. So, even organisms that don't use oxidative phosphorylation, like anaerobic fermenters, have ATP synthases so that they can still create a proton motive force to drive things like flagellar movement or ion transport.

Turning the citric acid cycle

The citric acid cycle is a key pathway in catabolism in all cells. As mentioned earlier, all the steps in the respiration of glucose up to the formation of pyruvate are the same as in fermentation. After this, the citric acid cycle is responsible for the oxidation of pyruvate to CO_2. For each pyruvate used, three molecules of CO_2 are made (see Figure 5-5).

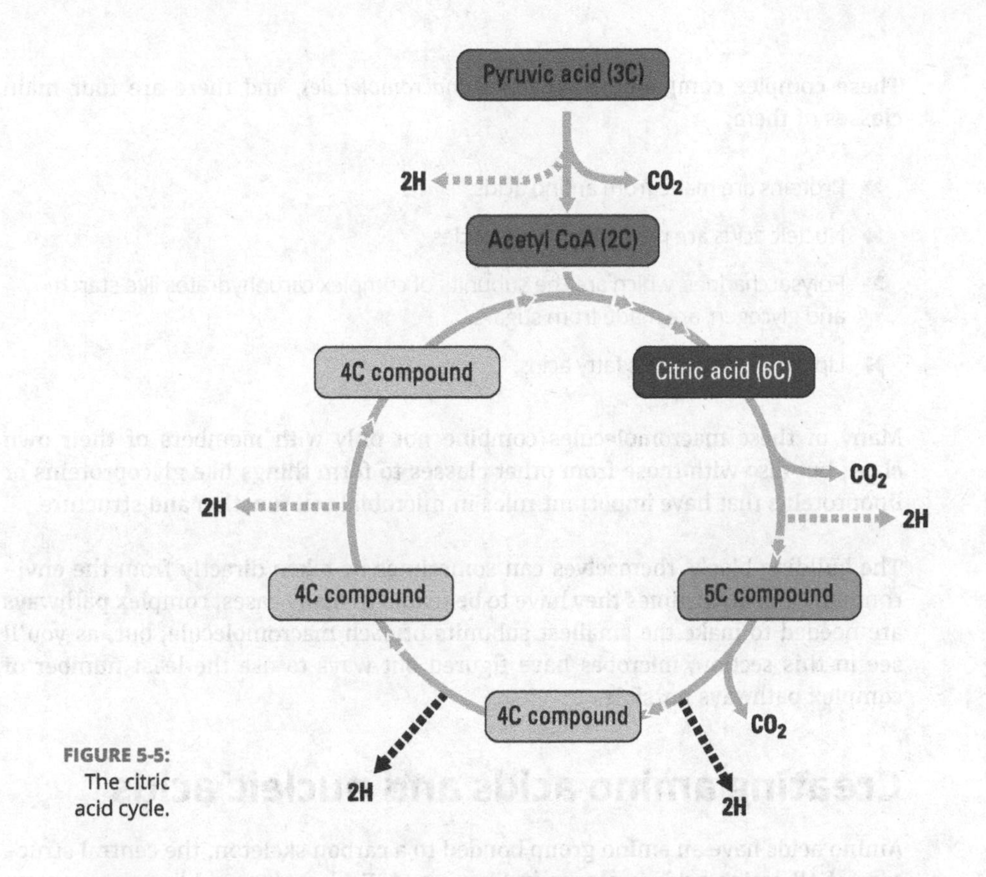

FIGURE 5-5: The citric acid cycle.

REMEMBER

In respiration, glucose is completely oxidized to CO_2 and much more energy is made than in glucose fermentation. Glucose is fermented to alcohol or lactic acid products and two ATP are made. On the other hand, glucose is respired to CO_2 and H_2O and 38 ATP are made.

Stacking Up with Anabolism

We've covered the many ways that a cell can use to generate energy by breaking down food, like glucose, and obtaining reducing power to make ATP. Now let's look at what the cell does with this energy.

The cell needs many compounds for life, including enzymes to perform its many functions, structural molecules to give it shape, as well as nucleic acids to store its genetic material. If it can, the cell will obtain some of these things from its environment, but when it has to make them itself, it uses *anabolism*.Anabolism is essentially making the complex things needed from smaller building blocks.

These complex compounds are called *macromolecules*, and there are four main classes of them:

- Proteins are made from amino acids.
- Nucleic acids are made from nucleotides.
- Polysaccharides, which are the subunits of complex carbohydrates like starch and glycogen, are made from sugars.
- Lipids are made from fatty acids.

Many of these macromolecules combine not only with members of their own class, but also with those from other classes to form things like glycoproteins or lipoproteins that have important roles in microbial cell function and structure.

The building blocks themselves can sometimes be taken directly from the environment, but other times they have to be made. In many cases, complex pathways are needed to make the smallest subunits of each macromolecule, but, as you'll see in this section, microbes have figured out ways to use the least number of complex pathways possible.

Creating amino acids and nucleic acids

Amino acids have an amino group bonded to a carbon skeleton, the central structure of all amino acids is shown in Figure 5-6. Each amino acid has one or more special side groups or chains (R) that give it a specific structure or function. Making amino acids from scratch is very expensive in terms of energy, so microbes try their best to get them from their environment. When they can't get them from outside, however, they use a kind of template method to reduce the amount of energy spent on different biosynthesis pathways.

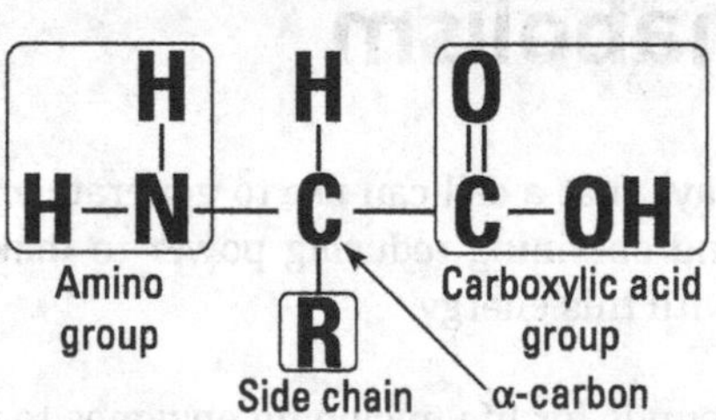

FIGURE 5-6: The common structure of all amino acids.

Amino acid biosynthesis starts with a carbon skeleton, made from intermediates of the citric acid cycle or glycolysis. A few different skeletons can be used in the production of several different amino acids, and what you get are classes of amino acids where the members of each class have a similar structure.

The next step is attachment of an amino group to the carbon structure. The amino group can be taken from ammonia (NH_3) in the environment and used to make glutamate or glutamine. Depending on the availability of ammonia, the enzymes used are either

- **Glutamate dehydrogenase:** Used when NH_3 is plentiful. It's the less energy expensive of the two.
- **Glutamate synthase:** Used in NH_3-poor conditions. It requires more energy than glutamate dehydrogenase.

Here's the second frugal thing that microbes do: After glutamine and glutamate are made, the amino group can be used in all the other 20 amino acids.

Nucleotides are the subunits of nucleic acids like ribonucleic acid (RNA) and deoxyribonucleic acid (DNA), and like amino acids they're expensive and time consuming to make. In fact, nucleotides are even more cumbersome to make because atoms of carbon and nitrogen have to be added one at a time! When it can't be avoided, microbes make two different nucleotide templates (one for purines and one for pyrimidines) that they modify to make other variants. This saves the expense of having a bunch of complex pathways for nucleotide biosynthesis at work in the cell.

The two classes of nucleotides that arise are the purines and the pyrimidines. The first key purine made is inosinic acid, from all kinds of carbon and nitrogen sources. Once made, it's modified to produce the main purines adenine and guanine. The first key pyrimidine made is uridylate, which is then modified into thyamine, cytosine, and uracil.

The structure of complete nucleotides is shown in Figure 5-7. They have a purine or pyrimidine ring attached to three phosphates (PO_3^-) and a ribose sugar backbone. After they're formed, they can either be used directly in RNA or the ribose can be modified, by reduction to deoxyribose, and be used in DNA.

Making sugars and polysaccharides

In the previous section, we talk a lot about the breakdown of sugars like glucose for energy, but cells can also use sugars for all sorts of other things. For example, the backbone of peptidoglycan, a major component of the bacterial cell wall, is made of sugars. *Hexoses* are six-carbon sugars like glucose, and *pentoses* are five-carbon sugars like ribose (see Figure 5-8). When hexoses need to be made, they're synthesized with *gluconeogenesis* using intermediates from glycolysis and the citric acid cycle. Pentoses can be made by removal of one carbon from a hexose.

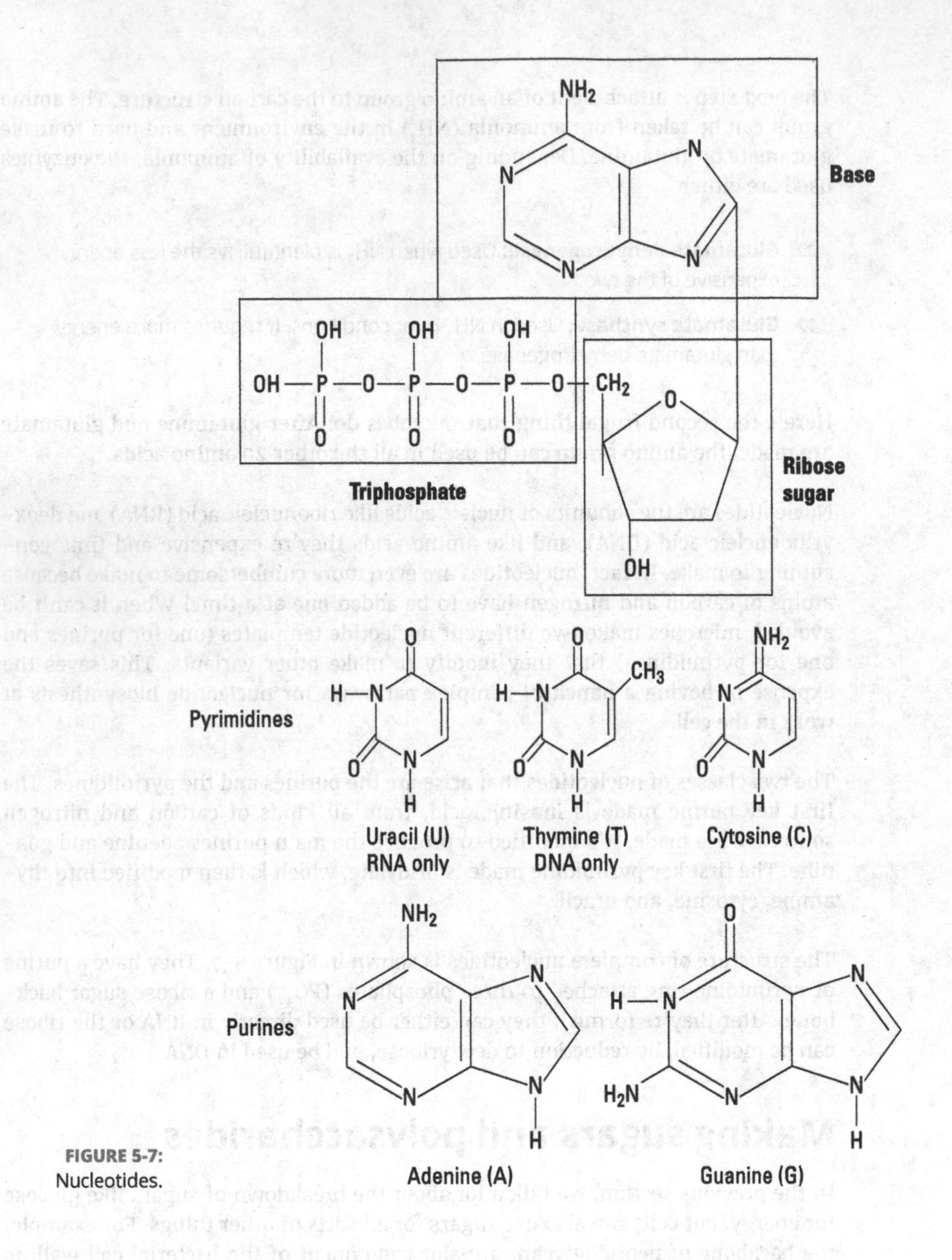

FIGURE 5-7: Nucleotides.

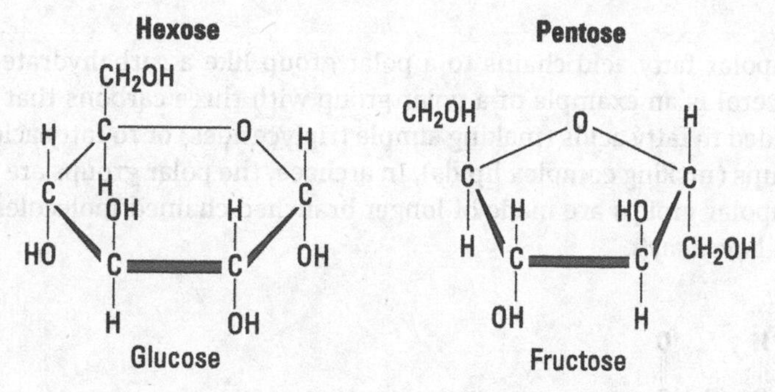

FIGURE 5-8: An example of a hexose and a pentose sugar.

Polysaccharides are molecules made up of many sugar subunits. In prokaryotes, they're synthesized from activated glucose. There are two forms of activated glucose; each is used to make different types of polysaccharides:

- **Uridine diphosphoglucose (UDPG)** is the precursor for the backbone of the cell wall component peptidoglycan, a precursor in parts of the Gram-negative outer membrane and a precursor of the storage molecule glycogen.
- **Adenosine diphosphoglucose (ADPG)** is the precursor of the storage molecule glycogen.

The process of making a polysaccharide involves adding an activated glucose to the growing chain of a polysaccharide.

Putting together fatty acids and lipids

Bacterial cells make a wide variety of lipids from subunits of fatty acids that have some important functions in the cell. Lipids are the main part of membranes and can be used as stores for longer-term storage of energy. Fatty acids are built by adding two carbons at a time to a growing fatty acid chain. The enzyme acyl carrier protein (ACP) is essential to this process because it holds onto the growing chain until all the carbons have been added.

Fatty acids can either be *straight-chained* or *branched;* in addition, they can either be *saturated* or *unsaturated.* Straight-chained saturated fatty acids are very straight, while unsaturated sites add kinks and bends in the chain (see Figure 5-9). Straight chains of lipids can be packed more densely next to one another (as in a membrane, for instance). These differences determine how a membrane will function (for example, under different temperatures).

Membrane lipids in particular need to have a polar side and a nonpolar side, meaning one side with a charge and one side without. This allows them to interact with other charged molecules (like water) on one side and to interact with other uncharged molecules (like fats) on the other. This is done by attaching the nonpolar fatty acid chains to a polar group like a carbohydrate or a phosphate. Glycerol is an example of a polar group with three carbons that can either all be bonded to fatty acids (making simple triglycerides) or to fatty acids and other side groups (making complex lipids). In archaea, the polar groups are the same but the nonpolar groups are made of longer branched chained molecules called *phytanyls* and *biphytanyls*.

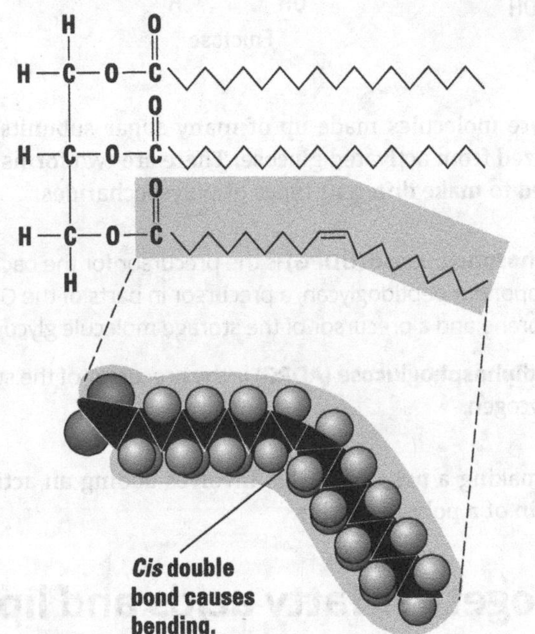

Cis double bond causes bending.

FIGURE 5-9: An example of bending in a fatty acid chain.

POLYSACCHARIDES GIVE BACTERIA AN EDGE

In addition to all the other great uses for polysaccharides, some bacteria use sugars to make compounds that they use to compete and survive. Two examples, in particular, are important because pathogenic bacteria use them to get an edge on us:

- *Streptococcus pneumoniae* uses a thick polysaccharide capsule, that surrounds the entire bacterial cell to hide from the human immune system.
- *Pseudomonas aeruginosa* excretes a sticky substance made of sugars around itself, forming what is called a *biofilm*, so that it can attach to the lungs of immunocompromized people. Once inside the biofilm, *Pseudomonas* is very hard to treat with antibiotics that can't get through the dense polysaccharide matrix.

IN THIS CHAPTER

» Seeing how genetic material is organized

» Finding out about transcription and translation

» Fine-tuning the amount and activity of proteins

» Keeping track of changes to the genome

Chapter 6

Getting the Gist of Microbial Genetics

The genetic material of the cell includes all its DNA, which acts as a copy of the instructions needed to make proteins. Because proteins are the machinery that perform all the functions in the cell — including metabolizing and providing structure — DNA is the central blueprint for how a cell is shaped and behaves.

In most other chapters of this book, we consider bacteria and archaea (often called prokaryotes) together, mostly because they share many aspects of cell structure, metabolism, and overall lifestyle. In those cases, it makes sense to talk about the differences between prokaryotes and eukaryotes. As discussed in Chapter 8, however, archaea and eukaryotes are more closely related than archaea and bacteria. It's no wonder then that the organization of their genetic material, as well as the mechanisms used to make protein, are less similar to those in bacteria than they are to those in eukaryotes. Viruses, on the other hand, are so different from other microbes that we save a discussion of viral genetics for Chapter 14.

Organizing Genetic Material

In all cells, the instructions within the DNA are first copied into RNA, which is then used to make protein. The central dogma of biology is then a flow of information from DNA to protein, through RNA, like this:

DNA → RNA → Protein

Instructions for individual proteins exist as a string of DNA called a *gene*, and all the genes in an organism are collectively called its *genome*. Within its genome, each organism has many more genes that it will use at any one time, because it will carry genes for proteins that it will need if or when conditions change. The genome is the genetic potential of an organism. All the proteins and products that the organism has made at the moment, which allow it to interact with its environment in a specific way, is called its *phenotype*.

In this section, we fill you in on the processes of microbial genetics mainly for bacteria, with notes on how things differ in archaea and eukaryotic microorganisms.

DNA: The recipe for life

The way that DNA encodes the instructions for proteins is through a set of four molecules called *bases*, each of which represents a letter of the genetic code (A = adenine, C = cytosine, G = guanine, and T = thymine). The bases are made of carbon and nitrogen rings and are bound to a sugar and a phosphate to form a nucleotide. The nucleotides are connected together to form a long chain with the bases pointing out. Because the nitrogenous bases can interact with each other — A binding with T and C binding with G — two such chains placed opposite to each other form the ladderlike structure of DNA, with paired bases making the rungs of the ladder (see Figure 6-1).

REMEMBER

Nucleotide bases will always pair in the same way, so each strand of DNA has the same sequence when read in the opposite direction to one another. The fact that each of the two DNA strands has the same sequence is called *complementarity*; it's essential to making sure that all cells get the same instructions during DNA replication and cell division.

Covalent bonds attach the subunits of the backbone together, whereas hydrogen bonds hold the paired bases together. Because these hydrogen bonds are much weaker than the rest of the bonds, the bases can be pulled apart, allowing things like DNA replication or RNA synthesis to occur.

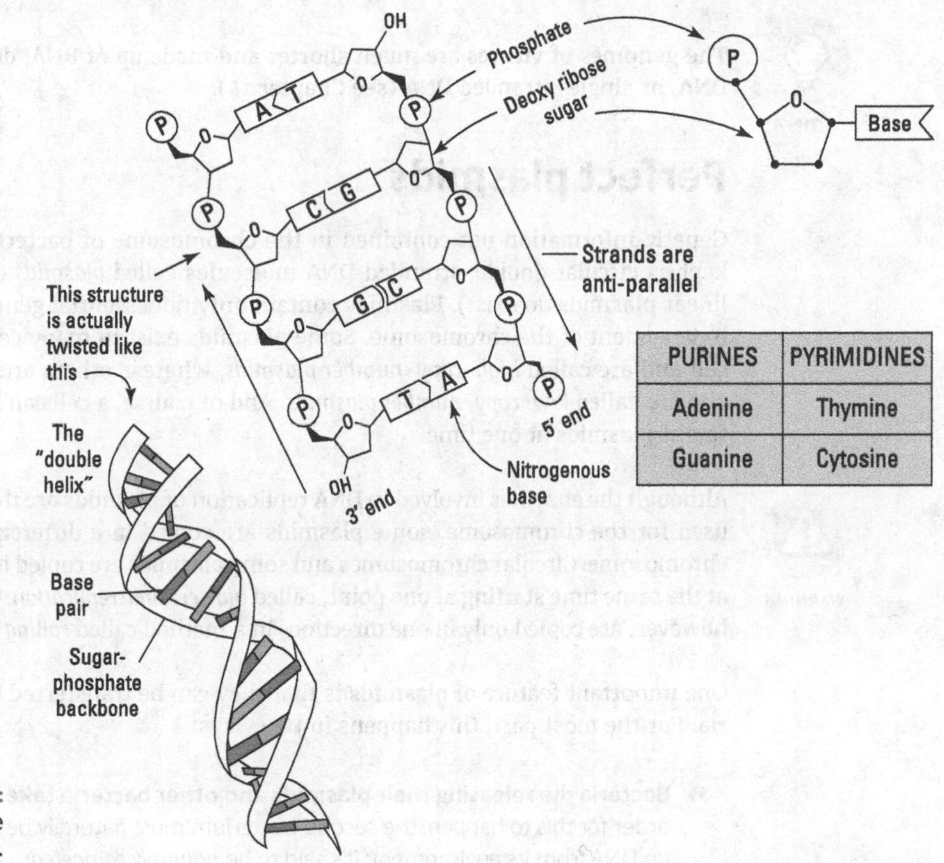

PURINES	PYRIMIDINES
Adenine	Thymine
Guanine	Cytosine

FIGURE 6-1: The structure of DNA.

The genomes of bacteria and archaea are, for the most part, arranged as a single circular chromosome and some extra-chromosomal genetic material, called *plasmids.* The chromosome contains all the essential genes required for life, whereas plasmids contain useful but not strictly essential genes. Eukaryotic genomes are usually contained in multiple linear chromosomes, although they can also have plasmids.

In both cases, the types of genes in the genome include

- Biosynthesis and metabolism genes
- Ribosomal RNA genes and transfer RNA genes
- DNA replication and repair genes

A bacterial genome is twisted up on itself to compactly fit inside a bacterial cell. The DNA for the genome of a eukaryote is wound around proteins called *histones* that help compact it without the DNA strand getting tangled. Archaea have a single circular chromosomelike bacteria that is wound with histones like eukaryotes.

The genomes of viruses are much shorter and made up of RNA, double-stranded DNA, or single-stranded DNA (see Chapter 14).

Perfect plasmids

Genetic information not contained in the chromosome of bacteria or archaea is kept as circular double-stranded DNA molecules called *plasmids* (although some linear plasmids do exist). Plasmids contain only nonessential genes and replicate independent of the chromosome. Some plasmids exist in many copies inside one cell and are called *high-copy-number plasmids*, whereas others are less numerous and are called *low-copy-number plasmids*. And of course, a cell can have many different plasmids at one time.

Although the enzymes involved in DNA replication of plasmids are the same as those used for the chromosome, some plasmids are copied in a different way than the chromosome. Circular chromosomes and some plasmids are copied in two directions at the same time starting at one point, called *bidirectional replication*. Other plasmids, however, are copied only in one direction, in a method called *rolling circle replication*.

One important feature of plasmids is that they can be transferred between bacteria. For the most part, this happens in two ways:

- **Bacteria die releasing their plasmids and other bacteria take them up.** In order for this to happen, the second bacterium must naturally be able to take up DNA from its environment; it's said to be *naturally competent*.
- **Plasmids, or other genetic material, are actively transferred from one bacterium to another by a process called *conjugation*.** Only some plasmids have the genes necessary to induce the transfer of genetic material between bacterial cells via conjugation; they're called *conjugative plasmids*. Conjugative plasmids can facilitate the transfer of themselves, other plasmids, and even chromosomal DNA between bacteria.

Aside from conjugative plasmids, the other main types of plasmids are

- **Resistance plasmids:** Carry genes for resistance to antibiotics, heavy metals, and other cellular defenses.
- **Virulence plasmids:** Carry virulence factors.
- **Colicin plasmids:** Carry bacteriocins used to inhibit or kill other bacteria. Bacteriocins have a narrower range than antibiotics, so they're specifically targeted to particular bacteria.

Plasmids are a useful tool for biotechnology, as explained in Chapter 16.

Doubling down with DNA replication

Before genes can be passed from parent cells to their progeny, a copy of the genome has to be made in a process called *replication*. For circular chromosomes, like those in bacteria and archaea, replication begins at the origin of replication and proceeds in two directions away from that point, simultaneously.

The steps involved in DNA replication, shown in Figure 6-2, must happen in a precise order:

1. Supercoiled double-stranded DNA is relaxed by an enzyme called *topoisomerase* (or *gyrase*) and then unwound by an enzyme called *helicase,* which opens up the two strands in one area at a time.
2. Nucleotides matching the bases exposed by the unwinding base pair with their match. The enzyme *DNA polymerase* then joins these bases together by catalyzing the formation of a bond between the phosphate of one nucleotide and the sugar (deoxyribose) of an adjacent nucleotide. This enzyme has a proofreading function, correcting any mistakenly added nucleotides.
3. The enzymes move farther along, unwinding the next section of DNA so that more nucleotides can join the growing chain of the new DNA strand.

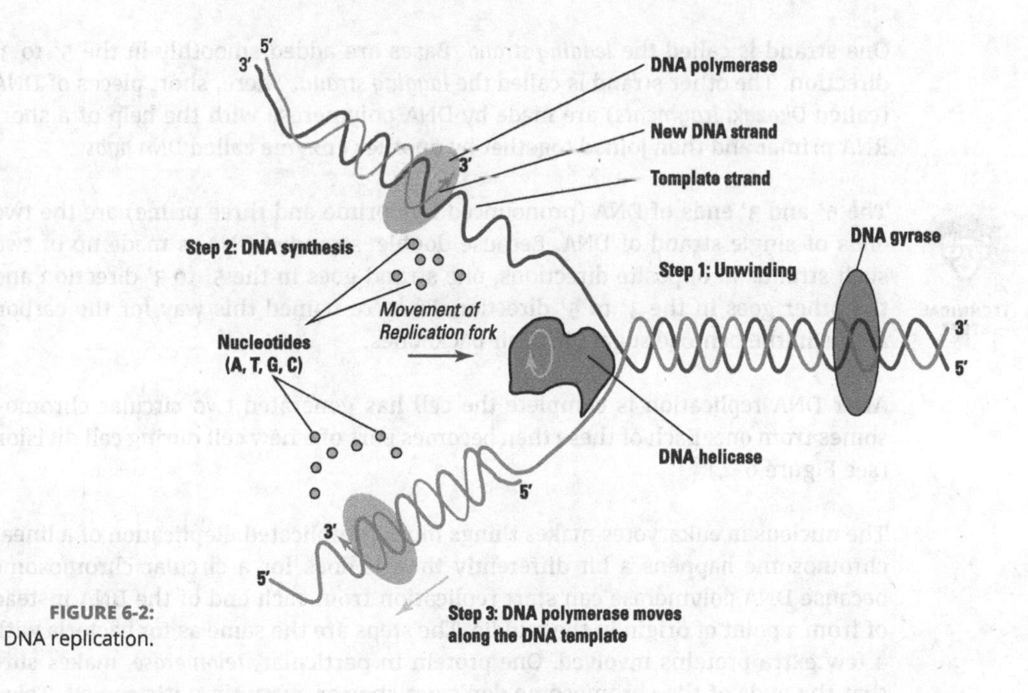

FIGURE 6-2: DNA replication.

The site where all this is happening is called the *replication fork.* Because each strand of a double-stranded DNA molecule gets incorporated into one of the two final copies of new DNA molecules, the process is called *semi-conservative replication.*

DNA polymerase can only add bases in the 5' to 3' direction, so replication proceeds differently on the two strands of DNA in the replication fork, as shown in Figure 6-3.

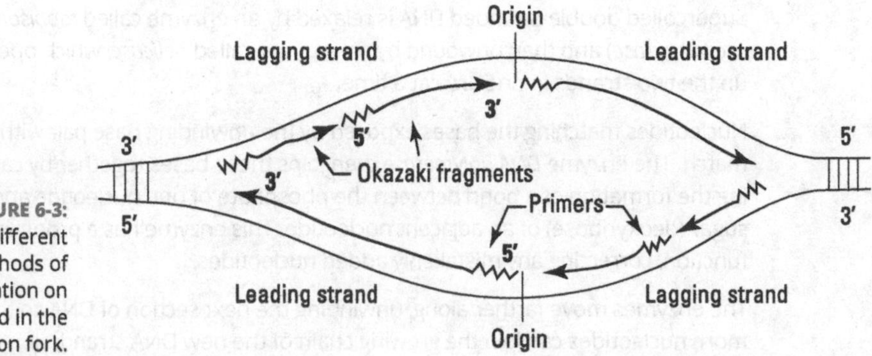

FIGURE 6-3: Different methods of replication on each strand in the replication fork.

One strand is called the *leading strand.* Bases are added smoothly in the 5' to 3' direction. The other strand is called the *lagging strand.* There, short pieces of DNA (called *Okazaki fragments*) are made by DNA polymerase with the help of a short RNA primer and then joined together by another enzyme called *DNA ligase.*

TECHNICAL STUFF

The 5' and 3' ends of DNA (pronounced five prime and three prime) are the two ends of single strand of DNA. Because double-stranded DNA is made up of two such strands in opposite directions, one strand goes in the 5' to 3' direction and the other goes in the 3' to 5' direction. They're named this way for the carbon atoms in the pentose sugars in their backbones.

After DNA replication is complete the cell has generated two circular chromosomes from one. Each of these then becomes part of a new cell during cell division (see Figure 6-4).

The nucleus in eukaryotes makes things more complicated. Replication of a linear chromosome happens a bit differently than it does for a circular chromosome because DNA polymerase can start replication from each end of the DNA instead of from a point of origin in the middle. The steps are the same as for bacteria with a few extra proteins involved. One protein in particular, *telomerase,* makes sure that the ends of the chromosome don't get shorter every time it's copied. Telomerase extends the ends of the chromosome by adding repeated sequences; these repeated sequences are called *telomeres.*

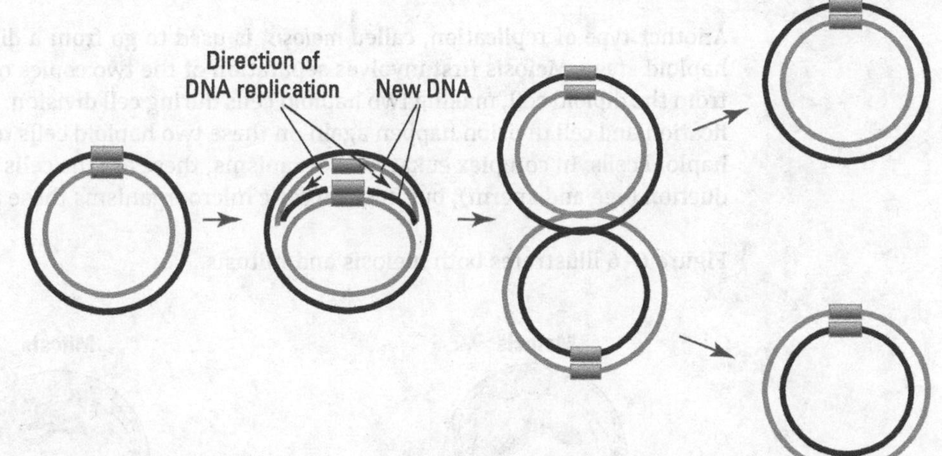

FIGURE 6-4: Replication of a circular chromosome.

After replication but before cell division, the chromosomes become very condensed into a form called *heterochromatin,* which can be seen with a microscope. Next, the nucleus disassembles so that copies of the chromosomes can be separated in two directions. Another important feature of eukaryotic chromosomes are the *centromeres,* areas of the chromosome where spindle fibers attach during cell division. Neither telomeres nor centromeres contain genes; they're structural parts of the chromosome only (see Figure 6-5). After cell division has occurred, two nuclei form around the chromosomes in each daughter cell.

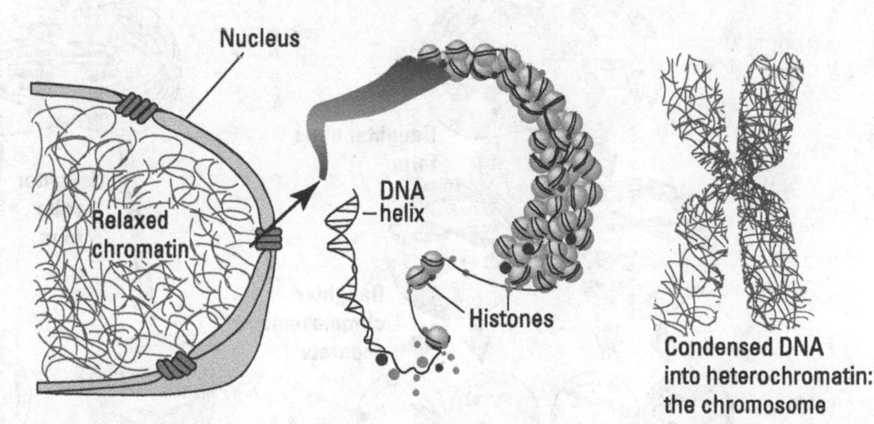

FIGURE 6-5: Eukaryotic chromosomes.

Eukaryotic organisms can have one or two copies of their chromosomes within the cell. When carrying two copies, the organism is said to be *diploid;* when carrying one copy, the organism is said to be *haploid.* When eukaryotic cells replicate their DNA prior to cell division it's called *mitosis.* Both haploid and diploid organisms undergo mitosis and cell division, during which they make sure to have the same number of chromosomes in each resulting cell that they started with.

Another type of replication, called *meiosis*, is used to go from a diploid stage to a haploid stage. Meiosis first involves separation of the two copies of chromosomes from the diploid cell, making two haploid cells during cell division. Next, DNA replication and cell division happen again on these two haploid cells to give four final haploid cells. In complex eukaryotic organisms, these are the cells used for reproduction (egg and sperm), but in eukaryotic microorganisms these are the spores.

Figure 6-6 illustrates both meiosis and mitosis.

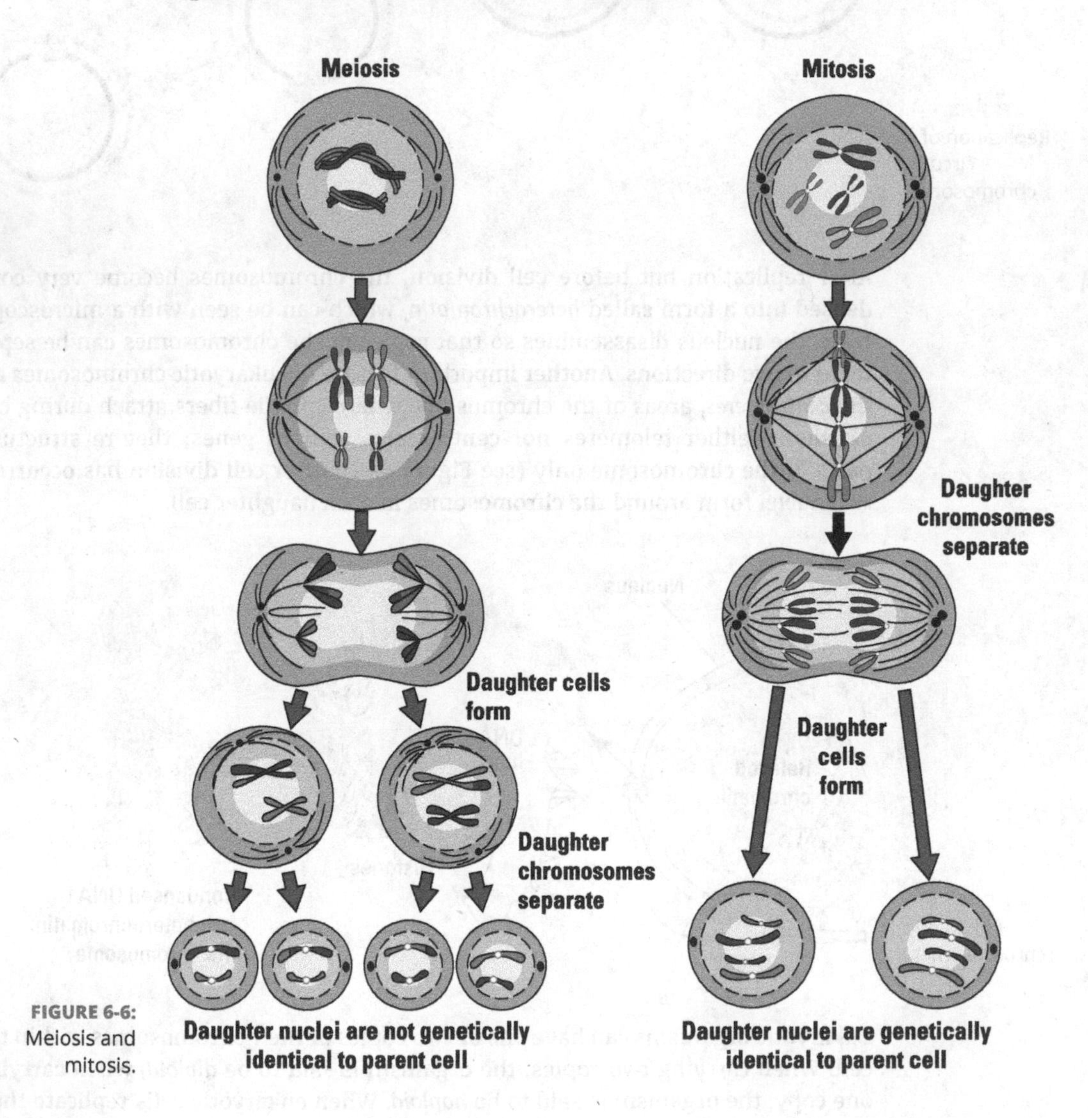

FIGURE 6-6: Meiosis and mitosis.

Assembling the Cellular Machinery

In order for the information carried within a cell's DNA to be useful, it must be turned first into ribonucleic acid (RNA) through transcription and then into protein through translation. Messenger RNA (mRNA) is used to ferry the message for a gene to the ribosome where it is made into protein. Because mRNA doesn't last forever in the cell, these short messages are a way of only making the protein that is needed at that time.

Making messenger RNA

RNA is made through transcription, where an enzyme called RNA polymerase transcribes the DNA sequence into a complementary version with the use of free RNA nucleotides. Three of the bases (adenine, guanine, and cytosine) are the same as in DNA, but the fourth (thymine) is replaced by uracil in RNA. Also, the backbone is slightly different, containing a ribose instead of a deoxyribose sugar.

Transcription begins at the promoter, which is a location in the DNA that signals the beginning of a gene. In bacteria, RNA polymerase binds to the promoter and transcription proceeds in the 5' to 3' direction until it encounters the terminator sequence, where it stops. The newly made mRNA molecule is then bound by the ribosome to begin translation.

TIP

In fact, protein synthesis can begin immediately in bacteria once part of the mRNA is available for binding. This is because in bacteria transcription and translation, both happen in the cytoplasm and also because the mRNA doesn't have to be processed prior to translation. Ribosomes bind to and start translation of mRNA before it's even finished being transcribed.

REMEMBER

In bacteria and archaea, many genes are organized into groups that are transcribed together, called *operons*. The operon has one promoter, which signals the start of transcription, followed by several genes next to one another and ending with a transcriptional termination sequence. Transcription of an operon results in an mRNA, called *polycistronic mRNA*, which codes for several proteins, each of which are translated in sequence. Grouping genes into operons is a way for the cell to coordinate the expression of genes that will be needed at the same time.

Transcription in eukaryotes is similar to transcription in bacteria except that after the mRNA is made, it has to be processed and then transported out of the nucleus to the cytoplasm for translation (see Figure 6-7).

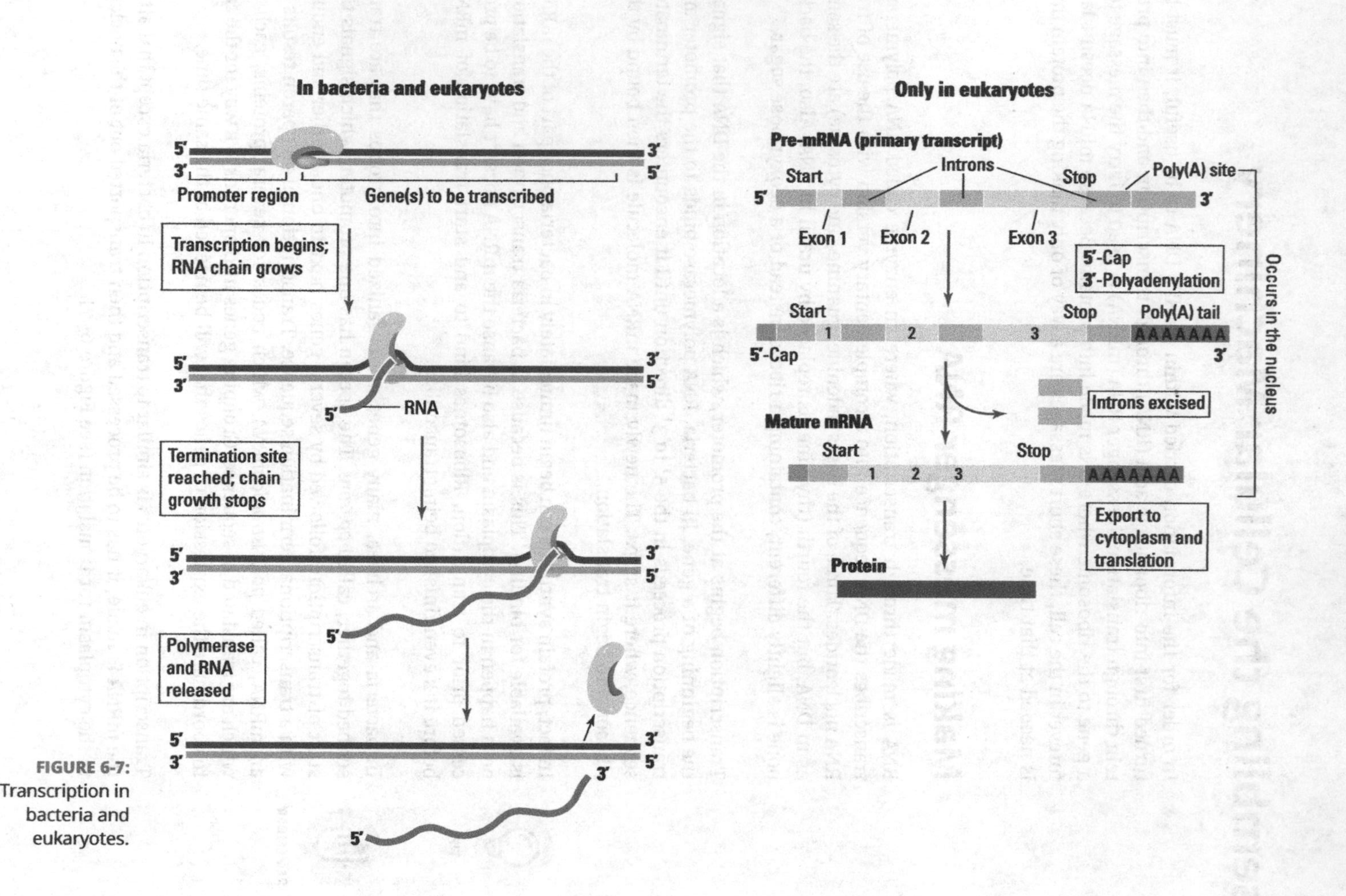

FIGURE 6-7: Transcription in bacteria and eukaryotes.

Here are the mRNA processing steps:

1. A cap is attached to the mRNA so that it can be recognized for translation by the ribosome.
2. The protein coding sequence of eukaryotic genes is interrupted by non-coding regions called *introns* that are transcribed by RNA polymerase but are later removed, or *spliced*, during RNA processing.
3. The mRNA is trimmed and a string of A's is added at the end, called a *poly A tail*. The poly A tail makes the mRNA stable and helps identify it to the ribosome as a sequence to be translated.
4. Mature mRNA is exported from the nucleus to the cytoplasm to be made into protein by ribosomes.

Archaea have promoters and RNA polymerase that are similar to those in eukaryotes, but transcription is regulated as in bacteria. Many archaeal genes are found in operons; some, but not all, have introns.

Remembering other types of RNA

The cell has a few other types of RNA, two of which are essential to protein synthesis: ribosomal RNA (rRNA) and transfer RNA (tRNA). These are called *non-translated RNAs* because they're never made into protein but perform their function as RNA. For rRNA and tRNA, this means first being folded into a three-dimensional structure that is held together by bonding between complementary bases in the sequence (see Figure 6-8). Folded rRNAs and tRNAs can then interact with proteins and with DNA.

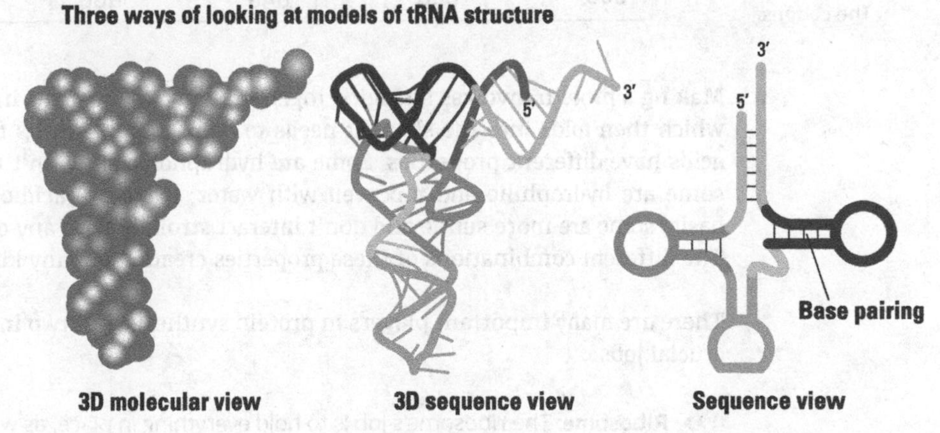

FIGURE 6-8: The three-dimensional structure of tRNA.

Synthesizing protein

The message contained within an mRNA is converted to protein through translation, where the genetic code is deciphered into amino acids. The bases in mRNA are decoded in threes into codons, each of which encodes an amino acid (see Figure 6-9); there are 20 amino acids. Several different codons encode the same amino acid.

First nucleotide	Second nucleotide: U	C	A	G	Third nucleotide
U	UUU Phe	UCU Ser	UAU Tyr	UGU Cys	U
U	UUC Phe	UCC Ser	UAC Tyr	UGC Cys	C
U	UUA Leu	UCA Ser	UAA STOP	UGA STOP	A
U	UUG Leu	UCG Ser	UAG STOP	UCG Trp	G
C	CUU Leu	CCU Pro	CAU His	CGU Arg	U
C	CUC Leu	CCC Pro	CAC His	CGC Arg	C
C	CUA Leu	CCA Pro	CAA Gln	CGA Arg	A
C	CUG Leu	CCG Pro	CAG Gln	CGG Arg	G
A	AUU Ile	ACU Thr	AAU Asn	AGU Ser	U
A	AUC Ile	ACC Thr	AAC Asn	AGC Ser	C
A	AUA Ile	ACA Thr	AAA Lys	AGA Arg	A
A	AUG Met	ACG Thr	AAG Lys	AGG Arg	G
G	GUU Val	GCU Ala	GAU Asp	GGU Gly	U
G	GUC Val	GCC Ala	GAC Asp	GGC Gly	C
G	GUA Val	GCA Ala	GAA Glu	GGA Gly	A
G	GUG Val	GCG Ala	GAG Glu	GGG Gly	G

FIGURE 6-9: The codons.

Making a protein involves stringing together many amino acids into a long chain, which then folds into the shape it needs to be in to perform its function. Amino acids have different properties. Some are hydrophobic and don't mix with water; some are hydrophilic and mix well with water; some are acidic and others are basic; some are more subtle and don't interact strongly with any other molecules. The different combinations of these properties create the many kinds of proteins.

There are many important players in protein synthesis, but two in particular have crucial jobs:

» **Ribosome:** The ribosome's job is to hold everything in place, as well as form the bonds between amino acids. All cells have ribosomes. Ribosomes are made of RNA and associated proteins, with a small subunit and a large subunit coming together during translation to catalyze protein synthesis.

» **Transfer RNAs (tRNAs):** Transfer RNAs are small RNA molecules that are folded into a specific shape necessary for fitting into ribosomes, carrying an amino acid and reading a codon. The way that each tRNA recognizes a codon is through base paring with a complementary sequence on the tRNA called the *anticodon*.

The start of translation is signaled by the codon AUG, which also codes for the amino acid methionine. The end of translation is signaled by one of three stop codons (UAA, UGA, or UAG), none of which codes for an amino acid. In prokaryotes, the process works like this (see Figure 6-10):

1. *Intiation* is the beginning of protein synthesis and involves assembly of the ribosome, the tRNA that recognizes the start codon, and the mRNA molecule itself, as well as other accessory proteins. A second tRNA for the next codon enters the ribosome, and the two first amino acids are joined with a peptide bond.
2. *Elongation* happens as the ribosome moves along the mRNA so that tRNAs can enter and add the appropriate amino acids to the growing peptide chain.
3. *Termination* occurs once the ribosome has reached the stop codon. At this point, the ribosome separates into its two subunits, and the mRNA molecule and the peptide chain are released.

REMEMBER

A peptide chain is a newly formed protein made up of amino acids covalently bonded by peptide bonds.

In eukaryotes the process is similar with a few key differences:

» Eukaryotic mRNAs are recognized by the ribosome by the mRNA's methylated cap and its poly A tail.

» The ribosomes are larger and use different accessory proteins for each step of translation. The ribosomes of archaea also use some of the same accessory proteins as those in eukaryotes.

The peptide chain then folds properly either on its own or with the help of other proteins. After it's sent to the proper location in the cell, the freshly made protein will be ready to perform its function in the cell. Some bacterial proteins need to be secreted to the *periplasm* (the space between the inner membrane and the outer membrane in Gram-negative bacteria) or inserted into the membrane. Proteins to be secreted have a signal peptide that is around 10 to 15 amino acids long. The signal peptide is bound by other proteins that will shuttle them to the area in the membrane where they can be exported from the cytoplasm.

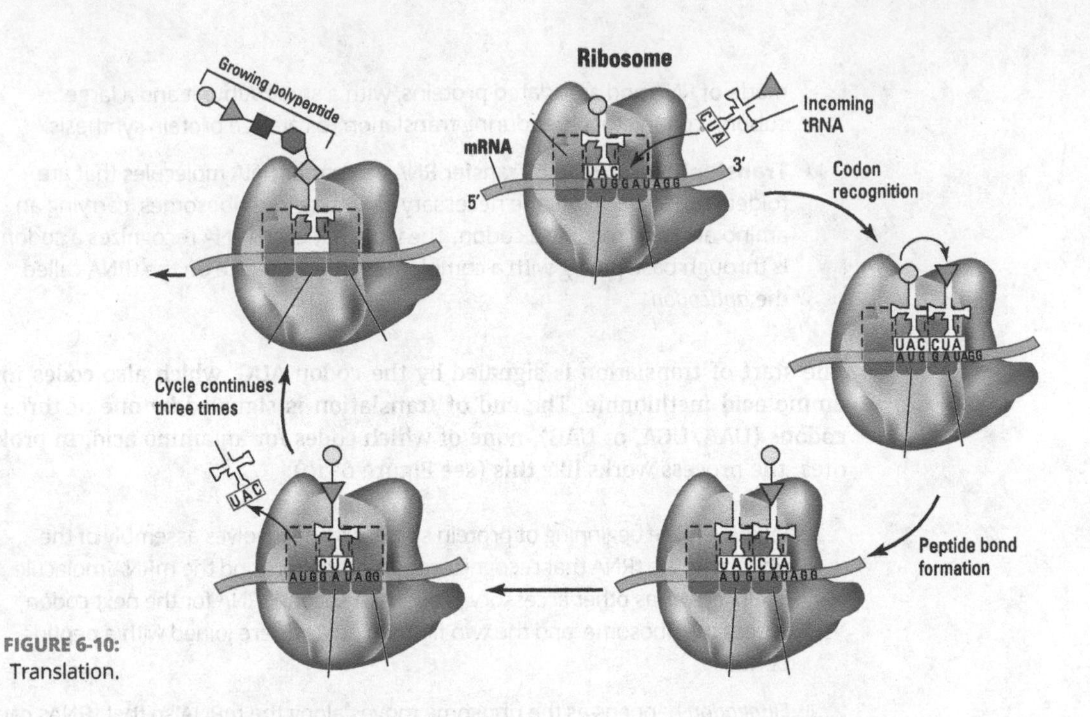

FIGURE 6-10:
Translation.

Making the Right Amount: Regulation

Of the many proteins at work in a cell, some are required all the time, whereas others are needed only under certain circumstances. For example, ATPase is always needed, whereas an enzyme that breaks down a specific sugar is needed only when that sugar is present.

Cells have genes for many more proteins than they need at a given time, and the process of turning those genes on and off or of changing the activity of those genes is called *regulation*. Protein synthesis is expensive for the cell, so it's important that it makes only what it needs.

Why have two main ways of regulating metabolism? Because neither of the two by itself could manage the resources necessary to keep a cell running efficiently. For example, when a protein is needed, it takes time to transcribe and translate the gene for it. Also, when the protein is no longer needed, it takes time for the amount of that protein in the cell to get old and stop working. So, although turning on and off transcription of a gene is important, it's also important to turn on and off the function of a protein (like an enzyme) so that the cell can react quickly to its environment.

Turning the tap on and off: DNA regulation

Regulation that occurs at the transcriptional level involves proteins that bind to DNA and either enhance or repress transcription. This form of regulation controls the amount of a protein that's made.

DNA-binding proteins, as their name suggests, are proteins that interact with DNA. There are two kinds of DNA-binding proteins: those that are sequence specific and those that are nonspecific. An example of a nonspecific DNA-binding protein is the histone than interacts with all the DNA in the cell in the same way. Sequence-specific DNA-binding proteins bind when they recognize a short region of the DNA sequence.

TECHNICAL STUFF

Histones interact with negatively charged DNA because they're very positively charged themselves. The association between the two neutralizes the charge on both and allows DNA to be compacted more than it if the negative charges were repelling each other.

Negative control of gene expression uses a *repressor protein* that, when active, binds to DNA and turns off expression of the gene. For some genes, the repressor is inactive until a *co-repressor* molecule is present. The corepressor binds to the repressor, activating it and causing expression of that gene to be turned off (see the arginine [arg] example in Figure 6-11). For other genes, the repressor is naturally bound to the gene to keep its expression turned off until an *inducer* molecule is present that binds to the repressor and inactivates it, which turns on transcription of that gene (see the lactose [lac] example in Figure 6-11).

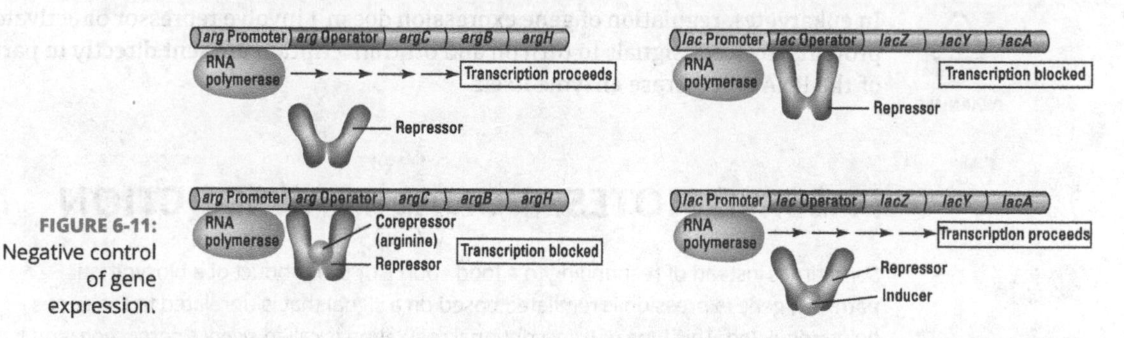

FIGURE 6-11: Negative control of gene expression.

REMEMBER

The process of transcription and translation of a gene is called the *expression* of that gene.

Positive control of gene expression involves a DNA-binding protein called an *activator* that binds to DNA and activates transcription. Activators usually need to first bind an inducer molecule that then allows them to bind DNA. When all three are bound, RNA polymerase can attach and begin transcribing the gene. An example of this is expression of the genes for breaking down maltose (a sugar) that require maltose to bind the activator, which then turns on the genes for maltose breakdown (see Figure 6-12).

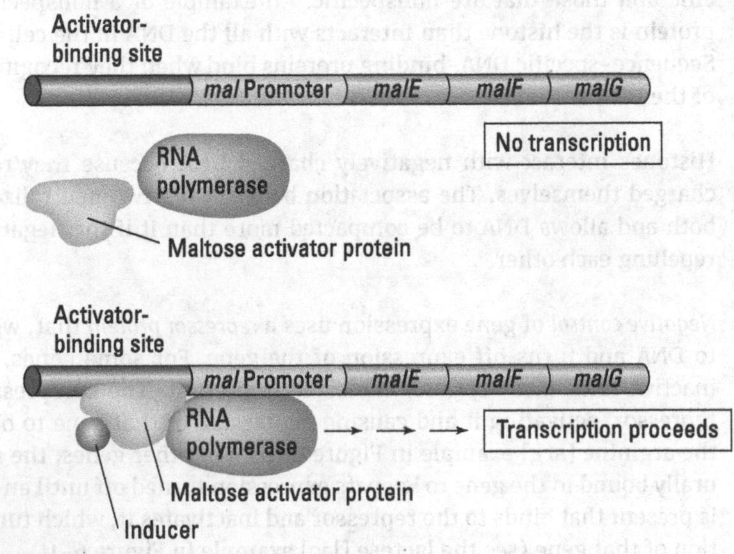

FIGURE 6-12: Positive control of gene expression.

In eukaryotes, regulation of gene expression doesn't involve repressor or activator proteins. Instead, signals to turn on and off transcription are sent directly to parts of the RNA polymerase enzyme itself.

PASSING NOTES: SIGNAL TRANSDUCTION

Sometimes instead of responding to a food source or the product of a biosynthetic pathway, gene expression is regulated based on a signal that is unrelated to the genes being regulated. This type of transcriptional regulation is called *signal transduction,* and it involves two parts: sensing the signal and responding to the signal. Rarely the same protein senses and responds to a signal, but more commonly two separate proteins work together in what is called a *two-component regulatory system*.

Regulating protein function

Regulating the activity of an enzyme is a way of fine-tuning metabolism to limit any waste of energy, to stop the accumulation of toxic waste product, or to avoid overusing a valuable resource.

When products build up in the cell, they can act to turn off the activity of the enzyme that made them. This is called *feedback inhibition*, and it acts to switch an enzyme from its active form to an inactive form, or from "on" to "off."

The activity of the enzyme can also be controlled in amore subtle way. Sites on the enzyme can be modified to incrementally decrease the amount of product an enzyme makes. This gives the cell precise control over a metabolic pathway.

Changing the Genetic Code

Genetic diversity is the basis of evolution and natural selection is the force driving it.

If a perfect copy of the genome were always passed on, there would be no genetic variety in the world. Genetic change is what drives evolution and has led to the great variety of life we know today. Changes to a genome can be in the form of a mutation or recombination, each of which occurs a few different ways.

In eukaryotes, DNA exchange is linked with cell division — two cells, each with one copy of the genome, fuse to form one cell with two copies of the genome. The genetic material can then recombine and split again, in meiosis (see "Doubling down with DNA replication," earlier in this chapter), to form haploid cells. In bacteria and archaea, however, DNA exchange is not always linked with cell division. Cells can take up genetic material that gets combined with their genome and separately reproduce by cell division.

Slight adjustments

Sometimes errors occur during DNA replication that alter the sequence by one or a small number of bases by adding too many nucleotides, too few nucleotides, or a wrong nucleotide. The result is called a *point mutation*. Point mutations can have a negative effect, a positive effect, or no effect on the protein. Most of the time, base substitutions have no effect on the final protein, so they're called *silent mutations*. Other times, however, changes in one or a few bases can alter the function of a protein in a negative or positive way by decreasing the function of the expressed protein or improving it.

Three types of mutations are possible due to base substitutions (see Figure 6-13):

» **Missense mutations:** A missense mutation happens when the wrong nucleotide is added within the coding region of a protein gene and that changes the amino acid that is incorporated into the resulting protein.

» **Nonsense mutations:** A nonsense mutation is the result of the wrong nucleotide being added within the coding region of a protein gene, creating a stop codon. The resulting protein will then be shorter than it is supposed to be, or *truncated*.

» **Frameshift mutations:** A frameshift mutation is caused by the loss or addition of a nucleotide that changes how the pattern of three nucleotides per codon is read. The result is that, from this point forward, completely different amino acids may be incorporated into the protein.

Point mutations can also be caused by mutagens, things like chemicals or radiation that damage DNA. Microorganisms have the ability to repair some of the DNA after damage from a mutagen, the process involves removing the incorrect or damaged bases and adding the correct ones.

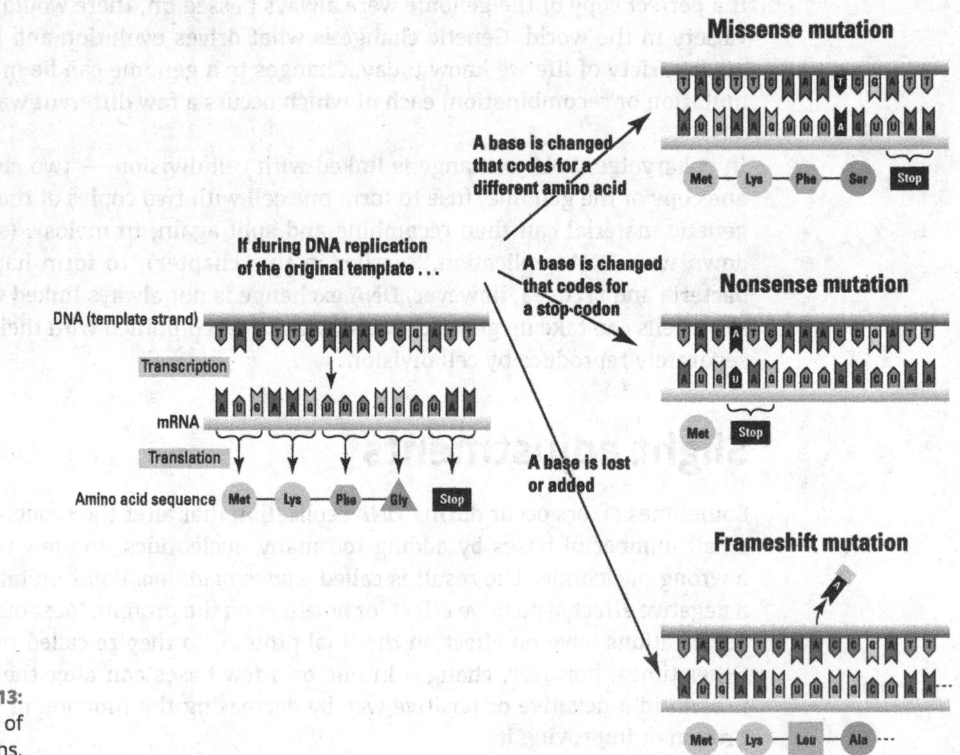

FIGURE 6-13: The effects of point mutations.

CODONS WOBBLE, BUT THEY DON'T FALL DOWN

The DNA code is degenerate — it has redundancy built in to protect its message from base substitutions. More than one three-base codon in the DNA codes for the same amino acid, and often the difference is in the third nucleotide of the codon, called the *wobble position.* Because of this base substitutions that change the wobble, nucleotides don't change which amino acid is added to the protein during translation. It's a way of allowing some natural variation so organisms can evolve over time without having a high number of negative mutations in any population. Base substitutions happen naturally at a rate of about 1 per 10^{10} bases copied, depending on the individual organism and gene. Mutation rates in viruses are much higher, which works to their advantage because they're constantly adapting to avoid their host's defenses.

Major rearrangements

Changes to an organism's DNA can also happen on a larger scale than with point mutations, where regions of DNA from two different sources get combined. The process of incorporating sequences from different sources into the same chromosome is called *recombination* (see Figure 6-14); it occurs in different contexts in eukaryotes and bacteria.

Eukaryote microorganisms have a specific genetic recombination step built into the formation of the sex cells, where parts of each pair of diploid chromosomes get exchanged, thus increasing the genetic diversity of the resulting spores. In bacteria, genetic recombination is not linked to cell division; instead, it's part of a few different strategies aimed at increasing genetic diversity, including transformation, conjugation, transduction, and transposition.

Transformation and conjugation

Bacteria come upon DNA in their environment all the time because the DNA of previous individuals hangs around long after they've died. Bacterial cells that are able to take up foreign DNA are said to be *competent*. When cells are competent, large molecules of DNA are able to pass through the cell wall into the cytoplasm, after which recombination is necessary if the DNA is to be incorporated into a bacterial genome.

When competent cells take up DNA from their environment, it's called *transformation.* When cells transfer genetic material with the help of a conjugative plasmid, it's called *conjugation.* Conjugation requires cell-to-cell contact, whereas transformation does not.

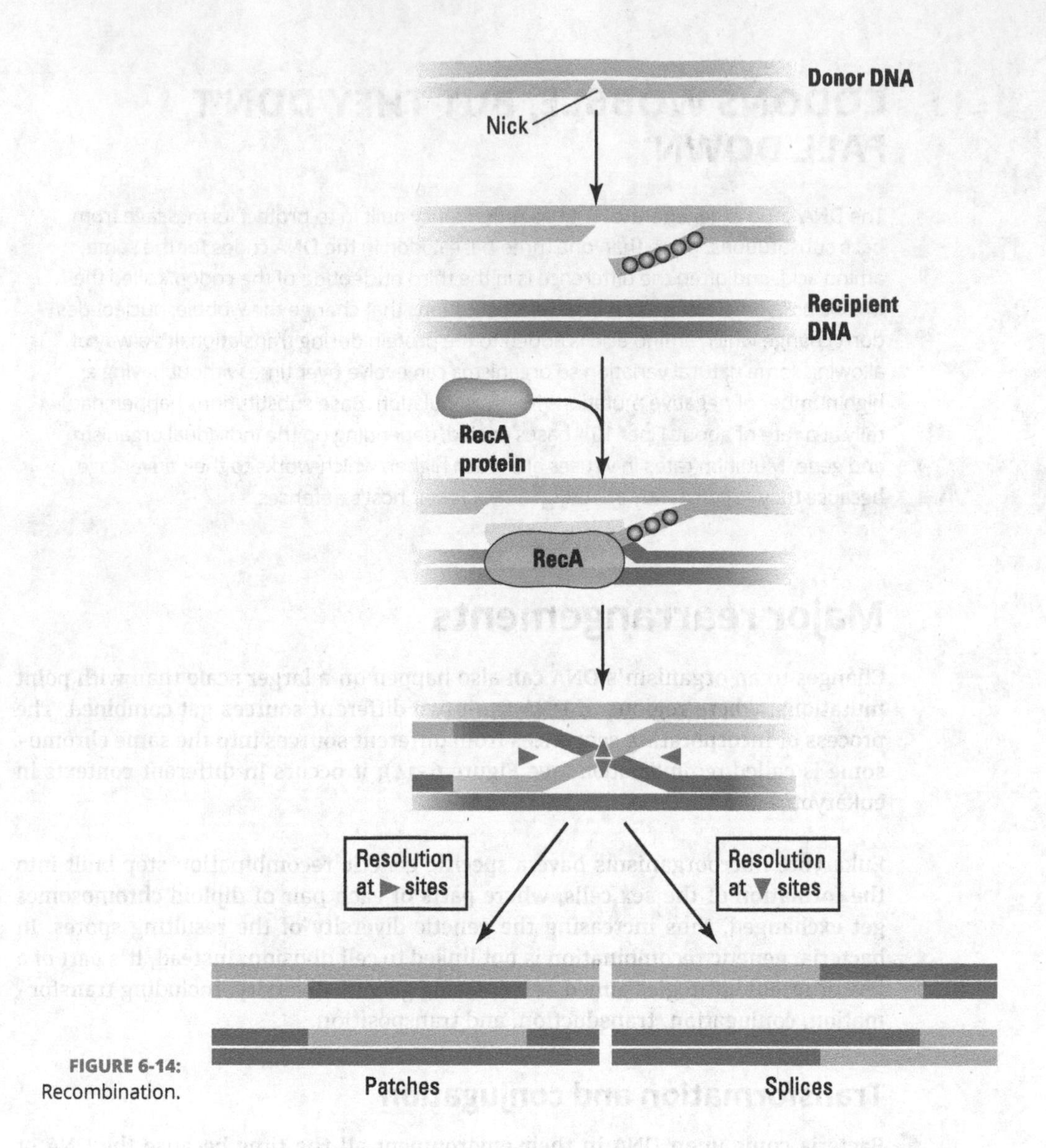

FIGURE 6-14: Recombination.

Gram-negative bacteria produce a sex pilus (a hairlike appendage on the surface of the bacteria) used to make contact between cells, whereas Gram-positive bacteria come into close proximity with one another and make a bridge across cells using sticky substances on the cell surface.

Transduction

Another way that genetic material can be transferred between bacteria is through the action of a virus that infects one bacterial cell and transfers genes to another bacterial cell, a process called *transduction*. Viruses that infect bacteria are called *bacteriophage* or simply *phage*. There are many kinds of phage, and we discuss them more completely in Chapter 14, but in general there are two types of transduction by phage:

- **Generalized transduction:** Generalized transduction happens when a phage packages any genes present in a bacterial cell, whether they're chromosomal, plasmid, or viral.
- **Specialized transduction:** Specialized transduction occurs when only some bacterial genes can be packaged into a phage, such as specific toxin genes or other virulence factors.

Transposition

Transposition is when a small piece of DNA, called a *transposon*, moves from one location to another within the DNA of an organism. For this to happen, there has to be an enzyme capable of cutting and resealing DNA and recognition sites where the enzyme acts. There are simple and complex transposons but all cause DNA modifications.

IN THIS CHAPTER

» Creating the right factors for microbial growth

» Seeing and counting microorganisms

» Reducing microbial numbers

Chapter 7 Measuring Microbial Growth

Your perspective on growing microbes will depend on whether you're more concerned with encouraging microbes to grow (for instance, to study them) or discouraging microbial growth (for example, for food preservation or sterilization). Growing some microorganisms in the lab has been very challenging, and microbiologists put a lot of effort into getting the conditions just right so that they can encourage these shy microbes to grow. Other microbes are easily grown in the lab and are used frequently as model organisms to study different aspects of microbial life. If you're trying to get rid of microbes, some are easy to eradicate and others, like spore-forming bacteria, are quite challenging to kill.

In this chapter, we walk you through the ins and outs of microbial growth. We start with the topic of encouraging microbial growth, because we love microbes!

Getting Growth Requirements Right

For each microorganism, there is a set of conditions (both physical and chemical) under which it can survive. Many factors combine to create these conditions. Within the range of conditions that a microbe can survive, there is a narrower range that is optimal for growth.

Physical requirements

Microbes have a variety of physical requirements for growth, including temperature, pH, and water stress. We explain these requirements in the following sections.

Temperature

Microbes can be separated into groups based on the range of temperatures at which they can survive. At the edges of each range, microbes can usually survive but not thrive, whereas the perfect conditions for growth are usually somewhere in the middle. There is also a lot of gray area between these groups because not all microbes in each group are the same. For instance, some psychrophiles can survive at 0°C but prefer 15°C, and others prefer 30°C, bringing them almost into the mesophile group.

Here are the various groups of microbes based on their physical requirements:

- **Psychrophiles** are microbes that can grow at 0°C. Some are inhibited at higher temperatures, preferring to live in cold climates, whereas others can survive in conditions above 20°C. This latter group are called *psychrotrophs* because they prefer colder temperatures but can live just fine in higher ones. The range for psychropiles is –10°C to 20°C, with an optimum at about 10°C. The range for psychrotrophs is from 0°C to 40°C with an optimum of about 20°C.
- **Mesophiles** like it best between 25°C and 40°C but can survive between 10°C and 50°C. Microbes that live within animals grow optimally at a temperature that matches that of their host. For instance, microbes that live in the human body grow between 34 and 37°C, which is body temperature.
- **Thermophiles** can tolerate temperatures up to 70°C and like it best between 50°C and 60°C. This group contains a subset considered *hyperthermophilic*, or extreme heat loving. All the known microorganisms in this category are archaea and some can even grow in temperatures above 120°C, deep in the sea where the pressure stops water from boiling at that temperature.

pH

Another physical growth condition important to microorganisms is pH. The pH is the measure of how acidic or alkaline a solution is, with values from 0 to 14. Acidic environments include acid mine drainage, iron lakes, and the jar of pickles in your cupboard, with ranges between 1 and 6. Neutral pH is around 7. Alkaline, or basic pH, is 8 to 14. Most bacteria prefer a pH range of 6.5 to 7.5, whereas fungi can grow in more acidic conditions, preferring pH 5 to 6. Some bacteria and archaea are *acididophilic* (acid loving); they grow in conditions far too acidic for other species.

Water stress

The last physical condition to consider is water stress, either from the concentration of solutes in the microbe's surroundings or from drying. As more solutes such as salts or sugar are dissolved in water, the concentration of water to solutes goes down. A microbial cell is permeable to water, so if the concentration of water is lower outside the cell than inside the cell, water will move out in order to balance the inside and outside solutions.

Too much water leaving the cell will kill it. This fact has been used to preserve meats and other foods, by either drying or curing them with salt or adding a lot of sugar (for example, honey and jam). On the other hand, environmental microbes have adapted to salty conditions; some grow quite well in low-salt environments, like seawater, or high-salt environments, like brine ponds.

One way bacteria have developed to deal with bad conditions is to transform themselves into *endospores*. The endospore is a dormant form of the original bacterial cell surrounded by a tough coating that makes it resistant to drying, as well as toxic compounds in its environment.

Chemical requirements

Unlike the physical requirements where a specific range or concentration is necessary for optimum growth, the chemical requirements just need to be present in the environment and a microbe will use what it needs. Microbes use compounds containing the following elements and vitamins to make everything in the cell including membranes, proteins, and nucleic acids:

- **Carbon:** Carbon is necessary for all life. In the microbial world, *chemoorganotrophs* get their carbon from organic matter, whereas *chemoautotrophs* get it from carbon dioxide in the atmosphere.
- **Nitrogen, sulfur, and phosphorus:** Nitrogen, sulfur, and phosphorus are necessary for protein and nucleic acid biosynthesis. Most microbes get these elements by degrading proteins and nucleic acids, but some capture nitrogen from nitrogen gas or ammonia or get sulfur from other ions in the environment.
- **Trace elements:** Trace elements such as iron, copper, molybdenum, and zinc are needed as cofactors for enzymes and must be obtained, in tiny amounts, from the environment.

- **Vitamins and amino acids:** Unlike humans, microbes can make vitamins, which also act as enzyme cofactors. Some microbes, however, lack the ability to make one or several vitamins and have to get them from their environment. The same is true of amino acids and these along with the vitamins needed are called *growth factors.* Although most bacteria can make all the amino acids they need, some can't quite make them all; these bacteria are called *auxotrophs.*
- **Oxygen:** The presence of oxygen affects a few different aspects of microbial growth (see Chapters 9 and 10). Different microbes respond differently to oxygen (see Chapter 10).

Culturing microbes in the lab

If you're interested in studying a particular microbe in a laboratory, you'll need to create an environment that will accommodate its growth well. *Culture media* contains all the things that a microbe needs for growth. *Culture conditions* are conditions that are within the microbe's *growth optima* (external environmental conditions that allow for the optimal growth of the organism).

The culture medium used for a particular microbe must contain all the chemical requirements for that organism, as well as have the right pH and solute concentration and be incubated at the right temperature and oxygen conditions. Here are some of the different kinds of culture media:

- **Chemically defined media** are those where the concentration of all the components within it are known. Often, chemically defined media are used to study specific aspects of microbial growth and metabolism.
- **Complex media** are those in which the concentration of nutrients is unknown. These media are used most commonly in microbiology labs because they work well to culture a variety of microorganisms. Complex media contain ingredients such as animal or plant products, for which the exact chemical ingredients are not known. They're often used to grow enough microbial biomass to be used for research.
- **Anaerobic growth media** contains reducing agents that remove the oxygen dissolved in them.
- **Selective media** are used to isolate particular microbes of interest. The idea behind selective media is to discourage or completely inhibit the growth of all but the microbe that you want to culture. This is accomplished through the use of additives that are harmful to all but the microbe of interest and conditions that favor the desired microbe, which then outgrows the rest.

- **Enrichment media** is a type of selective media specifically designed to encourage the growth of microbes that are either present in a sample in very small numbers or are easily outcompeted by the other microbes in the sample. If the microbes that you want to grow metabolize something rare, like phenol, it can be added as the sole carbon source and only allow microbes that degrade it to grow.
- **Differential media** contains compounds that change color depending on the metabolism of the microbes present. An example of this is blood agar on which bacteria form halos around their colonies as they *lyse* (break open), red blood cells.

Many media types exist that are both selective and differential so as to quickly grow and identify particular microbes.

Each of these types of media can be used as a liquid or as a solid. To make the solid version of each medium, *agar* (a substance extracted from algae) is added to the medium and then it's heated (usually during sterilization). The mixture is poured into a container. Upon cooling, the agar sets the medium into a semisolid form on which microbial cultures can grow.

Obtaining a pure culture of a microbe of interest is accomplished by inoculating a plate of solid media in a specific way, shown in Figure 7-1. In the figure, the black lines show what you would actually draw on an agar plate using a stick or a loop as a "drawing tool" after having touched it to some bacteria. The bottom part of the figure shows where the bacteria would grow (the circles) following along the lines you "drew." As the microbes are spread across the plate in the way shown, they are diluted out so that single colonies can be isolated. Colonies that look similar are likely from the same organism.

For microbes that require anoxic (free of oxygen) conditions, there are a couple ways to exclude oxygen from the growing conditions. The first involves a sealed jar containing a chemical combination that produces hydrogen (H_2) and carbon dioxide (CO_2). The hydrogen combines with the oxygen (O_2) in the presence of the catalyst to form water. The other method is with a large chamber that is purged with an inert gas, usually nitrogen gas (N_2), which replaces regular air.

Some microbes — namely, the ones that live in animals — can grow only with elevated levels of CO_2. The range of CO_2 needed is usually between 3 percent and 5 percent and can be achieved by lighting a candle in a sealed jar or using more modern growth chambers specifically designed to regulate the amount of CO_2.

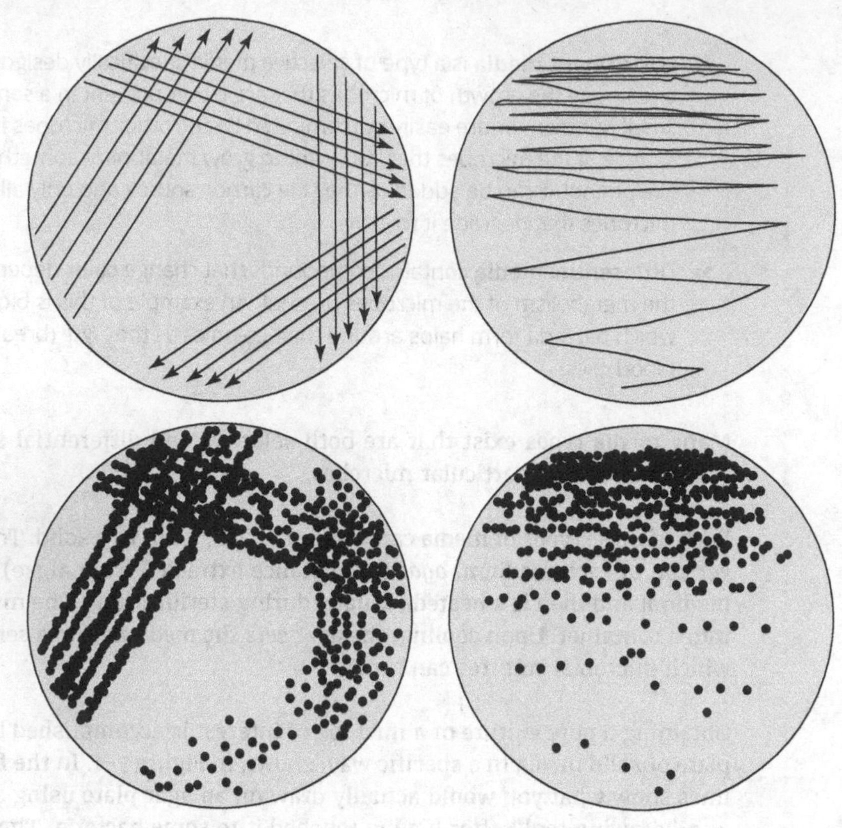

FIGURE 7-1: The method for streaking out isolated bacterial colonies.

The biosafety level of a microorganism is the level of precaution needed when working with it and is based on its known ability to cause human disease. Biosafety levels start at 1 (safest) and go to 4 (hot zone) and have been decided for all microbes on which research is performed. To work with a level 3 pathogen, special containment rooms and equipment are required that keep everything under negative pressure and filter all air coming out of them. Level 4 pathogens require the same precautions, plus personnel have to wear protective suits that have their own air supply.

Observing Microbes

When studying microorganisms it's often helpful to be able to see them. This may be in order to count them, tell the different kinds apart from one another, or study their structures. To deal with their extremely small size, two very different strategies are used.

The first strategy is to increase their numbers to a level that can be observed with the naked eye. This usually requires that the microbe of interest be in a pure culture that when grown up to sufficient numbers will form a characteristic type of *colony* (a large number of microbial cells that came from a single individual microbial cell) on solid media.

The second strategy is to magnify the cells many times with the use of microscopy. Often this method gives good detail about individual cell structures but requires that the sample be prepared with the use of dyes and other chemicals.

Counting small things

Microbial experiments often require that we know just how many bacteria, for instance, are present. Knowing the number of microbial cells helps to indicate whether cells are growing or dying and it helps experiments to be consistent from day to day. Methods for deciphering the number of cells present easily and accurately are used in microbiology labs every day. These include direct methods where colonies are directly counted and indirect methods where an estimate of the number of cells is made.

Direct methods

Determining the number of bacterial cells that are alive in a sample is done using *viable counts*. Each method for calculating viable numbers of organisms takes advantage of the fact that when a suspension of bacteria is plated on solid media, each live bacterial cell will grow to form a colony, which is large enough to be seen. When the concentration of bacteria in a sample is high, it must be diluted enough so that the colonies are distinct from one another, which is done with *serial dilution*, illustrated in Figure 7-2.

In Figure 7-2, a culture of bacteria is repeatedly diluted by adding a fixed volume to the first tube (the first dilution) and then taking a fixed volume from the first tube and adding it to the second tube (the second dilution). This is repeated several times (which is why it's called *serial* dilution) in a effort to dilute the original bacterial culture. A small volume of each dilution is then plated on agar and individual bacterial cells will each form one colony. Increasing dilutions will result in a decreasing number of bacteria present and, hence, a decreasing number of colonies on each agar plate. The number of colonies is then used to calculate backward to determine the exact number of cells that were present in the original sample.

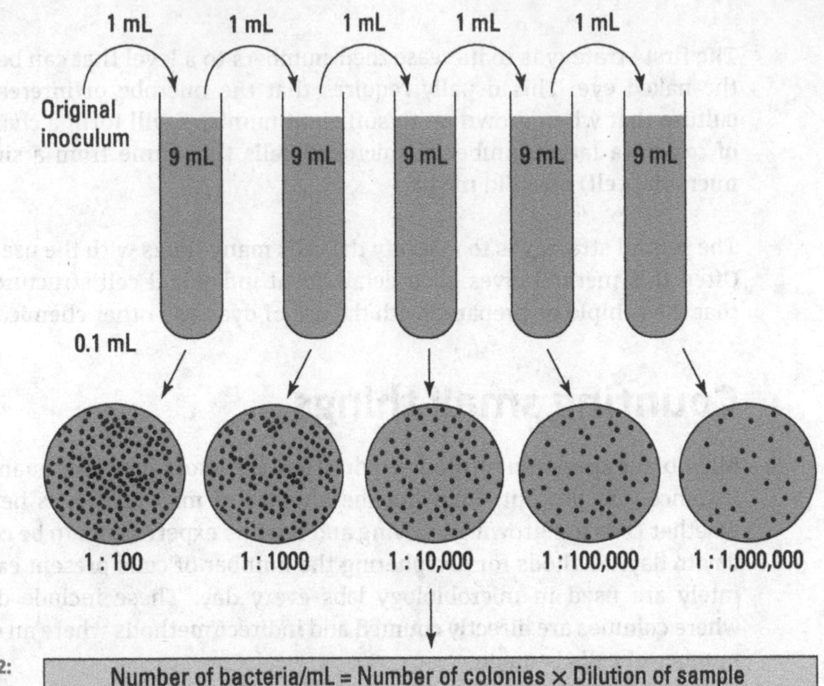

FIGURE 7-2: Serial dilution.

REMEMBER

To calculate the number of viable cells in the original sample, the number of colonies present after the plates have been incubated is multiplied by the dilution factor. The number of cells in a sample is usually expressed as colony forming units (CFU) per unit of volume (in this case mL) or CFU/mL.

When the numbers of cells in a sample are very low, such as in lake water, a large volume of the sample can be passed through a filter, which is then cultured with the proper medium to produce colonies of the microbe of interest. In this case, the number of colonies is divided by the volume of sample filtered. For example, if 5 liters (L) of lake water were filtered and 100 colonies appeared after incubation, then the original number of bacteria in the water was 100 colonies/5 L = 20 CFU/L.

Another way of counting the number of cells in a sample is by *direct microscopic counts*. With this method, all the cells in the sample — whether alive or dead — are counted, but it gives you an immediate estimate of bacterial numbers without the need to incubate a sample for 24 hours or more.

Indirect methods

Although not as precise, indirect methods are faster and more convenient than direct methods at estimating the number of microbes in a sample.

Turbidity is the amount of light scattering caused by the cells in a liquid culture. A machine called a *spectrophotometer* passes a beam of light through a clear tube containing a bacterial culture. As the number of cells in a culture increases, more light is scattered and less of it passes through the tube to be recorded on the other side. Turbidity is expressed as absorbance and as optical density (OD). The higher the number (between 0 and 2), the more concentrated the cells in the solution.

Another way to indirectly estimate the number of cells is to measure the metabolic activity of the sample for a specific substrate. In this way, the amount of waste produced — either a gas that can be captured and measured or a metabolite that can be used with an indicator in the media — can be used to calculate how many growing cells are in the culture.

The last method to indirectly estimate the number of cells is by calculating the dry weight of the organism in culture. This method is often used with fungi that form long filamentous colonies that aren't easily broken up by dilution. The sample is concentrated and the liquid is removed. Then the cells are dried and weighted.

Seeing morphology

Scientists have been peering at microorganisms through microscopes for centuries. For some, the shape of their cells can offer clues to their identity, but it's often necessary to use stains that tell us a bit more about their cellular structure.

Simple stains contain a single dye that can bind to microbial cells and show off their basic structure. Examples of simple stains include crystal violet, safranin, methylene blue, and carbolfuchsin.

Differential stains distinguish one type of microbe from another. An example of a differential stain that tells apart Gram-negative from Gram-positive cells is the Gram stain, shown in Figure 7-3. It takes advantage of the fact that Gram-positive bacterial cells have a much thicker cell wall than Gram-negative cells do, which stops them from being washed clean of the first stain by the alcohol-washing step.

The acid fast stain takes advantage of the fact that bacteria with a waxy substance in their cell walls will hold onto a stain even after being washed with alcohol. Bacteria that remain colored after washing with alcohol are called *acid-fast* and include human pathogens from the *Mycobacterium* group.

Other structures in the cell can be observed after being stained with special techniques. These include bacterial capsules, endospores, and flagella.

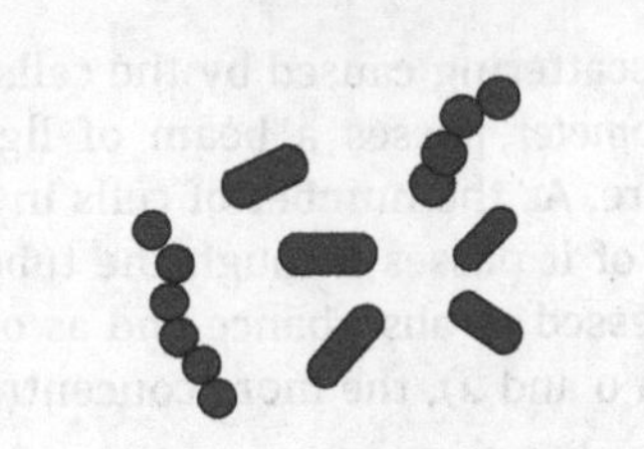

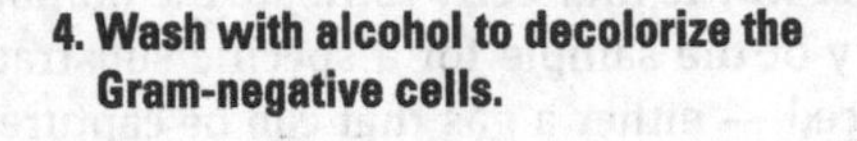

3. Fix the stain with a mordant.

4. Wash with alcohol to decolorize the
Gram-negative cells.

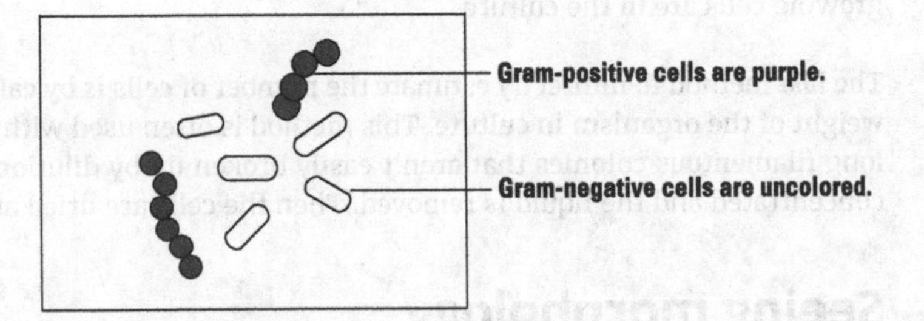

5. Stain with a counterstain like safranin.
This makes the Gram-negative cells easier to see.

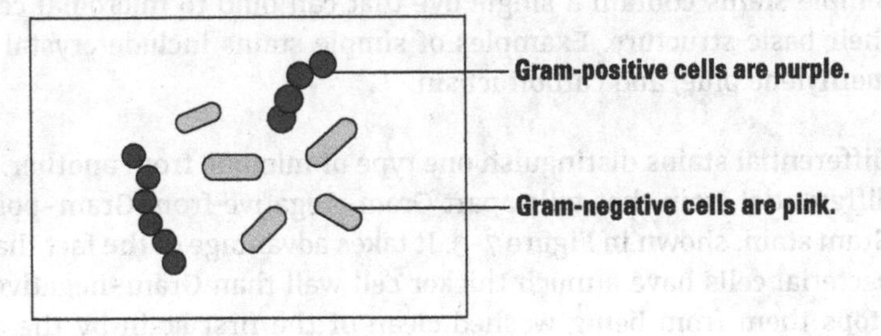

FIGURE 7-3: The Gram stain.

Calculating Cell Division and Population Growth

Unlike animal growth, which is measured both in the size and number of individuals, microbial growth is all about the population size. Only the number of cells matters when calculating the size of the population. Because cells are usually grown in solution the level of growth is referred to as *culture density* or *concentration of microbial cells.*

Dividing cells

When microbial cells are growing happily on a food source, they'll increase in numbers by cell division. Not all microbes use the same method for increasing the number of cells in the population. Here are the ways that cells divide:

- Binary fission, which is used by many bacteria, is a process in which the growing cell first replicates its DNA and then the cell wall constricts, dividing the cell into two (see Figure 7-4).
- Some bacteria and fungi, like yeast, form new cells through budding. This is where an area on the cell's surface starts to grow outward and is eventually pinched off when big enough.

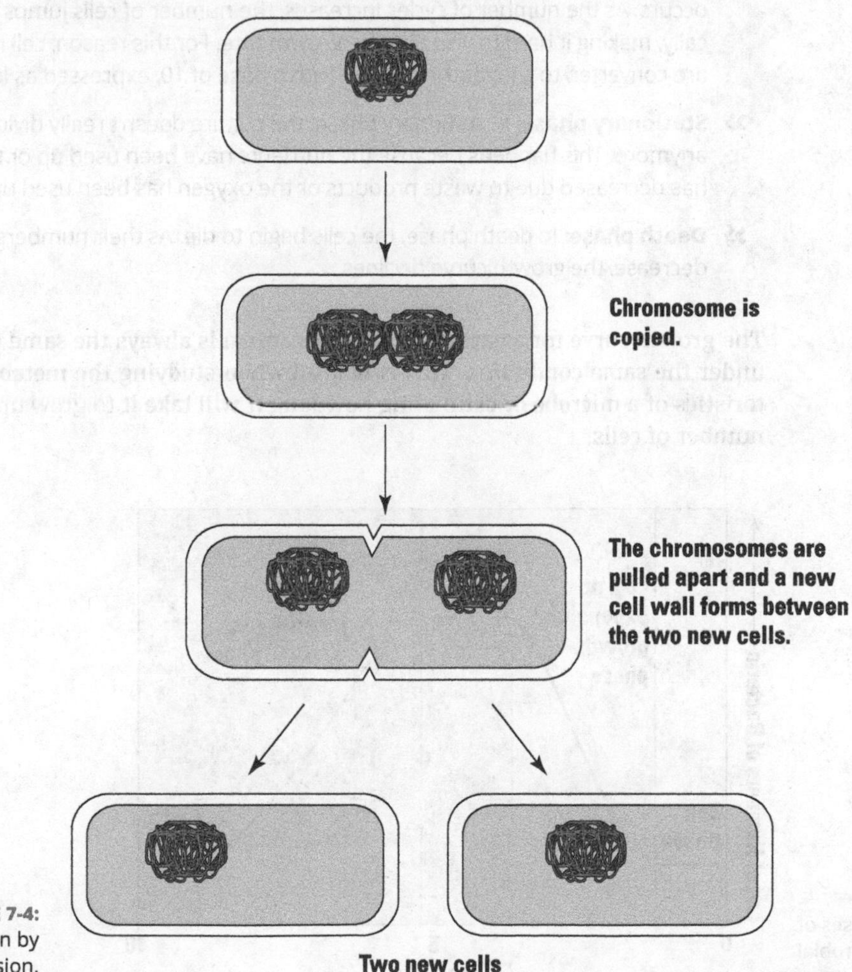

Chromosome is copied.
The chromosomes are pulled apart and a new cell wall forms between the two new cells.
Two new cells

FIGURE 7-4: Cell division by binary fission.

» Filamentous bacteria and fungi form long chains that don't completely separate. To make new colonies, they form a special structure called a *conidiospore,* which then separates, or a part of the chain will simply break off.

Following growth phases

Microbes in culture follow a typical pattern called a *growth curve* (see Figure 7-5), which can be broken down into a few phases:

» **Lag phase:** The lag phase is where the cells are metabolizing but not increasing in numbers.

» **Log phase:** The log phase is when the greatest increase in cell numbers occurs. As the number of cycles increases, the number of cells jumps drastically, making it hard to visualize the growth rate. For this reason, cell numbers are converted to a logarithmic value with a base of 10, expressed as $\log_{10}$.

» **Stationary phase:** In stationary phase, the culture doesn't really divide anymore. This happens because the nutrients have been used up or the pH has decreased due to waste products or the oxygen has been used up.

» **Death phase:** In death phase, the cells begin to die. As their numbers decrease, the growth curve declines.

The growth curve for a particular microorganism is always the same when grown under the same conditions. This is helpful when studying the metabolic characteristics of a microbe or estimating how long it will take it to grow up to a certain number of cells.

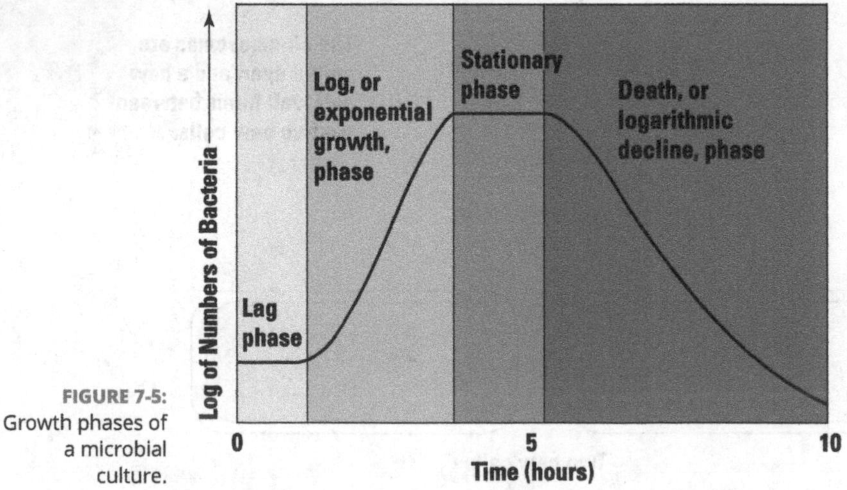

FIGURE 7-5: Growth phases of a microbial culture.

If a microbial culture is given infinite nutrients and the waste products of its metabolism are taken away, then much higher cell densities are possible. This technique is often used in industry in a process called *continuous culture.*

Inhibiting Microbial Growth

The terminology used to describe methods for reducing or removing microbes from a surface can sometimes be confusing. There are different reasons for wanting to get rid of microbes, but not all of them require *sterilization* (the complete eradication of all living things), which is needed for surgical equipment. For the purposes of food preparation, for example, steps are used to ensure that food-borne pathogens that could make people sick have been removed. An example of this is *commercial sterilization* used on canned food, where the conditions are used to reduce the chances of including spores that cause botulism, but the food is not completely sterile. When cleaning a place where food is served, like a restaurant, the process of *sanitation* gets rid of microbes that can be passed between people. *Disinfection* is the removal of growing microbial cells from a surface. When that surface is skin, the disinfecting agent is milder and is called an *antiseptic.*

Another distinction to make is whether a chemical is *bacteriostatic* (which stops the growth of bacteria but doesn't kill it) or *bacteriocidal* (which does kill bacteria).

In this section, we tell you about some of the compounds known to have activity in inhibiting microbial growth.

Physical methods

These physical methods disrupt proteins, killing microorganisms, and some have been used specifically to destroy bacterial endospores.

Heat, although useful for destroying microbes, isn't enough to sterilize most things. It's only in the combination of heat with pressure that resistant spores are killed. This is accomplished with special equipment like an autoclave or pressure cooker that uses steam under pressure.

In the absence of pressure, heat can kill most microbes, which is why boiling things in water for ten minutes is recommended to be fairly safe, although it won't be free of the spores of some bacteria, if present.

If the object to be sterilized can withstand high heat, like metal can, then flaming can be used. With this method, the object is placed directly into a flame or is first dipped into alcohol that is then ignited. Needless to say, this is an effective way to sterilize an object.

Pasteurization is the heating of things like milk and beer at lower temperatures than what would be needed to sterilize them in order to kill pathogens and lower microbial numbers while still protecting the flavor of the product. Two technological advances to the food industry include high-temperature short-time (HTST) pasteurization and ultra-high-temperature (UHT) treatment, the latter of which actually sterilizes the product without severely impacting its flavor.

Liquid solutions can be sterilized by filtration through a membrane with a pore size smaller than the microbes to be removed. For bacteria and fungi, pore size is 0.22 μm to 0.45 μm, but for viruses and some sneaky bacteria it's closer to 0.1 μm.

Low temperature serves to slow or halt the growth of most microbes. Some, however, can grow at low temperatures. If you want proof of this, just think of how leftovers can spoil in the refrigerator if left for too long.

There are several examples of radiation used to kill microorganisms. The first example is ionizing radiation such as X-rays and gamma rays that either directly damage cells or indirectly damage them via the production of free radicals. The second kind is nonionizing radiation that comes from ultraviolet (UV) light, which causes irreversible changes in a microbe's DNA that make it useless.

Disinfectants

Examples of disinfectants are all around us. Some are easily identified and others are harder to spot. Here are some of them:

- A type of phenol that comes from coal tar is called a *cresol* and is the main ingredient in Lysol. A bisphenol called *triclosan* has been used in the manufacture of kitchen utensils like cutting boards to deter microbial growth.
- Halogens like iodine and chlorine have been used for centuries and are still used today to treat water (iodine tablets) and in the household disinfectant Clorox.
- Alcohol is a convenient disinfectant that works for many microbes but isn't able to get rid of endospores or unenveloped viruses.
- Heavy metals like silver, copper, and mercury are very effective against microorganisms and along with their historical use in medicine and industry have recently been incorporated into sporting equipment and clothing to reduce the microbes that cause odor.

3

Sorting Out Microbial Diversity

IN THIS PART . . .

Get a glimpse of the early earth and see how the evolution of life on earth began with microorganisms that changed the climate to what it is today.

Understand the pathways microbes use to get energy and the ways in which they get the carbon they need to make their cellular material, from either carbon dioxide or organic matter.

See how microbes drive the carbon and nitrogen cycles and how great of an impact this has on all other life on earth.

Examine the many places where microbes live, from the bottom of the sea to inside the human body.

Unravel the interconnected communities of microbes and see how they cooperate and compete with one another.

IN THIS CHAPTER

» Finding the origins of life

» Getting an overview of evolution

» Understanding the phylogeny of microorganisms

Chapter 8
Appreciating Microbial Ancestry

The earth has undergone major geologic and climatic changes since it was formed 4.5 billion years ago. When microbes first appeared, the earth was very different than it is today. As the earth changed, it provided microbes with new opportunities for diversification. As microorganisms evolved, they had a significant impact on the earth's atmosphere, leading to even more profound changes that made possible the diversity of life we see today. Organizing our understanding of microbial diversity involves learning about how to tell microbes apart and how to name them. Measuring the diversity of microorganisms today lets us read the history of microbial evolution, which, as it turns out, is also our evolutionary history.

Where Did Microbes Come From?

The earth was formed around 4 billion years ago, but before life could spring forth, the earth had to cool to below 100°C, making it possible for water to exist as a liquid. After the oceans formed, the surface of the planet was still extremely volatile, with asteroid crashes and cosmic radiation bombarding the surface. For these reasons, scientists think that life began not on the surface in warm ponds, but deep in the ocean near *hydrothermal vents* (openings where water heated by the earth's core escaped to the surface). In the mud near hydrothermal vents, energy and elements were likely plentiful and conditions for the first life forms may have been right.

Tracing the origins of life

Within this ancient mud at the bottom of the ocean, the following processes likely happened:

1. The formation of organic molecules was catalyzed spontaneously and without live cells. These organic molecules included lipids, amino acids, nucleic acids, and sugars. No one was around to consume this first organic material, so it just accumulated.
2. Ribonucleic acids (RNA) formed, some with *catalytic activities* (meaning that they helped chemical reactions to happen). These catalytic RNAs are called *ribozymes* because they're made of RNA but have activities like enzymes. One of the activities of an early ribozyme was replication of RNA, making self-replicating RNA.
3. Another activity of early ribozymes was protein synthesis. Proteins with enzyme activities started catalyzing biochemical reactions, including nucleic acid replication.
4. Deoxyribonucleic acid (DNA) evolved and was more stable than RNA, so it took over the role of storing genetic information. The flow of genetic information in the molecular biology of life was then set:

   DNA → RNA → Protein
5. Last, lipid membranes were formed. They encircle things and have embedded within them proteins that can move molecules from one side to the other. Lipid-enclosed biochemical processes were protected from the randomness of the outside world and the cell was born!

The cell has undergone many improvements in the billions of years since it arose, but it remains fundamentally the same today. Exceptions include some viruses that store their genetic material as RNA and contain the enzyme reverse transcriptase, the discovery of which added an arrow from RNA to DNA to the above equation. In Chapter 14, a discussion of viruses highlights some other exceptions to these steps.

Diversifying early prokaryotes

Early on, primitive cells diverged in two different directions, eventually giving rise to bacteria and archaea. Each of these branches was likely well suited to different environmental conditions, so each gave rise to different structural and metabolic specializations. Bacteria and archaea would have had to develop processes that were similar to each other to survive, yet the exact details would be different in each. Table 8-1 gives some examples.

TABLE 8-1 Differences in the Fundamental Structure between Bacteria and Archaea

	Bacteria	Archaea
Lipid membrane	Ester-linked	Ether-linked
Cell wall	Peptidoglycan	No peptidoglycan
Enzymes for transcription of nucleic acids	One enzyme with four subunits	One enzyme with eight to ten subunits
Enzymes for protein synthesis	70S ribosome	70S ribosome
Specialized metabolism	Chlorophyll-based photosynthesis	Can grow above 100°C
	Endospores	

REMEMBER

Organisms need a source of carbon and a source of energy. The carbon is used to make the carbon containing molecules used for cell structure and as energy storage. The energy is used to power metabolic and other cellular processes.

Different types of metabolism evolved, each with its own substrates and waste products. As early microbes consumed one substrate, others evolved to use a new energy or carbon source. The products of their metabolism built up and had a great impact on the chemical composition of the earth.

The impact of prokaryotes on the early earth

Early phototrophs used sunlight as energy, but it wasn't until oxygenic phototrophs (which produced oxygen as a waste product) evolved that the diversity of life really started to take off. At first, the oxygen produced merely reacted with all the reduced compounds around. But after a couple hundred million years, oxygen eventually accumulated to levels that were high enough for aerobic microorganisms to evolve.

OZONE, WE LOVE YOU

As the earth's atmosphere became filled with oxygen (O_2), some of it reacted with ultraviolet (UV) radiation to form ozone (O_3). It formed a thick layer in the upper atmosphere that still covers many parts of the earth today. Because ozone absorbs radiation much better than plain air does, it protected the earth's surface from strong UV and cosmic radiation. Up until that point, the levels of radiation bombarding the surface were too high to allow cells to survive. With protection from the ozone layer, life could now leave the relative protection of the sea to start colonizing the land, leading to an explosion of organisms adapted to terrestrial and freshwater habitats.

It was another 2 billion years before oxygen levels in the atmosphere would get up to the 21 percent that they are today.

Why did the presence of oxygen in the atmosphere allow such an explosion of different forms of life? The amount of energy gained from reducing O_2 to H_2O is very high, so aerobic organisms can grow much more quickly than anaerobic organisms, producing many more cells from the same amount of a resource. Many more cells means many more mutations capable of adapting to new niches. What the rising oxygen levels meant for anaerobic microorganisms is that their reduced substrates were limited as oxygen spontaneously reacted with them. For many anaerobic microorganisms, oxygen was toxic, so they could either develop mechanisms to deal with it or be restricted to locations without oxygen.

As oxygen levels rose, microorganisms with organelles, the eukaryotes, evolved. This was the beginning of what is now the domain Eukarya. It started with algae that diversified extensively in the oceans and eventually gave rise to large multicellular organisms. Within 600 million years, these large multicellular organisms gave rise to the many forms of plants and animals that have lived and are alive today.

Hitching a ride: Endosymbiosis

The fundamental difference between eukaryotic and prokaryotic cells is the presence of a nucleus and membrane-bound organelles, which made many early microbiologists assume that the two had different evolutionary beginnings. In fact, many aspects of eukaryotic biology are more similar to members of the Archaea than to members of the Bacteria. It turns out that eukaryotes evolved from an archaeal ancestor that engulfed but didn't destroy a bacterial cell, resulting in a symbiotic relationship called *endosymbiosis* (see Figure 8-1). This symbiotic relationship gave rise to the Eukaryotes.

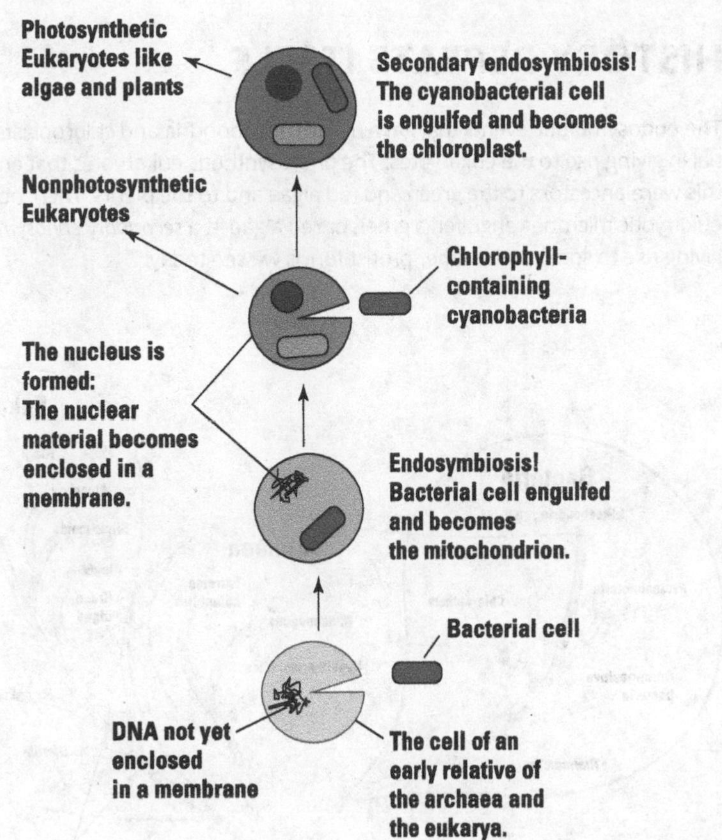
Photosynthetic
Eukaryotes like
algae and plants
Secondary endosymbiosis!
The cyanobacterial cell
is engulfed and becomes
the chloroplast.
Nonphotosynthetic
Eukaryotes
Chlorophyll-
containing
cyanobacteria
The nucleus is
formed:
The nuclear
material becomes
enclosed in a
membrane.
Endosymbiosis!
Bacterial cell engulfed
and becomes
the mitochondrion.
Bacterial cell
DNA not yet
enclosed
in a membrane
The cell of an
early relative of
the archaea and
the eukarya.

FIGURE 8-1: Endosymbiosis.

There is still some question about the exact order in which this happened, but evolutionary biologists generally agree that the following things happened within a common archaea-eukaryote ancestor cell:

1. An aerobic bacterial cell was engulfed to eventually become the mitochondrion.
2. A nuclear membrane formed around the chromosome of the cell.
3. A cyanobacteria was engulfed to eventually become the chloroplast, giving rise to the algae and eventually plants.
4. Membrane-bound organelles formed.

The result is an evolutionary tree like the one shown in Figure 8-2.

HISTORY REPEATS ITSELF

The endosymbiotic events that gave rise to mitochondria and chloroplasts were essential in giving rise to the eukaryotes. The photosynthetic eukaryotes that emerged from this were ancestors to the green and red algae and to the plants. Then, nonphototropic eukaryotic microbes engulfed a green or red algae in a *secondary endosymbiosis* event, giving rise to some of the other protist forms we see today.

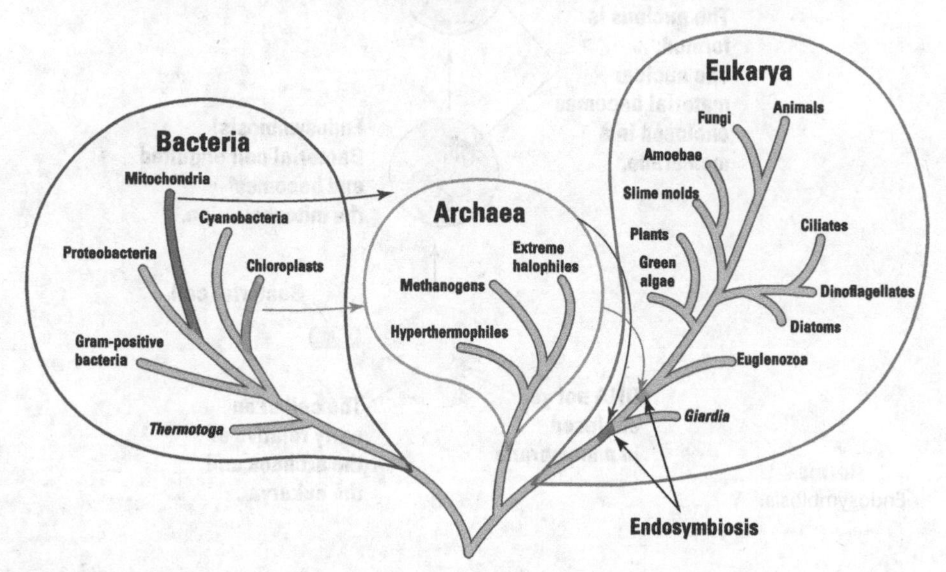

FIGURE 8-2: The tree of life, including endosymbiotic events.

The advantages of endosymbiosis were obvious for the bacteria. Bacteria could give up having to defend themselves from being engulfed and having to acquire their own food. The host cell could use some of the hydrogen and the energy produced by the bacteria for the small price of some substrates and having to carry around extra cargo. Cells at that time were likely experimenting with increased size and could use the extra energy to power cellular processes.

Photosynthetic eukaryotes show up in the evolutionary record later so scientists think that this early eukaryotic cell, with mitochondria and a nucleus, engulfed a cyanobacteria so that it could use sunlight for energy as well. Next to the diversification of prokaryotes, the appearance of eukaryotes marks the second major explosion of diversity with much more complex cells that could take advantage of brand new habitats.

Understanding Evolution

Before we can talk about the study of evolution in microorganisms, we should first describe exactly what evolution is. Evolution is made up of a few parts:

- The descent of one organism from another, also known as *ancestry*. The ancestors of a species are called its *lineage*.
- Differences between individuals in different generations.
- Environmental pressures that favor the traits of one individual over another. This is the driving force of evolution called *natural selection*. The individual habitats with all their specific environmental pressures is called a *niche*.

For evolution to work, there has to also be *extinction*, where the survival of the more successful organisms provides the stock for the next round of diversification and selection. In this way, over time, lineages become more and more specialized for their niches. Figure 8-3 shows the theoretical evolution of different species over a time period.

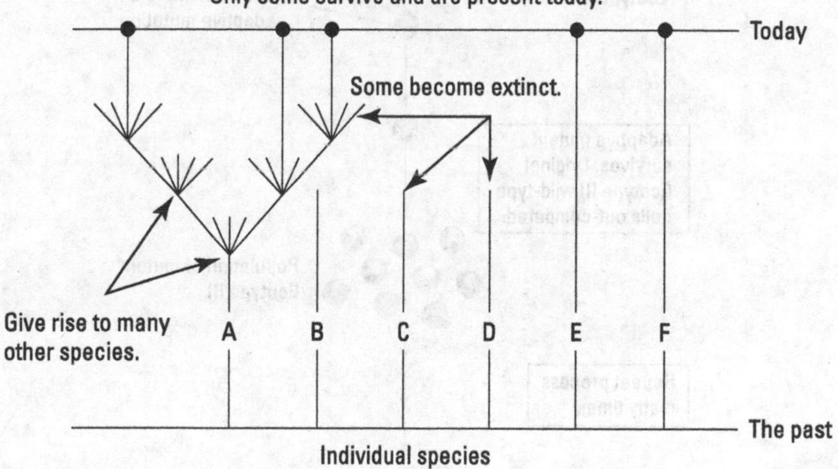

FIGURE 8-3: Lineages arise due to evolution.

Mutations (changes) between generations affect many different characteristics of the species, but only the ones that give it an advantage are kept. Some mutations are deadly or make the individual less successful, other mutations are neutral, and still others provide an advantage. Only the neutral and beneficial mutations remain throughout the evolutionary history of that species.

REMEMBER

Evolution of animals happens on time scales that are too long for us to observe. Even the evolution of organisms with short periods between generations is too slow to see in a lifetime. Bacterial generations are sometimes on the order of minutes, and evolution of lab strains can be measured, but it's tricky to measure in nature.

Genetically similar organisms living in the same habitat and having the same environmental pressures are called *ecotypes*. When changes accumulate that are beneficial enough to give one member of an ecotype an advantage over the others (see Figure 8-4), it will reproduce much more quickly than the others and eventually swamp them out until they're extinct. But if conditions change, different members of one ecotype can continue to exist, each taking advantage of different resources, making them into two new ecotypes.

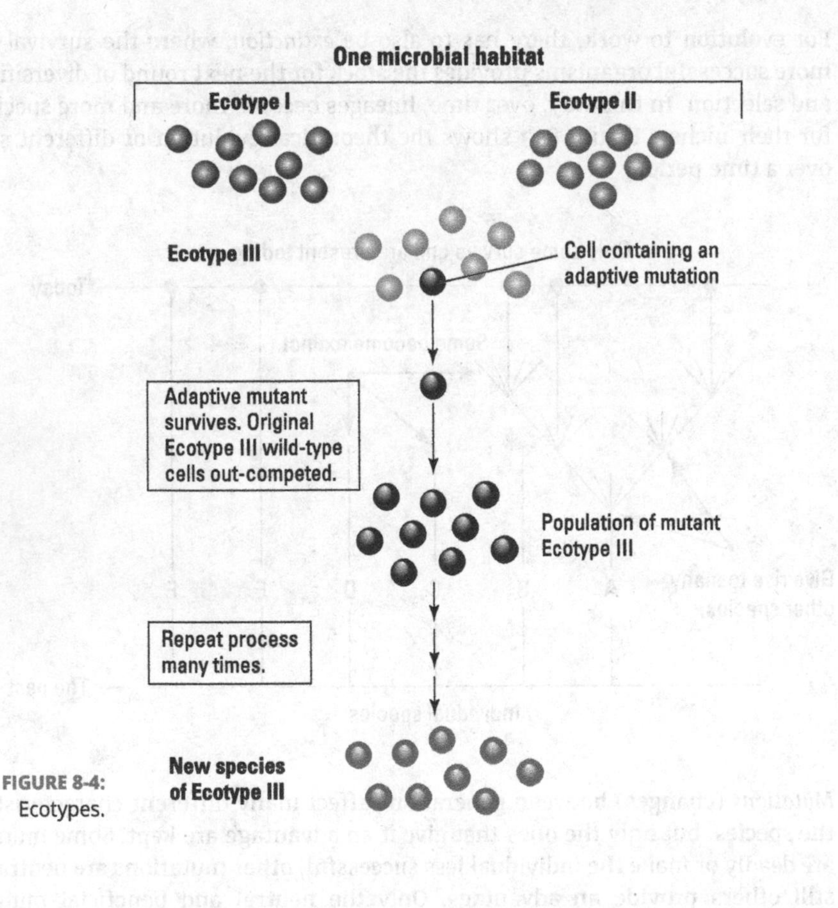

FIGURE 8-4: Ecotypes.

Studying Evolution

There are many different methods used to identify microbes. Some take advantage of metabolic differences between microorganisms by using biochemical tests to exclude groups one by one until a single species or strain is left. Others look at aspects that all microorganisms have — for instance DNA — and measure the differences between them.

Each of these approaches is important to identifying a microorganism. The first approach, based on looking at the unique aspects of each microbe, is used extensively in medical microbiology and is covered in Chapter 17. The second approach, based on the similarities between microorganisms, is used most often to study evolutionary relationships and is covered here and used in Chapter 18.

Phylogeny (the study of the evolutionary history) of the organisms alive today is possible by looking at the differences in their DNA. Notice in Figure 8-2 that all the species present now have common ancestors. If changes in the DNA are preserved from when the change happened until today, calculating the relatedness of organisms based on the degrees of difference between them should be possible.

Choosing marker genes

Changes from one generation to another are a result of mutations to an organism's DNA. Some mutations are beneficial, others are detrimental, and many are neutral. These neutral mutations don't affect evolution, so they build up in the genome and increase the number of genetic differences between species. If you consider that throughout evolution each organism carries with them all the changes that have occurred to the DNA of their ancestors, then an organism's genome contains a record of its evolutionary history.

To calculate the genetic changes within individuals, *marker genes* are used. These are genes that

- Are present in all organisms
- Are essential, so they aren't lost or inactive in some individuals
- Haven't changed too quickly, so they still have a lot of similarity overall

TECHNICAL STUFF

Examples of marker genes include the gene for the small subunit of the ribosomal RNA, elongation factor Tu gene, and the DNA gyrase gene, among others. The most widely used is the small ribosomal subunit gene. In bacteria and archaea, this is the 16S rRNA gene, and in eukaryotes it's the 18S rRNA gene. This rRNA gene in particular has been extremely useful for taxonomic assignments, as well as phylogenetic studies, but it isn't always ideal for distinguishing between very closely related organisms. To do that, either several different genes or the whole genome are compared.

Different genes accumulate changes at different rates. Some, like the 16S rRNA genes, change slowly, likely because their function is essential, so changes that affect function are quickly selected against. Other genes, like those for *luminescence* (the production of light), change more quickly over time because their function is more flexibly controlled. More slowly evolving genes are useful for telling the difference between distantly related microorganisms, whereas more quickly evolving genes are better for distinguishing between closely related species.

Seeing the direction of gene transfer in prokaryotes

One of the challenges in determining the evolutionary history of organisms based on the genetic code of bacteria and archaea is that they can transfer their genetic code in more ways than one, both *horizontally* and *vertically*.

Vertical gene transfer

Vertical gene transfer is the transfer of genetic material from the parent directly to the progeny. In bacteria and archaea, this is done through cell division. Errors during replication of the parent genome are transferred to each of the daughter cells, resulting in mutations that may persist through the next generations.

Horizontal gene transfer

Horizontal gene transfer, sometimes called *lateral gene transfer*, is the incorporation in the genome of whole genes or other large sections of DNA from an outside source. This happens through several mechanisms, including transformation and transfection (see Chapter 6).

The impact of horizontal transfer is that genes from completely unrelated species can be present within the same genome. This complicates phylogenetic analysis based on the differences in DNA sequence because it is assumed that mutations accumulated over time happen through vertical gene transfer.

Classifying and Naming Microbes

Taxonomy is the science of classifying living organisms for the purpose of

- Telling them apart
- Describing an individual
- Allowing microbiologists to speak a universal language
- Sorting out their evolution

Classification is the formal ranking that all described species are grouped into. Table 8-2 gives examples for a few different microbes. Ranks are ordered from the most inclusive at the top to the most specific at the bottom.

TABLE 8-2 The Classification of Four Microorganisms

	Bacteria from the Human Colon, Rarely Disease-Causing	Bacteria on Human Skin, Sometimes Disease-Causing	Archaea from Sulfur Acid Springs	Button Mushroom
Domain	Bacteria	Bacteria	Archaea	Eukaryota
Phylum	Proteobacteria	Firmicutes	Crenarchaeota	Basidiomycota
Class	Gammaproteobacteria	Bacilli	Thermoprotei	Agaricomycetes
Order	Enterobacteriales	Bacilliales	Sulfolobales	Agaricales
Family	Enterobacteriaceae	Staphylococcaceae	Sulfolobaceae	Agaricaceae
Genus	*Escherichia*	*Staphylococcus*	*Sulfolobus*	*Agaricus*
Species	*E. coli*	*S. aureus*	*S. acidocaldarius*	*A. bisporus*

Many unofficial ranks are added by taxonomists to help them make sense of complicated groups of organisms, but none of these is official. One helpful unofficial ranking for microbiologists is that of a *strain* (a subspecies label for bacteria that are within the same species but have different characteristics, called *phenotypes*). An example of a strain is *E. coli* strain 0H157:H7, which can make people sick, when the majority of the rest of the species is harmless.

Nomenclature is the naming system, based on Latin words, given to each species. All described species are given a two-part name that is considered its scientific name in contrast to its common name. The two-name system is important for having an accurate record of each individual species described to date. The complete name is made up of the genus and species — for example *Escherichia coli* in Table 8-2, is always italicized and is often shortened to an initial for the genus and the species name: *E. coli.*

The genus can be thought of as a noun and the species name as an adjective. For instance, *Escherichia* is a Latinized version of the last name of Theodor Escherich, who discovered this bacterium, and *coli* means "of the colon." Other examples include *Homo* ("man") *sapiens* ("wise") and *Staphylococcus* ("bacteria that look like a bunch of grapes") *aureus* ("golden").

Organizing and keeping track of all of the known prokaryotes is a big job, but several resources exist:

- **Guides for identification and classification:** These are essential when describing a new strain or species or when trying to identify a strain. The gold standard for this is *Bergey's Manual of Systematic Bacteriology,* which has both descriptions of and methods for identifying organisms based on growth and metabolism.
- **Culture collections:** Culture collections are an important resource because they keep live cultures of microorganisms that can be bought by microbiologists. Examples include: ATCC (`www.atcc.org/en/Products/Cells_and_Microorganisms.aspx`), BCCM (`http://bccm.belspo.be/catalogues`), and CIP (`http://www.pasteur.fr/ip/easysite/pasteur/en/research/collections`), among others.
- **Lists of currently described microorganisms:** These are easy to browse and search through. Some examples include: List of Prokaryotic Names with Standing in Nomenclature (`www.bacterio.cict.fr`), Bacterial Nomenclature Up-to-Date (`www.dsmz.de/bactnom/bactname.htm`), and Taxonomy Browser (`www.ncbi.nlm.nih.gov/taxonomy`).

With the use of modern phylogenetic techniques, the classification and naming of many microorganisms, especially bacteria, have changed as microbiologists have realized that some species within the same genus were actually distantly related to each other and more closely related to other species. Two important examples are the reclassification of many members of very large genera into their own separate genus. For example, some *Streptococcus species* were renamed as *Enterococcus sp.* and some *Pseudomonas species* as *Burkholderia species*. This has proven to be important especially in tracing the origins of a human infection.

CLEARING UP THE CONCEPT OF A SPECIES IN MICROBIOLOGY

In regular biology, a *species* is defined as group of organisms isolated from breeding with others either by genetics or circumstance, termed *reproductive isolation.* The concept of reproductive isolation only works for organisms that reproduce sexually, so a different approach had to be taken for prokaryotes.

The current species concept for prokaryotes is based on 16S rRNA gene identity and DNA-DNA hybridization. Together, these two methods can distinguish between most species. In medical microbiology, bacteria are identified by specific biochemical and physiological tests aimed at separating different disease-causing microbes, but of course this doesn't work for all organisms.

Another approach has been proposed for defining species: the *phylogenetic species concept,* where several conserved genes are used to build a better evolutionary history for bacteria or archaea. More complex phylogenies are both more useful and harder to make than phylogenies based on a single gene.

It's hard to even guess how many total bacteria and archaea species there are today, partly because of the difficulty in defining a species and partly because we just don't have the resources to describe them all. To date, 7,000 species of bacteria and archaea have been found for certain, with likely tens or hundreds of times more (maybe even a million total!) that haven't been identified.

Climbing the Tree of Life

It's natural to be confused about how many domains of life there are. Intuitively, it seems like there is such a variety of life around us that we can observe with our own eyes. Historically, scientists have gone through many versions of the tree of life before coming up with the one we use today. Until the advent of molecular phylogenetics, there were many different classification systems based on methods available at each time. A big problem was always where to put the bacteria.

Before microscopy became sensitive enough to see subcellular structure, bacteria were grouped with algae and fungi as part of the kingdom Planta. Next, the question was whether to have a separate kingdom for the protists and include in it all the microscopic organisms. The term *kingdom* has even been superseded by the higher ranking of domain. Within the domain Bacteria and Archaea there are no kingdom ranks, yet this rank is important within the domain Eukarya. Based on DNA sequencing of the 16S and 18S rRNA genes, we now have a universally accepted *three domain system*, which includes the Bacteria, the Archaea, and the Eukarya domains (see Chapter 3).

Phylum and lower-level reorganization happens often as we discover and describe new organisms. In fact, the fine level detail of the phylogenetic trees shown in this book will likely have changed by the time you're reading this!

WARNING

Viruses aren't classified in the tree of life because they aren't technically alive. They use the machinery of their host cell to reproduce, and their genomes can become incorporated into and mingled with that of their host.

CONTINUING THE DEBATE

Since 1978, taxonomists have enjoyed the certainty that there are three domains of life, but recently this concept has been challenged. New evidence for the existence of only *two* domains is causing controversy in the field. The debate involves how to organize the branches of Archaea and Eukarya.

If each domain represents a distinct splitting of evolutionary paths, then members of one domain should never be found to arise from another domain, but there is some evidence that this is the case with the Archaea and Eukarya. Because these branching events happened over 3 billion years ago, piecing together the history is very tricky. Also, because the Archaea were only discovered in the last 40 years, their evolutionary history is still being filled in as new members are found.

It's an exciting time for evolutionary biologists as new species are found that help answer questions about how life began!

IN THIS CHAPTER

» Understanding carbon fixation

» Seeing how light energy is used in photosynthesis

» Going through the processes of chemolithotrophy

Chapter 9 Harnessing Energy, Fixing Carbon

Living cells need energy, and they need carbon compounds that make up the bulk of the material in a cell. Microbes can get energy from the chemical bonds of complex compounds, from light or heat energy or by stripping electrons from molecules in their surroundings. Many microbes fix carbon by reducing CO_2 from their surroundings (plants do this, too); this provides not only for their carbon needs but also for the needs of organisms that can't fix carbon from CO_2. In order to fix carbon, an organism has to have a source of reducing power (provided by electron donors).

This chapter explains how microorganisms harvest energy from light or from inorganic compounds, how they get the reducing power they need, and the processes used to fix carbon.

Forging Ahead with Autotrophic Processes

For microbial cells, two things are at the top their list of priorities:

- Obtaining energy with which to create a proton motive force used to generate adenosine triphosphate (ATP)
- Converting carbon into cellular material

If a microorganism gets its carbon from organic material, the microorganism is a *heterotroph.* If the microorganism gets its carbon from inorganic carbon, like CO_2, then it's an *autotroph.* There are plenty of autotrophs in the microbial world, several of which are discussed here. One of the most common ways of assimilating CO_2 into cellular material is through the Calvin cycle.

Fixing carbon

The biological world — from microorganisms to humans — is made of carbon-based life. So, all forms of life have to obtain carbon for their cellular material. But whereas all other organisms have to use *organic* (reduced carbon) substrates for food, autotrophs have devised ways of reducing the CO_2 from the air, a process called *fixing carbon.*

There are a few different ways to fix carbon, which we cover in this section.

The Calvin cycle

This method of reducing atmospheric CO_2 is used by all green plants, purple bacteria, cyanobacteria, algae, and most chemolithotrophic bacteria, among others. It's no surprise then that the first enzyme in the cycle, ribulose bisphosphate carboxylase (better known as RuBisCO), is the most abundant protein on earth. RuBisCO makes two molecules of phosphoglyceric acid (PGA) from ribulose bisphosphate and CO_2 (see Figure 9-1). PGA then has a phosphate group (PO_3^-) added to it, after which PGA is reduced to glyceraldehyde 3-phosphate.

Glyceraldehyde 3-phosphate is one of the intermediates in the pathway of glucose breakdown called *glycolysis,* so using the fact that the enzymes for glycolysis can function in the reverse direction, as most enzymes can, glucose is then made from glyceraldehyde 3-phosphate. In this way, the cell has taken CO_2 from the atmosphere, put in ATP and NADPH, and made glucose, which can be used for building cellular material or as energy if needed.

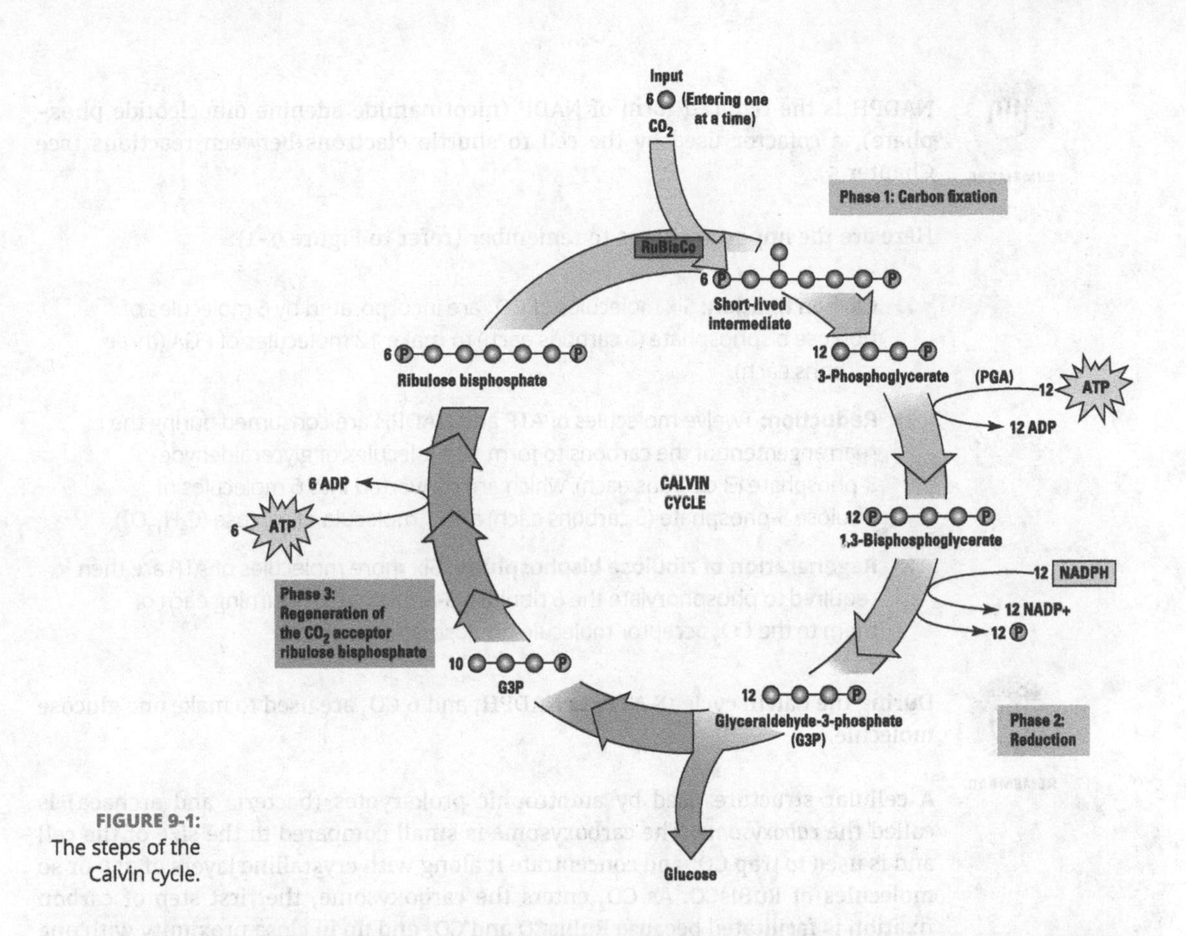

FIGURE 9-1: The steps of the Calvin cycle.

CARBON: A VERSATILE ATOM

The reason all life on earth is carbon based is that the material that makes up cells contains chains of carbon atoms. Carbon is one of the few versatile atoms with which this is possible because

- It's abundant on earth.
- It's able to form four bonds with other atoms.
- It's small enough that the strength of these bonds can support branching structures.

This makes it thermodynamically possible to use carbon building blocks to create all the many different shapes of molecules necessary in the cell.

REMEMBER

NADPH is the reduced form of NADP (nicotinamide adenine dinucleotide phosphate), a cofactor used by the cell to shuttle electrons between reactions (see Chapter 5).

Here are the important steps to remember (refer to Figure 9-1):

- **Carbon fixation:** Six molecules of CO_2 are incorporated by 6 molecules of ribulose bisphosphate (5 carbons each) to make 12 molecules of PGA (three carbons each).
- **Reduction:** Twelve molecules of ATP and NADPH are consumed during the rearrangement of the carbons to form 12 molecules of glyceraldehyde 3-phosphate (3 carbons each), which are converted into 6 molecules of ribulose 5-phosphate (5 carbons each) and 1 molecule of glucose ($C_6H_{12}O_6$).
- **Regeneration of ribulose bisphosphate:** Six more molecules of ATP are then required to phosphorylate the 6 ribulose 5-phosphates returning each of them to the CO_2 acceptor molecule ribulose bisphosphate.

REMEMBER

During the Calvin cycle 18 ATP, 12 NADPH, and 6 CO_2 are used to make one glucose molecule.

A cellular structure used by autotrophic prokaryotes (bacteria and archaea) is called the *caboxysome.* The carboxysome is small compared to the size of the cell and is used to trap CO_2 and concentrate it along with crystalline layers of 250 or so molecules of RuBisCO. As CO_2 enters the carboxysome, the first step of carbon fixation is facilitated because RuBisCO and CO_2 end up in close proximity with one another.

A second important function of the caboxysome is to keep the RuBisCO enzyme away from O_2. When oxygen is present, the enzyme will sometimes combine O_2 and ribulose bisphosphate (a process called *oxygenation*) instead of carboxylating it (combining it with CO_2). When this happens it becomes much more expensive, in terms of ATP and reducing power (electron donors), for the cell to use the oxygenated ribulose bisphosphate in further reactions.

Alternative pathways

Two other carbon-fixation pathways — which are used by the green sulfur and the green nonsulfur bacteria — are the *reverse citric acid cycle* and the *hydroxypropionate pathway.* The reverse citric acid cycle uses many of the enzymes of the citric acid cycle in the reverse direction. The hydroxypropionate pathway may be one of the earliest forms of autotrophy on earth.

THE REVERSE CITRIC ACID CYCLE

Two key steps of the reverse citric acid cycle are catalyzed by ferredoxin-linked enzymes that work to reduce CO_2 (see Figure 9-2). The first one carboxylates (adds a carboxyl group made of a carbon two oxygens and a hydrogen) succinyl-CoA to form α-ketoglutarate, and the second carboxylates acetyl-CoA to pyruvate. Other steps in this process simply reverse the action of most of the enzymes in the regular citric acid cycle (see Chapter 5).

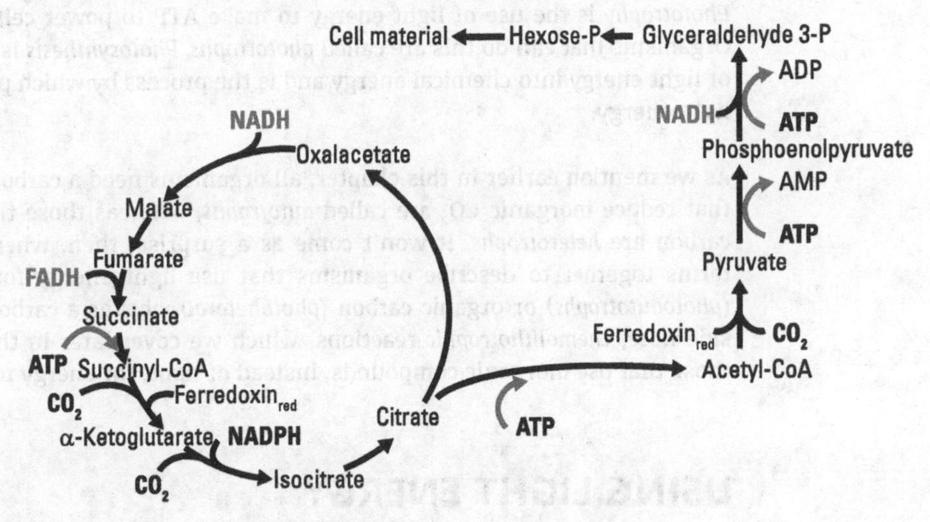

FIGURE 9-2: The reverse citric acid cycle.

REMEMBER

Ferredoxin is a nonheme iron-sulfur protein, one of the proteins important for the light reactions in these bacteria that acts as a strong electron donor.

Each turn of the reverse citric acid cycle uses three molecules of CO_2 to make pyruvate, which is then converted to glyceraldehyde 3-phosphate, which can be used to make cell material. This autotrophic pathway is common in the green sulfur bacteria, but it has been seen among a few other groups so it's likely widespread among prokaryotes.

THE HYDROXYPROPIONATE PATHWAY

Another autotrophic pathway that hasn't been seen widely among microbes is the hydroxypropionate pathway, which has been described mainly in *Chloroflexus*, a type of green nonsulfur bacteria. In this pathway, two molecules of CO_2 are converted to glyoxylate and on to cell material. Because *Chloroflexus* is thought to be the ancestor of most of the other bacterial groups that use sunlight for energy, and because this pathway has only been found here and among very ancient archaea, this pathway is thought to be one of the oldest forms of autotrophy for photosynthetic organisms on earth.

Using the Energy in Light

Phototrophy is the use of light energy to make ATP to power cellular processes. Organisms that can do this are called *phototrophs. Photosynthesis* is the conversion of light energy into chemical energy and is the process by which phototrophs use light energy.

As we mention earlier in this chapter, all organisms need a carbon source; those that reduce inorganic CO_2 are called *autotrophs*, whereas those that use organic carbon are *heterotrophs.* It won't come as a surprise, then, when we put these terms together to describe organisms that use light energy for ATP and CO_2 *(photoautotrophs)* or organic carbon *(photoheterotrophs)* as a carbon source . As a side note, *chemolithotrophic* reactions, which we cover later in this chapter, are those that use inorganic compounds, instead of light, for energy to generate ATP.

USING LIGHT ENERGY

Sunlight is electromagnetic radiation with a wide range of wavelengths. The *visible spectrum,* or the part that we can see, is only a tiny part of the electromagnetic range and it's made up of all the colors that we see. Each color is measured in wavelengths and carries a different amount of energy. A *wavelength,* as its name implies, is the distance a photon of light travels along a wavelike trajectory. Light with a shorter wavelength has a higher frequency, and vice versa. Higher-frequency light has more energy than lower-frequency light.

For our purposes, the wavelengths we need to talk about are between around 300 nanometers (nm) and 900 nm in length because light in this range is absorbed by microbes for energy. The other cool thing about light is that when one particular wavelength is absorbed — say, blue light at 420 nm — the rest of the light in the spectrum will assume the complementary color — in this case, yellow — and be seen that way by our eyes. Just outside the visible range (400 nm to 800 nm) are wavelengths that can also be used by microbes, in the ultraviolet part (300 nm to 400 nm) and the infrared part (around 900 nm) of the spectrum. Light in the ultraviolet and infrared part of the spectrum are sometimes referred to ultraviolet or infrared *radiation*.

Harvesting light: Chlorophylls and bacteriochlorophylls

Phototrophs are able to capture the energy in light thanks to photosynthetic pigments, like *chlorophyll* and *bacteriochlorophyll*, which absorb light energy kicking off a process that eventually results in the production of ATP. There are two main types of photosynthesis: those that generate oxygen (called *oxygenic photosynthesis*) and those that don't (called *anoxygenic photosynthesis*). For the most part, oxygenic phototrophs have chlorophyll whereas anoxygenic phototrophs have bacteriochlorophyll.

The overall structure of these two pigments is very similar (see Figure 9-3). They both have the distinctive tetrapyrrole ring with a Mg^{2+} in the center and a long 20-carbon phytol tail that helps anchor them to the photosynthetic membrane. The differences occur in the substitutions around the ring (highlighted in Figure 9-3) and in the length and substitutions on the phytol tail.

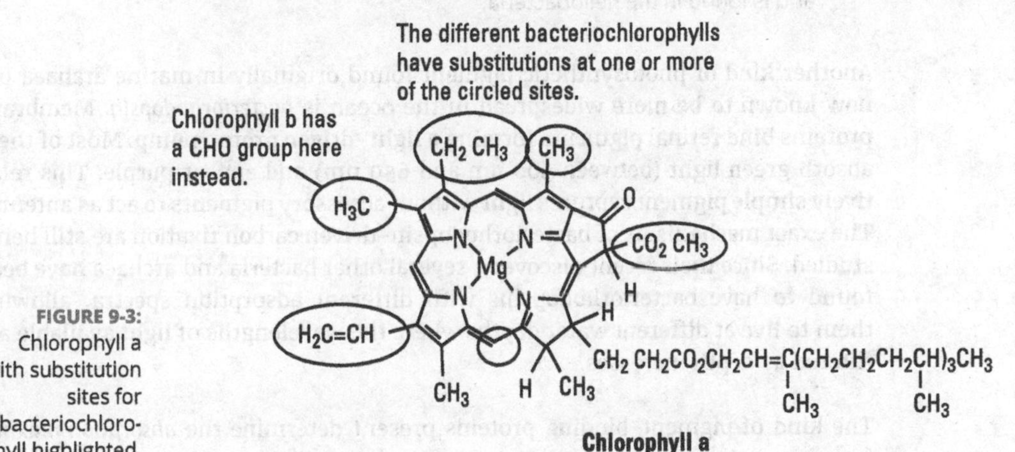

FIGURE 9-3: Chlorophyll a with substitution sites for bacteriochlorophyll highlighted.

There are four different types of chlorophyll; a and b are the most common. There are also seven known variants of bacteriochlorophylls. These types of chlorophyll and bacteriochlorophyll differ in structure, and those differences affect the specific wavelength of light that each can absorb, which allows several different species of microbes together to collect the full spectrum of light, each absorbing a different range of wavelengths. Here is a list of the known types of chlorophyll and bacteriochlorophyll:

» **Chlorophyll a** absorbs red light (around 680 nm) and is the main pigment in higher plants, many algae and the cyanobacteria.

» **Chlorophyll b** also absorbs red light (660 nm) and is found in all higher plants, as well as a group of bacteria called prochlorophytes.

» **Chlorophyll c** is found in eukaryotic microbes, like marine and freshwater algae, and absorbs red light (between 450 and 640).

» **Chlorophyll d** is found in a type of cyanobacterium that lives in areas lacking visible light, but containing infrared radiation (700 nm to 730 nm), like nestled underneath corals and algae.

» **Bacteriochlorophyll a and b** absorb infrared radiation (in the range of 800 to 1,040 nm) and are found in the purple bacteria.

» **Bacteriochlorophyll c, d, and e** absorb far red light (in the 720 nm to 755 nm range) and are found in the green sulfur bacteria.

» **Bacteriochlorophyll c_s** also absorbs far red light (720 nm) and is found in the green nonsulfur bacteria.

» **Bacteriochlorophyll g** absorbs red or far red light (at either 670 nm or 788 nm) and is found in the heliobacteria.

Another kind of photosynthetic pigment found originally in marine archaea but now known to be more widespread in the ocean is *bacteriorhodopsin*. Membrane proteins bind retinal pigments forming a light-driven proton pump. Most of them absorb green light (between 500 nm and 650 nm) and appear purple. This relatively simple pigment captures light without accessory pigments to act as antenna. The exact mechanisms of bacteriorhodopsin-driven carbon fixation are still being studied. Since their recent discovery, several other bacteria and archaea have been found to have bacteriorhodopsins with different adsorption spectra, allowing them to live at different water depths where the wavelengths of light available are filtered as depth increases.

The kind of pigment-binding proteins present determine the *absorption maxima* for an organism. The absorption maxima is the range of wavelengths of light that provide the most energy to that organism. There are many types of these pigment-binding proteins, and their position around the light-harvesting pigments can change the spectrum of light absorbed. The combination of light-harvesting pigments and pigment-binding proteins is called the *photocomplex* and always occur within a membrane. The arrangement within a membrane is essential for creating the proton motive force necessary to generate ATP.

The photocomplexes are organized so that there is a central reaction center (usually made of chlorophyll or bacteriochlorophyll) around which as many as 300 accessory pigments are arranged to gather up the light energy and pass it to the reaction center. When the surrounding pigments contain additional chlorophyll or bacteriochlorophylls, they're appropriately called *antennae*, because they absorb as much light as possible and funnel it to the reaction center. The pigments in the reaction center participate directly in the reactions involved in converting light energy to chemical energy. This setup is especially essential for phototrophs that live in low light conditions.

Photosynthetic membranes, which house the photocomplexes, occur in all phototrophs, but their structure can be very different in each organism. In eukaryotes, which have cellular compartments called organelles, it's common to find structures called *chloroplasts*. Chlorplasts contain *thylakoid membranes* arranged in stacks called *grana* (see Figure 9-4a) and participate in generating the proton motive force during photosynthesis. In microorganisms that don't have traditional organelles, like the bacteria and archaea, a variety of photosynthetic membranes can be found:

- The cyanobacteria have thylakoid membranes, which are not contained within a chloroplast.
- The purple bacteria use structures called *lamellae,* that are made by the inward folding of the cytoplasmic membrane (see Figure 9-4b) and *chromatophores* (see Figure 9-4c), that are vesicular structures made from the membrane.
- The cytoplasmic membrane itself is used by the heliobacteria.
- The *chlorosome,* a specialized structure in the green bacteria, allows growth in environments such as deep in lakes and in areas that have the lowest light intensities. Instead of the antenna pigments surrounding the reaction center within the photosynthetic membrane, they're arranged into dense arrays inside the chlorosome, which lies adjacent to the cytoplasmic membrane where the reaction centers are located (see Figure 9-4d).

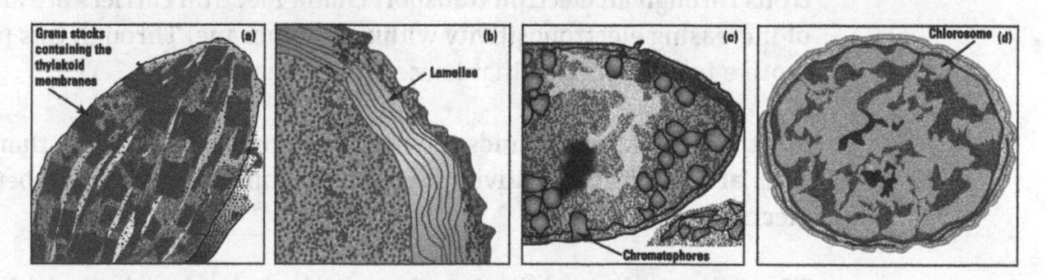

FIGURE 9-4: Types of phytosynthetic membranes: (a) grana, (b) lamellae, (c) chromatophores, and (d) chlorosomes.

Helping photosynthesis out: Carotenoids and phycobilins

Making a living from light energy is a double-edged sword — it means being constantly exposed to a source energy that can be harmful. Bright light causes the formation of singlet oxygen ($^1O_2$) through photo-oxidation reactions. Singlet oxygen, and free radicals in general, are toxic because they can randomly energize other molecules. When members of the photosynthetic complex are oxidized by singlet oxygen, they become nonfunctional, which can cause major problems. Carotenoids come to the rescue by absorbing high-energy blue light (440 nm to 490 nm) and quenching the toxic oxygen species (namely, $^1O_2$) before they cause damage.

Carotenoids are the reason many photosynthetic bacteria, other than the cyanobacteria, appear in bright colors of pink, red, brown, or yellow. This is because they absorb blue light and reflect red, brown, or yellow and because of the sheer number of them in the cell. Although carotenoids can act as accessory pigments transferring the light energy they've gathered to the reaction center, for the most part they're photoprotective.

Members of another class of pigments called *phycobilins* are important to cyanobacteria and red algae. They enhance the light-gathering ability of these organisms by absorbing energy and transferring it to the main chlorophylls involved in photosynthesis. Red phycobilins absorb green light (550 nm), whereas blue phycobilins absorb red light (620 nm). All phycobilins have a structure called a *bilin*, which is a pyrrole ring along an open chain (seen in the nearby sidebar), and are bound to a protein to make the a complex called a *phycobiliprotein*. Many phycobiliproteins combine to form a *phycobilisome*, a structure of tightly packed light-harvesting molecules that help cyanobacteria and algae to grow in places with very low light levels.

Generating oxygen (or not): Oxygenic and anoxygenic photosynthesis

The purpose of photosynthesis is to harness light energy and use it to move electrons through an electron transport chain. Electron carriers are arranged, in order of increasing electropositivity within a membrane. Through this process, a proton motive force is created that is used to produce ATP.

REMEMBER

Electronegative compounds are better at donating electrons than electropositive ones are. As electropositivity increases, a compound becomes better at accepting electrons.

The compounds used to carry electrons include pheophytin (chlorophyll without the magnesium ion (Mg^{2+}) center), quinones, cytochromes, plastocyanins (copper-containing proteins), nonheme iron sulfur proteins, ferredoxin, and flavoproteins (covered in detail in Chapter 5).

There are two main types of photosynthesis: *oxygenic* (the kind that generates O_2) and *anoxygenic* (the kind that doesn't generate O_2). Anoxygenic photosynthesis is used mainly by the purple bacteria, the green sulfur and nonsulfur bacteria, the heliobacteria and the acidobacteria. Oxygenic photosynthesis is used by the cyanobacteria, the algae, and by plants.

Oxygenic photosynthesis

Oxygenic photosynthesis occurs in, among others, eukaryotic microorganisms like algae and in bacteria such as cyanobacteria; the same mechanism is at work in both. Electron flow happens through two different electron transport chains that are connected; together, these electron transport chains are called the *Z scheme* (see Figure 9-5). The stars of each chain are photosystem I (PSI) and photosystem II (PSII), each containing chlorophyll reaction centers surrounded by antenna pigments.

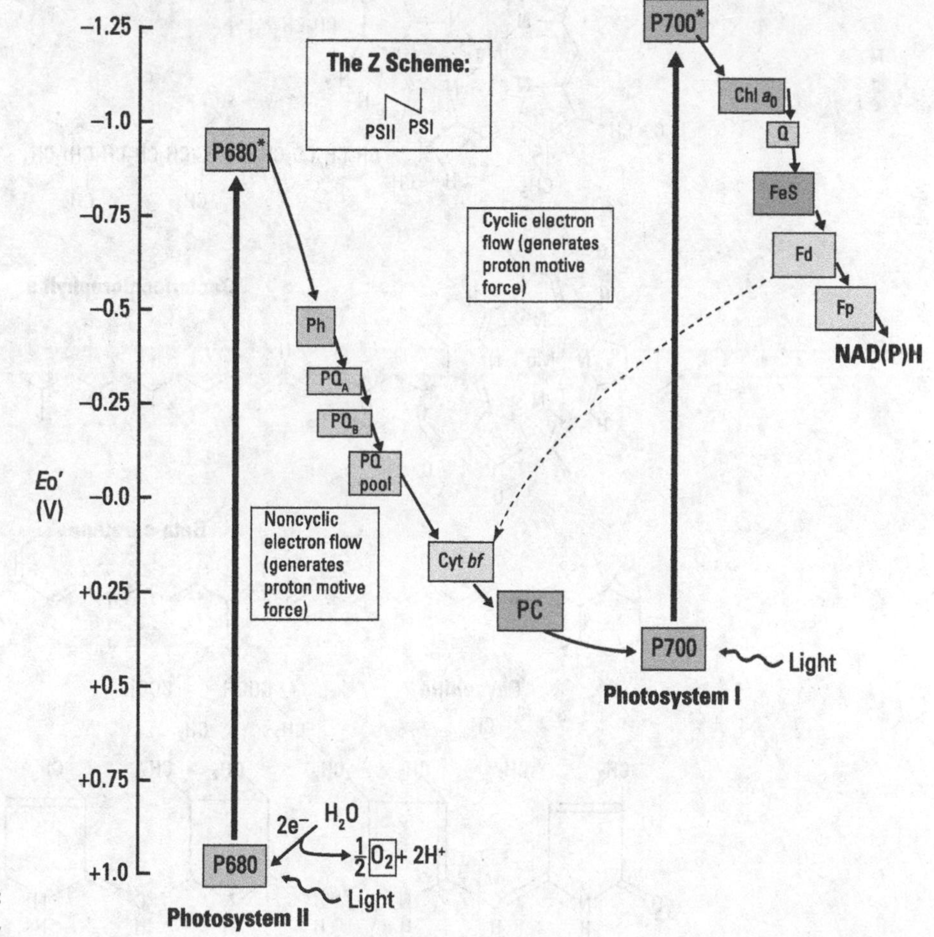

FIGURE 9-5: The Z scheme.

STRUCTURES OF PHOTOSYNTHETIC PIGMENTS

If you squint at the structure of some of the photoactive pigments discussed in this chapter, you'll see that they all have a few things in common: chains of carbons arranged into rings, a lot of double bonds, and side chains (refer to the following figure). The thing about double bonds and rings is that they're great at sharing electrons with each other so all the electrons along the structure of the molecule form a kind of cloud of electronegativity. When such a molecule absorbs a packet of light energy (called a *photon*), the electron cloud becomes excited, causing one electron to be transferred to another molecule rather easily.

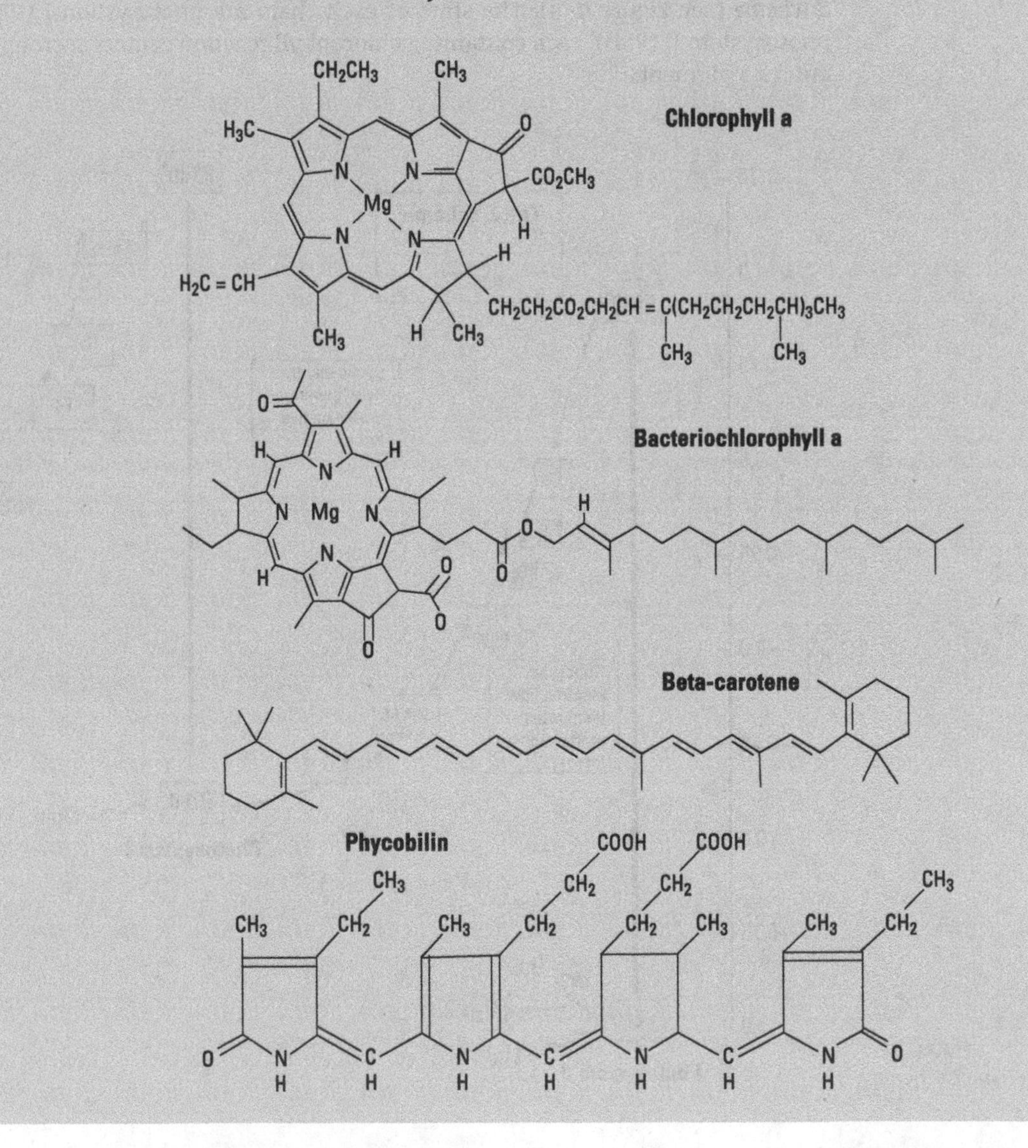

TIP

Don't let the names fool you, the flow of energy is from PSII to PSI.

The chlorophyll in PSI is called P700, and the chlorophyll in PSII is called P680, for the wavelengths of light each absorbs most efficiently. The steps involved are illustrated in Figure 9-5 and summarized here.

1. Light energy (a photon of light) is absorbed by PSII, exciting P680 and making it into a good electron donor that reduces the first member of the electron transport chain, pheophytin.

   PSII is normally very electropositive and it would just remain reduced unless excited by light.

2. Water is split to generate electrons used to reduce P680 back to its resting state. The protons (H^+) from water act to create the proton motive force, whereas the oxygen is released (giving the pathway its name).
3. The electrons travel through several electron carriers until eventually reducing P700 in PSI. P700 is already oxidized after having absorbed light and donated an electron to the next electron transport chain.
4. After passing through a series of electron carriers, the last step in the process is the reduction of $NADP^+$ to NADPH.

Aside from the production of NADPH, electron transport functions to create the proton motive force, which is used by ATP synthase to generate ATP.

REMEMBER

Because electrons don't cycle back to reduce the original electron donor, this pathway is called *noncyclic photophosphorylation*. If things are ideal and enough reducing power (extra electrons) is available, some of the electrons do travel back to reduce P700 and in the process add to the proton motive force that generates ATP (or *phosphorylation*). When this happens, it's called *cyclic photophosphorylation*.

The cool thing about microbes is how resistant they are to extenuating conditions. For example, when PSII is blocked, some oxygenic phototrophs can use cyclic photophosphorylation with PSI alone in a similar way to how anoxygenic phototrophs do it. Instead of oxidizing water, they use either H_2S or H_2 as the electron donor to provide the reducing power (the electrons) for CO_2 fixation.

Anoxygenic photosynthesis

Many of the steps in anoxygenic photosynthesis are the same as those for oxygenic photosynthesis (see the preceding section). For example, light excites the photosynthetic pigments, causing them to donate electrons to the electron transport chain and ATP is again generated from the proton motive force created by electron transport (see Figure 9-6).

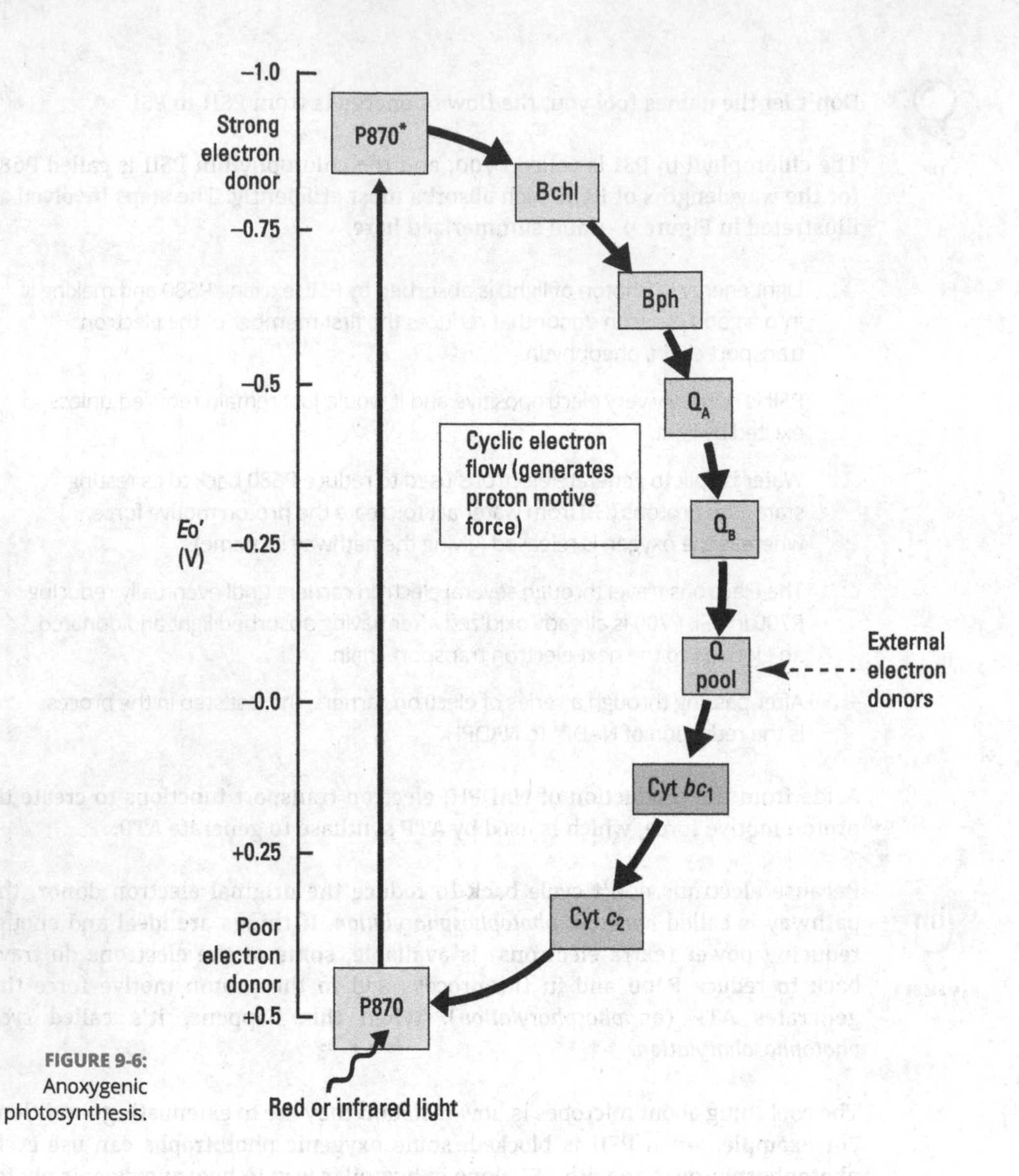

FIGURE 9-6: Anoxygenic photosynthesis.

Here are the main ways that anoxygenic photosynthesis differs from oxygenic photosynthesis:

- Oxygen is not released because P680 of PSII is not present. Water is too electropositive to act as the electron donor for the photosystem.
- Depending on the species, the reaction center can consist of chlorophyll, bacteriochlorophyll, or other similar pigments. The reaction center in purple bacteria is called P870.

- Some of the carriers within the electron chain are different, including bacteriopheophytin, which is bacteriochlorophyll without its Mg^{2+} ion.
- Electrons cycle back to reduce P870, so this is a cyclic electron transport chain leading to generation of ATP through cyclic photophosphorylation.
- Unlike in oxygenic photosynthesis, where NADPH is the terminal electron acceptor, no NADPH is made because electrons are cycling back into the system.

Without NADPH, cells have to come up with another way of generating the reducing power necessary to drive the Calvin cycle for carbon fixation. This is accomplished through oxidization of things like inorganic compounds. The electrons donated are added to either the quinone pool (purple bacteria) or donated to iron-sulfur proteins (the green sulfur and nonsulfur bacteria, and the heliobacteria).

When the electron acceptor is not sufficiently electronegative (as in the case of quinone), then *reverse electron flow* is needed to get the necessary reducing power. Reverse electron flow uses the proton motive force to push electrons to reduce $NADP^+$. This mechanism is used frequently in other situations, where several turns of the electron transport cycle are necessary to generate enough power to reduce one molecule of NAD^+ or $NADP^+$.

In some phototrophs, both ATP and reducing power (that is electron donors like NADH or NADPH) are produced from the light reactions, whereas in others (like the purple bacteria) the light reaction producing ATP but reducing power has to be obtained in separate reactions (like oxidizing inorganic compounds).

Getting Energy from the Elements: Chemolithotrophy

Chemolithotrophy is the use of inorganic compounds for energy. Most chemolithotrophs are autotrophs (reduce CO_2 to make cellular material), but some are *mixotrophs,* meaning that they use inorganic chemicals for energy and organic compounds as carbon sources. In a nutshell, inorganic compounds are oxidized so that NAD^+ can be reduced, and reverse electron transport is used only if the source of electrons is more electropositive (has a higher affinity for electrons) than NADH.

Inorganic substances used for chemolithotrophy include but are not limited to

- Hydrogen sulfide (H_2S) and elemental sulfur (S^0) from volcanoes
- Ammonia (NH_3), nitrate (NO_2^-) and ferrous iron (Fe^{2+}) from mining, fossil fuels, agriculture, and industrial waste.
- Hydrogen (H_2), as well as ammonia and hydrogen sulfide, all produced by other organisms.

REMEMBER

Only microorganisms are able to perform these chemolithotrophic reactions, which are at the heart of most nutrient cycling in nature.

Harnessing hydrogen

Oxidation of hydrogen (H_2) can happen in both the presence and absence of oxygen (in oxic and anoxic environments). The main difference in these two situations is the terminal electron acceptor used. When oxygen is present, it's the obvious choice because it has a high affinity for electrons and when reduced it forms water (H_2O). The oxidation of H_2 allows ATP to be made by the proton motive force. The enzyme involved is hydrogenase and it comes in two forms: a membrane-bound form for making ATP and a soluble form for reducing NAD^+.

The H_2 these microbes use is the result of fermentation reactions, which occur for the most part anaerobically. This means that there is little H_2 in oxic environments. Also, the H_2 present in anoxic conditions is snapped up quickly by anaerobic hydrogen oxidizers, which are plentiful in these environments. Because of this, aerobic hydrogen bacteria have adapted to live in microaerobic environments. Here the small amount of oxygen inhibits the growth of anaerobic bacteria and as H_2 is continuously available it floats up from the anoxic space below.

Most hydrogen bacteria can also be chemoorganotrophic, meaning that they can use organic carbon when it's available. This ability to obtain carbon in an alternate way makes them *facultative* chemolithotrophs. When glucose is present, for instance, they repress CO_2 fixation and hydrogenase.

Securing electrons from sulfur

The oxidation of sulfur compounds is performed by the sulfur bacteria. Sources of sulfur include hydrogen sulfide (H_2S), elemental sulfur (S^0), and thiosulfate ($S_2O_3^{2-}$); the oxidized form of these is sulfate (SO_4^{2-}). There are two parts to complete sulfur oxidation; some microbes perform both, whereas others only use one part.

The first part is as follows:

$$H_2S + ½\,O_2 \rightarrow S^0 + H_2O$$

The S^0 is deposited either outside or inside the cell. When stored inside the cell, it's used as an energy reserve. When deposited into the environment, S^0 forms crystals that are insoluble. Microbes have to attach to the crystals directly in order to oxidize it.

The second part is as follows:

$$S^0 + 1½\,O_2 + H_2O \rightarrow SO_4^{2-} + 2H^+$$

Bacteria that perform this last stage of the reaction have to be acid tolerant because pumping protons (H^+) out of the cell lowers the pH of the surrounding environment. One of the enzymes used is *sulfite oxidase*, which transfers electrons directly from thiosulfate to cytochrome c. A regular electron transport chain is involved, and ATP is made as a result. Another pathway uses an enzyme called *adenosine phosphosulfate reductase*, which is usually found in sulfur-reducing bacteria, but here its action is reversed to oxidize sulfur and generate one phosphate bond in ATP (which has three). A third type of reaction uses the sulfur oxidation (or Sox) system, which is made up of more than 15 gene products. This complex system is used by several chemolithotrophs and even some phototrophs that oxidize sulfur in order to reduce CO_2, as discussed earlier.

Reduced sulfur compounds donate electrons that travel through the electron transport chain to O_2. As always, this creates the proton motive force and ATP. The electrons for carbon fixation, however, come from reverse electron flow (where part of the proton motive force generated is used) in order to reduce NAD^+ to NADH. When oxygen isn't present, however, some use nitrate as an electron acceptor.

Pumping iron

Aerobic oxidation of ferrous iron (Fe^{2+}) to ferric iron (Fe^{3+}) happens in acidic conditions. The reaction creates insoluble iron in water in the form of ferric hydroxide or $Fe(OH)_3$ and acidifies the environment further. The bacteria that do this get so little energy from the reaction that they have to oxidize massive amounts of iron to survive, creating red iron deposits where they live.

In these acidic conditions, Fe^{2+}/Fe^{3+} is very electropositive so the route to reduction of O_2 is necessarily very short. Another aspect of this system is that the transfer of electrons from the first electron acceptor (cytochrome c) to rusticyanin is actually energetically unfavorable. The reaction proceeds, however, because removal of Fe^{3+} from the system by the formation of $Fe(OH)_3$ pulls the whole thing forward.

Other bacteria make a living from catching ferrous iron that comes up from groundwater. As groundwater flows over iron deposits, it collects ferrous iron and keeps it in an anoxic environment. When not in acidic conditions, ferrous iron oxidizes spontaneously when it comes into contact with oxygen.

One of the most ancient forms of iron oxidation happens anaerobically. This type of reaction happens at neutral pH and is likely how the large iron deposits in the earth's surface were laid down way before we had an oxygenated environment. The way this works is that NO_3^- acts as the electron acceptor, and the energy is used either for energy (ATP) or for reducing carbon.

Oxidizing nitrate and ammonia

The oxidization of nitrogen compounds is called *nitrification.* The inorganic compounds ammonia (NH_3) and nitrite (NO_2^-) are oxidized by nitrifying bacteria and archaea. Most nitrifiers are also autotrophs, getting their carbon from CO_2. These types of electron donors are very electropositive and hang on to their electrons for dear life. This means that the electron acceptors have to be very, very electropositive and only a small amount of energy is released. Because of this, the proton motive force is weak and growth rates are slow.

There are two steps in nitrification. The first step is being catalyzed by ammonia monoxygenase (AMO) and the second step is being catalyzed by nitrite oxidoreductase:

$$NH_3 \rightarrow NO_2^- \rightarrow NO_3^-$$

Bacteria and archaea are known to have similar AMO genes, yet to date only the purple phototrophic bacteria have been found to perform the second step. They do this in anoxic conditions and use the energy for reducing power in carbon fixation.

Nitrifiers are especially important for the nitrogen cycle. In nature, decomposing organic material releases ammonia (NH_3), which when oxidized to NO_3^- provides an essential nutrient for plant, algae, and cyanobacteria growth.

Anammox

Another fascinating route for ammonia oxidation is through anammox or *anoxicammoniaoxidation.* This reaction is done under strict anaerobic conditions by members of the *Planctomycete* group of bacteria. These bacteria lack the kind of cell wall of other bacteria and have a structure called the *anammoxosome* (see Figure 9-7), which takes up most of the cell. This structure has an impermeable membrane made of unique lipids and functions to stop any diffusion of the products of ammonium oxidation into the cytoplasm, which is important since they are toxic. This precaution is necessary because an intermediate of the reaction is hydrazine (N_2H_4), an ingredient in rocket fuel. The full reaction sees the oxidization of ammonium (NH_4^+) with nitrite as the electron acceptor. These bacteria are autotrophs, but they don't use the Calvin cycle for CO_2 fixation. Instead, they use the strong reducing power of hydrazine in the acetyl CoA pathway to fix carbon.

The nitrite for anammox comes from aerobic ammonia oxidizing organisms in nitrogen-rich environments like sewage treatment plants and other areas where there is wastewater. Here, suspended particles can contain both oxic and anoxic areas acting as habitats for both aerobic and anaerobic microbes. Because they're inhibited by O_2, anammox can only exist in strictly anoxic environments where ammonium and nitrite are both found.

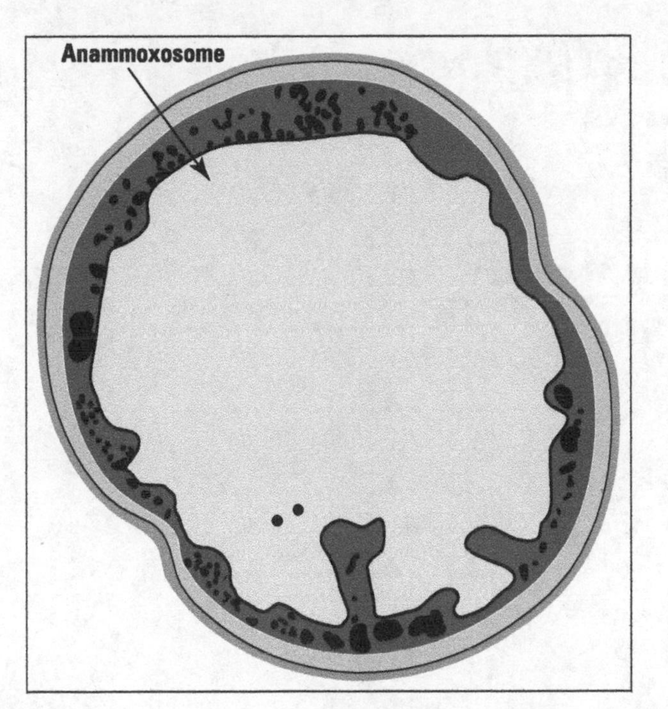

FIGURE 9-7: A *Planctomycete* cell with the anammoxosome.

IN THIS CHAPTER

» Using glucose for energy

» Digesting the steps in respiration

» Understanding the different types of fermentation

Chapter 10 Comparing Respiration and Fermentation

Every cell, whether microbial or not, must continuously produce adenosine triphosphate (ATP) to power cellular reactions. ATP, the currency of the cell, is a short-term energy-storing molecule and is around for only a few seconds to a minute at a time. Organisms that get energy from the organic compounds in their environment are called *chemoorganotrophs*. There are two main strategies for extracting energy from organic compounds: respiration and fermentation. In this chapter, we show you how these two strategies differ and compare the amount of ATP made from glucose by each. You see that respiration is also possible in the absence of oxygen and that most of the metabolic diversity in prokaryotes is due to fermentation pathways. We also talk about how some organisms can metabolize in more than one way and why that can be helpful for survival when conditions change.

Lifestyles of the Rich and Facultative

The presence, or absence, of oxygen can cause profound responses from many microorganisms. These can be put into four groups based on their relationship with molecular oxygen (O_2). The term *obligate* refers to the absolute need for something, whereas the term *facultative* means that they can function in conditions that are not their preferred conditions.

Aerobes use oxygen in their metabolism as the terminal electron acceptor for the electron transport chain in respiration. Different aerobes can grow in a range of oxygen concentrations for example:

- Some microbes, simply called **aerobes,** grow well at full oxygen levels (about 21 percent in air).
- **Microaerophiles** prefer to grow in conditions where the concentration of oxygen is very low.
- **Facultative aerobes** either grow either in the presence or the absence of oxygen. When oxygen is present these organism use it during respiration and when oxygen isn't present they use other pathways for their metabolism.

Anaerobes don't use oxygen for their metabolism but instead these type of microbes use fermentation or anaerobic respiration. Different anaerobes respond differently to oxygen in their environment:

- **Obligate anaerobes** are either inhibited or killed by oxygen. They live in environments that are devoid of all oxygen, like aquatic sediments or the colon of animals.
- **Aerotolerant anaerobes** essentially ignore oxygen in their environment and can grow well in it's presence or absence. Because they lack the citric acid cycle and/or an electron transport chain, they don't switch to aerobic respiration when oxygen is present.

In addition to the machinery needed to use oxygen in metabolic processes, another set of enzymes is needed if a microbe is going to survive in *oxic environments* (environments with oxygen).

Oxygen is one of those molecules that can be converted easily to a high-energy ion, called a *free radical*, through a series of steps, usually as part of its use as an electron acceptor. Free radicals are a problem because their unstable high-energy state can readily cause damage to sensitive molecules in the cell like DNA and proteins.

Free radicals from oxygen are a reality that every cell in an aerobic environment has to live with, so enzymes such as superoxide dismutase, catalase, and peroxidase have evolved to stop them before they even get started. All the groups of microbes listed in this section have a form of one of these enzymes, except for obligate anaerobes, which can't even tolerate oxygen.

Seeing the Big Picture

Sugars such as glucose, and organic substance in general, are great sources of energy and are used by many microorganisms for fuel.

REMEMBER

Organic substances are those that contain at least two carbons bonded to one another; *inorganic* substances are everything else.

For the most part, the breakdown of an energy source can happen in one of two ways: by respiration or by fermentation. But before respiration or fermentation, glucose is first broken down by glycolysis.

Respiration and fermentation are fundamentally different, mostly in the way they handle the substrate pyruvate. With exceptions, pyruvate is the product of glycolysis that is the main input into both pathways. These two metabolic strategies also differ in the products they release and how much total energy they can extract in the process. Figure 10-1 provides an overview of the respiration and fermentation processes.

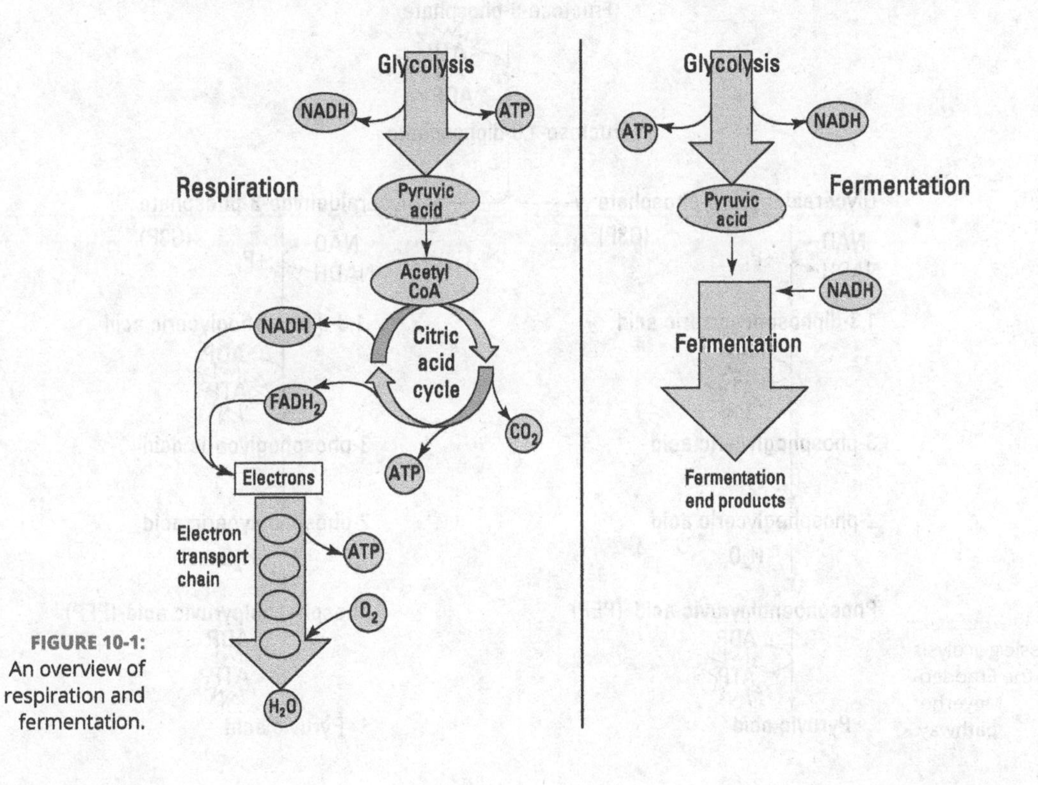

FIGURE 10-1: An overview of respiration and fermentation.

TECHNICAL STUFF

Glycolysis, discussed briefly in Chapter 5, is the breakdown of glucose and happens by way of three possible pathways. The most common, which is present in all eukaryotes and some prokaryotes is the Embden-Meyerhof pathway or classic glycolysis. The other two glycolytic pathways are phosphoketolase (sometimes called the heterolactic pathway) and the Entner-Doudoroff pathway. Because it often precedes either respiration or fermentation, classic glycolysis is discussed here. The other pathways are discussed in the "Figuring Out Fermentation" section, later in this chapter.

The point of glycolysis is to split a six-carbon sugar such as glucose into two molecules of pyruvic acid, which has three carbons. The steps are shown in Figure 10-2. Although this looks a bit scary, it's actually fairly straightforward.

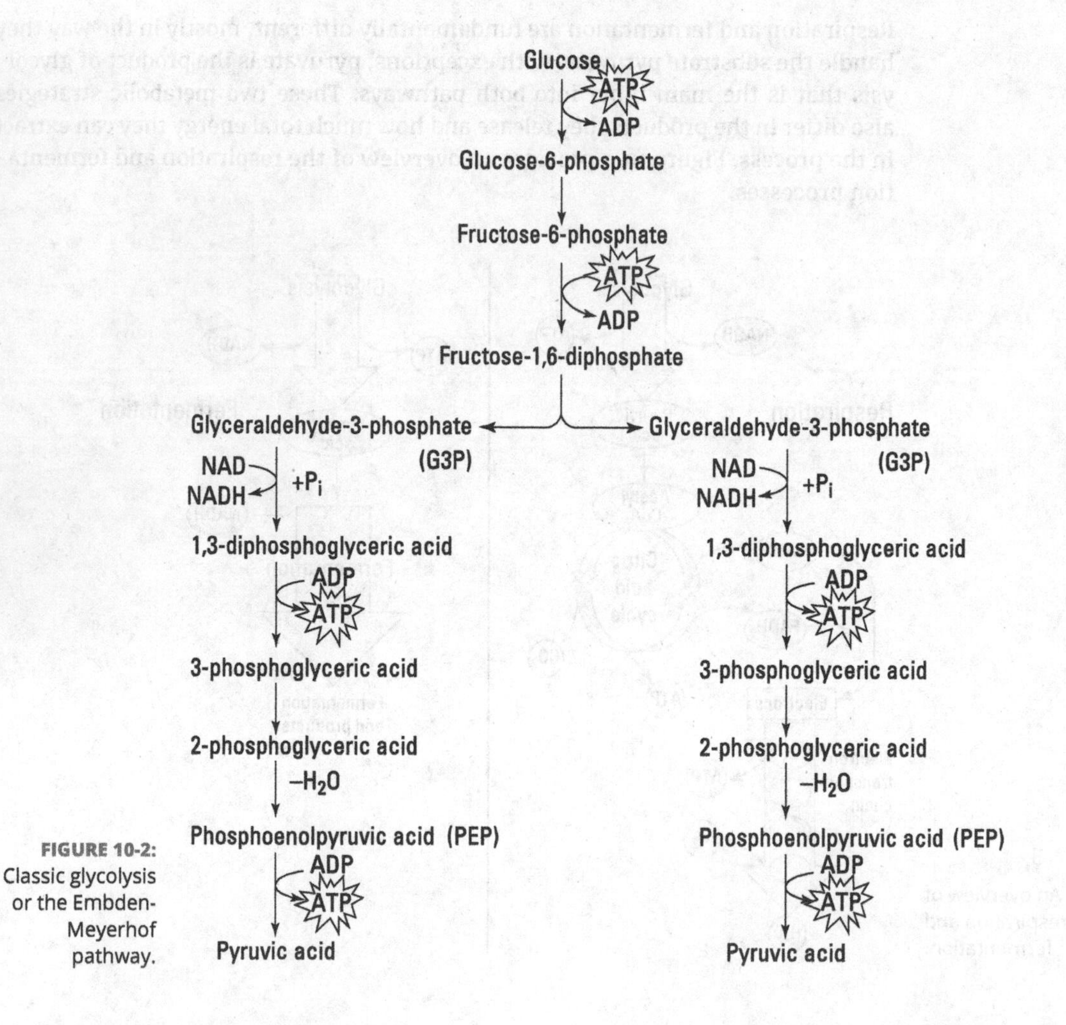

FIGURE 10-2: Classic glycolysis or the Embden-Meyerhof pathway.

In the first stage of this process, called the *preparatory stage*, some energy has to be put in to get the molecules ready. Two ATPs are used during this preparatory stage, each donating one phosphate (P_i) that will be attached to the intermediate products. At the end of the prep stage, the six carbon sugar fructose-1,6-diphosphate is cleaved into two molecules of glyceraldehyde-3-phosphate (G3P) with three carbons each that go on to the *energy conservation stage*.

TECHNICAL STUFF

Technically, in the first stage of glycolysis, one molecule of G3P and one molecule of dihydroxyacetone phosphate (DHAP) are produced, but because DHAP is easily converted to G3P and used in the next stage of glycolysis, we can say that two G3P were made.

TIP

Following the number of carbons in each step helps when trying to account for all the products of glycolysis and respiration. It's only when we get to fermentation that you want to pay more attention to the other atoms bonded to the sugar because the products of fermentation still have energy in them and also have important properties.

In the second stage of glycolysis, the energy conservation stage, the cofactor NAD^+ is reduced to NADH and ATP is formed along with pyruvic acid, which can then be used for either fermentation or respiration. Here's how this happens: Each molecule of G3P is first oxidized when it donates two protons to two molecules of NAD^+ to form two molecules of NADH. Next, another P_i is added to it so that it now has two phosphate bonds. One of these P_i is transferred to ADP to form ATP. Next, a rearrangement happens: A water molecule (H_2O) is lost and a second ATP is formed by the transfer of the last P_i to ATP from the three-carbon intermediate phosphoenolpyruvic acid (PEP). What's left after this step is pyruvic acid, a three-carbon molecule with no phosphates.

So, glycolysis breaks glucose into two molecules of pyruvic acid and produces a net amount of two ATP molecules and two NADH molecules. This may seem off, but remember that two ATP and one NADH were produced for each molecule of G3P, of which there were two. The net amount of ATP and NADH made in glycolysis is as follows:

-2 ATP (put in at the beginning)

+4 ATP and 2 NADH produced

2 ATP and 2 NADH net

The ATP is used to power cellular processes that require energy. NADH is considered a store of reducing power (a cofactor used in redox reactions) that can either go into the electron transport chain in respiration or be used to reduce pyruvate in fermentation. Both of these scenarios are discussed in this chapter.

Digging into Respiration

Compared to fermentation, respiration is a very efficient, but slow way of getting energy from glucose. It requires pyruvic acid, a terminal electron acceptor, and an electron transport chain, and it generates ATP, NADH and $FADH_2$ (another cofactor used in redox reactions). When oxygen is present, it acts as the terminal electron acceptor and is reduced to water; in anoxic conditions, where no oxygen is present, another inorganic or sometimes organic compound is reduced.

Respiration, whether aerobic or anaerobic, involves the citric acid (TCA) cycle and an electron transport chain. Unlike the citric acid cycle, which uses the same enzymes in almost all organisms, the electron transport chain employs slightly different electron carriers in different microbes. Despite these details, the end result is the same for all electron transport chains: Energy is released and used to create the proton motive force across a membrane, which in turn is used to make ATP.

So, now that we have pyruvic acid made from glucose through glycolysis, we can go straight to talking about the citric acid cycle, right? Not so fast. First, each pyruvic acid has to be oxidized to form acetyl-CoA (a two-carbon molecule), during which NAD^+ is reduced to NADH and CO_2 is released. Acetyl-CoA then enters the citric acid cycle where more ATP is made and reducing power is stored in cofactors.

Spinning the citric acid cycle

Imagine the citric acid cycle, also called the Krebs cycle, as a merry-go-round with kids jumping on and off as it turns. Acetyl-CoA, NAD^+, and FAD are the kids jumping on, and CO_2, NADH, $FADH_2$, and ATP are the kids jumping off.

Figure 10-3 shows the citric acid cycle. The important steps are that acetyl-CoA first combines with oxaloacetic acid. Then, through a series of steps, it's converted to succinyl-CoA that is used to make ATP by *substrate-level phosphorylation*, leaving succinate that is then returned to oxaloacetate so that it can combine with another acetyl-CoA.

ATP can also be made without the use of the electron transport chain using a process called *substrate-level phosphorylation*, that involves converting an organic compound from one form to another. During the conversion enough energy is released to make one molecule of ATP. Figure 10-4 gives examples of substrate-level phosphorylation and how they generate ATP without the enzyme ATP synthase. In the "Figuring Out Fermentation" section of the chapter, this way of generating ATP will be very common.

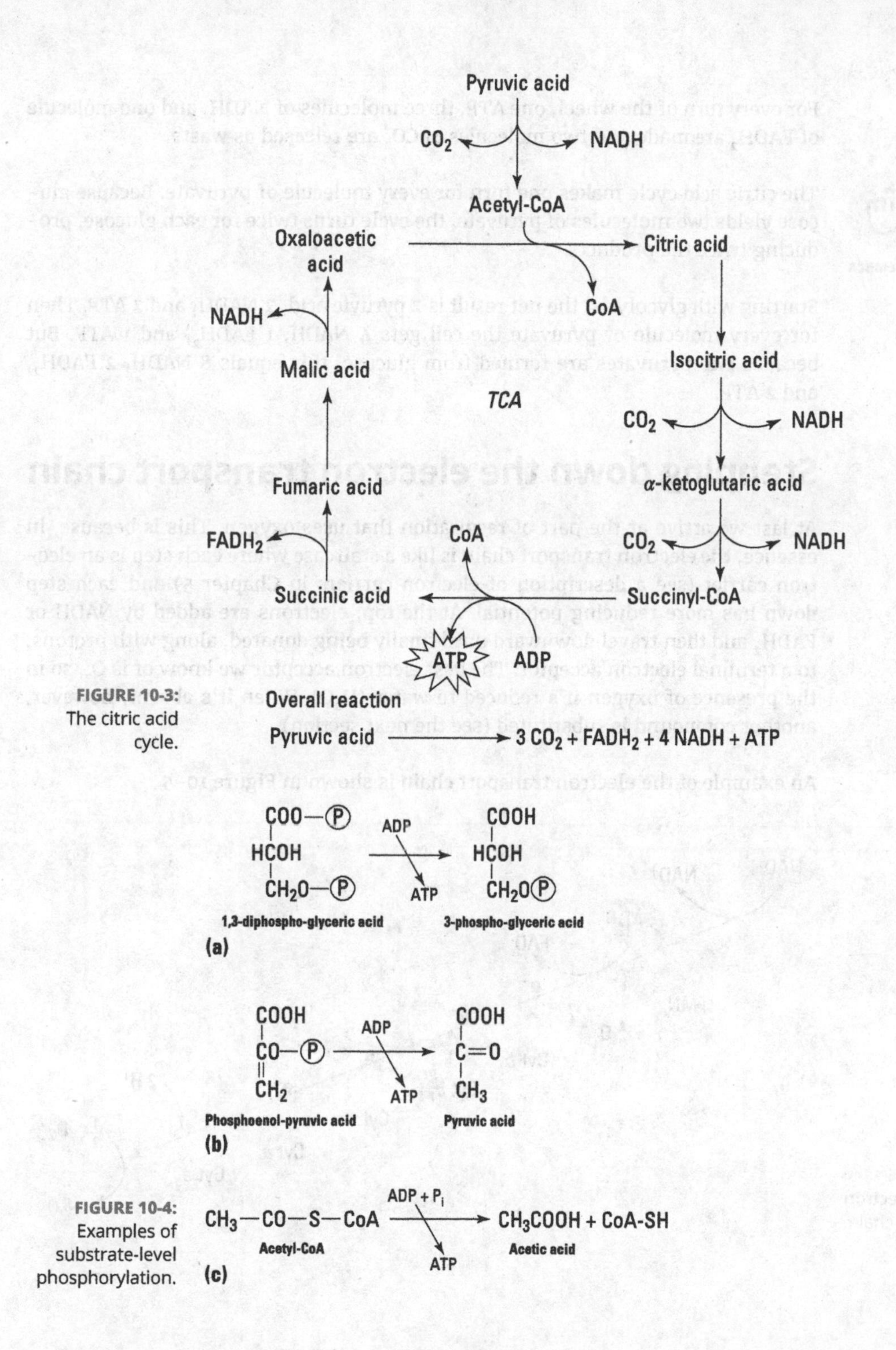

FIGURE 10-3: The citric acid cycle.

FIGURE 10-4: Examples of substrate-level phosphorylation.

For every turn of the wheel, one ATP, three molecules of NADH, and one molecule of $FADH_2$ are made and two molecules of CO_2 are released as waste.

REMEMBER

The citric acid cycle makes one turn for every molecule of pyruvate. Because glucose yields two molecules of pyruvate, the cycle turns twice for each glucose, producing twice the products.

Starting with glycolysis, the net result is 2 pyruvic acid, 2 NADH, and 2 ATP. Then for every molecule of pyruvate the cell gets 4 NADH, 1 $FADH_2$, and 1 ATP. But because two pyruvates are formed from glucose, this equals 8 NADH, 2 $FADH_2$, and 2 ATP.

Stepping down the electron transport chain

At last we arrive at the part of respiration that uses oxygen. This is because, in essence, the electron transport chain is like a staircase where each step is an electron carrier (see a description of electron carriers in Chapter 5) and each step down has more reducing potential. At the top, electrons are added by NADH or $FADH_2$ and then travel downward until finally being donated, along with protons, to a terminal electron acceptor. The best electron acceptor we know of is O_2, so in the presence of oxygen it's reduced to water (H_2O). When it's absent, however, another compound is substituted (see the next section).

An example of the electron transport chain is shown in Figure 10-5.

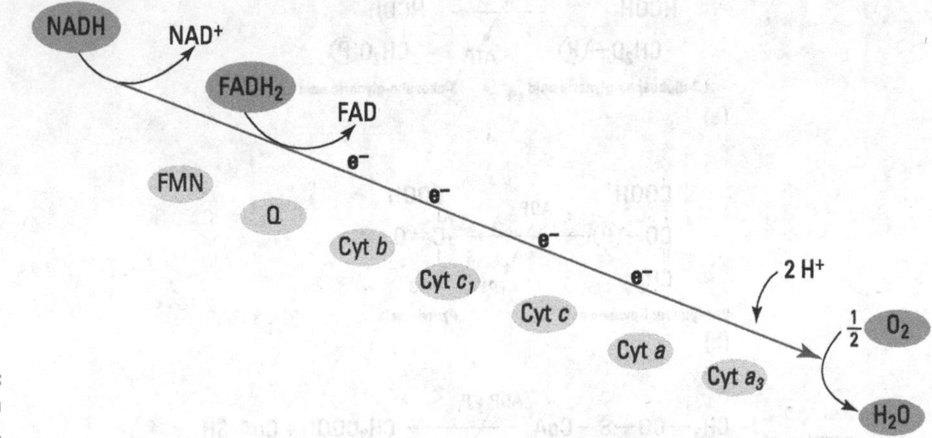

FIGURE 10-5: The electron transport chain.

When electrons go from a higher energy state (top of the stairs) to a lower energy state (bottom of the stairs), the energy they carry can be used to power enzymes. In this case, the enzymes are embedded in a membrane and use the energy gained to pump protons (H^+) across the membrane. As more H^+ accumulate on the outside of the membrane, a difference in charge and pH between the inside and the outside of the cell builds up. Because of the difference in acidity and electrical charge across the membrane (called the *proton motive force*), protons naturally want to re-enter the cytoplasm and do so through the enzyme ATP synthase, which is embedded in the membrane, in the process powering the formation of ATP from ADP.

The exact order and kind of electron carriers in the electron transport chain vary among microorganisms, but the overall result is the same: Electrons release energy that is used to move protons across a membrane, creating a difference in charge and pH.

Respiring anaerobically

Microbes that have an electron transport chain can still use it even if oxygen isn't present. For this they use other inorganic compounds as the terminal electron acceptor. Examples of this include nitrate (NO_3^-), sulfate (SO_4^{2-}), and carbonate (CO_3^{2-}) among others. Reduction of these compounds by microorganisms is actually very important for the natural cycle of nitrogen, sulfur, and carbon in the environment. Some organisms are able to respire both aerobically and anaerobically, whereas others are only able to do one or the other. Besides the three main substrates used as terminal electron acceptors (nitrogen-, sulfur-, and carbon-containing compounds) other compounds used include ferric iron, manganese, and inorganic and organic compounds.

Microbes can use inorganic substances in two different ways. When inorganic substances are incorporated into cellular material (for example, amino acids), the process is called *assimilative metabolism*. When inorganic substances are used as sources of energy, it is called *dissimilative metabolism*. There's a big difference between these two types of metabolism: Dissimilative uses large quantities of the substrate to gather the energy needed, whereas only the amount of each compound needed for biosynthesis is used in assimilative metabolism. The other major difference is where many organisms use inorganic substrates assimilatively only a small number of specialist microbes use dissimilative metabolisms of inorganic substances.

Nitrate and denitrification

Nitrate (NO_3^-) is a very common terminal electron acceptor. When it's completely reduced, it forms nitrogen gas (N_2). The process of reducing nitrogen compound to nitrogen gas is called *denitrification*. Nitrogen gas makes up about 78 percent of air and is completely unavailable to plants and animals as a nitrogen source.

The nitrogen gas can be recaptured by microorganisms and returned into a form that is available to plants in a process called *nitrogen fixation* (see Chapter 11). Denitrification is important for removing fixed nitrogen from processes like sewage treatment so that the water released at the end isn't full of nitrates because these nitrates cause algae to grow, fouling up rivers and streams.

There are several steps involved in complete denitrification (see Figure 10-6), but not all microbes go all the way. Some only convert nitrate to nitrite. Each step uses a different enzyme, but the result is the reduction of each compound during steps in the electron transport chain.

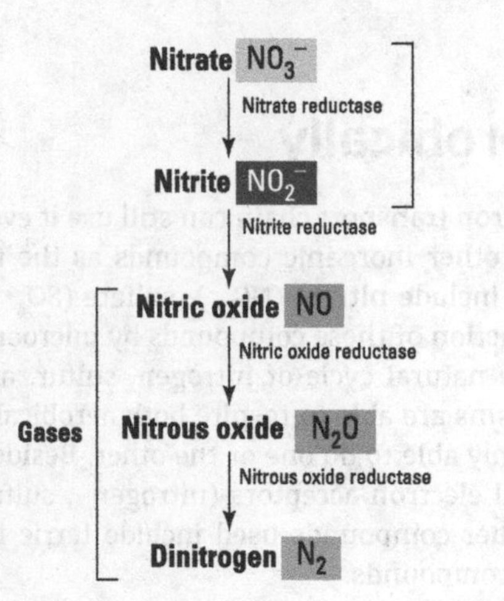

FIGURE 10-6: The steps in denitrification.

Sulfate and sulfur reduction

Sulfate (SO_4^{2-}) is an inorganic sulfur-containing ion that is common in the ocean. Because it's the most oxidized form of sulfur, bacteria such as *Desulfovibrio* use it as the terminal electron acceptor in the electron transport chain, reducing it to sulfite (SO_3^{2-}) and eventually hydrogen sulfide (H_2S) anaerobically:

$$SO_4^{2-} \rightarrow SO_3^{2-} \rightarrow H_2S$$

Although many plants and microbes can use sulfate for biosynthesis of sulfur-containing molecules in assimilative sulfate reduction, only the sulfate-reducing bacteria can use it for energy with dissimlative sulfate reduction. The hydrogen sulfide (H_2S) produced is a gas with a distinctive rotten egg smell, that can be used by sulfur-oxidizing microorganisms or can react with metals forming metal sulfides such as cadmium yellow (CdS), used as a paint pigment, or iron pyrite (FeS_2), commonly known as fool's gold.

Sulfur-reducing bacteria produce hydrogen sulfide from elemental sulfur (S^0) in the following way:

$$S^0 + 2\,H \rightarrow H_2S$$

The sulfate-reducing bacteria are fairly well studied, whereas the sulfur-reducing bacteria are less well understood.

Acetogenesis and methanogenesis

Another group of organisms that use anaerobic respiration for energy conservation are the acetogens and the methanogens.

These organisms reduce carbon dioxide (CO_2) and, as their names suggest, form acetate (CH_3COO^-) and methane (CH_4). They have one major thing in common: their use of the acetyl-CoA pathway to reduce CO_2. Unlike the autotrophs in Chapter 9, strict anaerobes like these don't use the Calvin cycle, the reverse citric acid cycle, or the hydroxypropionate cycle. Instead, they produce acetyl-CoA from two molecules of CO_2.

Acetogens are interesting because they both produce ATP from an electron transport chain, with CO_2 as the terminal electron acceptor, and use acetate to form more ATP through substrate-level phosphorylation. Although both bacteria and archaea are known to produce acetate, all known methanogens are archaea. Methanogens are abundant in the cow *rumen* (first part of the cow's stomach important for digesting cellulose) even in the human intestine, producing large amounts of methane, a smelly gas that is released as, well, you can guess.

Oxidizing hydrocarbons and other compounds

Hydrocarbons are molecules containing only carbon and hydrogen. They're chains made up of carbons, and they can be pretty much any length, but the longer the chains, the more insoluble they are in water and the less available they are for microbial breakdown. Hydrocarbons occur mainly in crude oil where the abundance of carbon from the decomposition of plant material has been bonded into chains and covered with hydrogen atoms.

In both the presence and absence of oxygen, hydrocarbons can be oxidized for energy by microorganisms. Anaerobically, it's the denitrifying and the sulfate-reducing bacteria that are responsible for the breakdown of hydrocarbons, using the hydrocarbon as the electron donor and reducing either NO_3^- or SO_4^{2-} at the end of the electron transport chain. Many microbes can degrade hydrocarbons aerobically; the process is much quicker than it is anaerobically. In both pathways, long chains are broken into shorter chains, which are then oxidized to CO_2.

Compounds other than glucose are broken down through respiration or fermentation and many of the pathways used are essentially the same (see Figure 10-7).

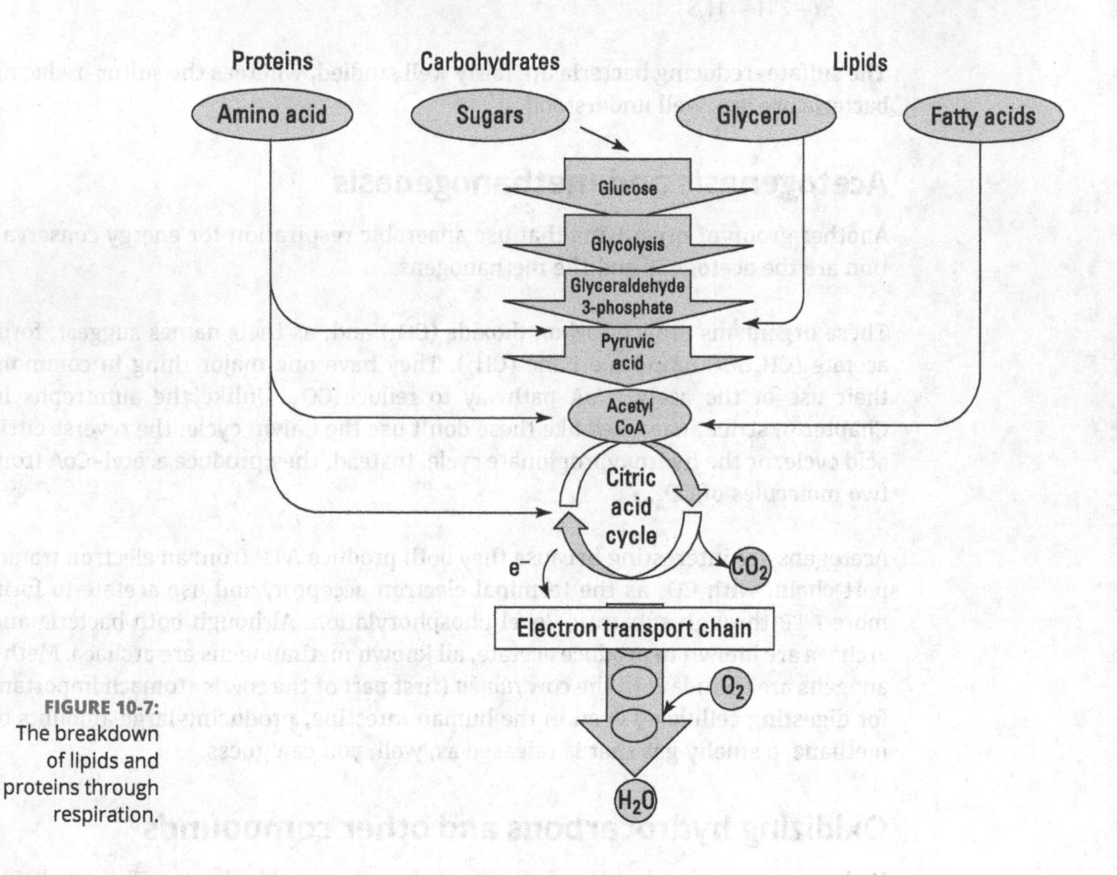

FIGURE 10-7: The breakdown of lipids and proteins through respiration.

Figuring Out Fermentation

Both respiration and fermentation are ways of generating ATP, the energy currency of the cell, but they're very different. In fermentation, ATP is produced directly from the products of glycolysis, so the amount of ATP produced is much lower. But fermentation is much faster than respiration and is a clever way to extract energy from a variety of substances in the absence of oxygen, although it can still happen when oxygen is present. One important function of fermentation is to oxidize NADH back to NAD^+ so that it can be used again in glycolysis. This happens during the reduction of pyruvic acid into fermentation products.

REMEMBER

Fermentation doesn't use the citric acid cycle or an electron transport chain to produce ATP. In fact, little to no ATP is made from the fermentation reactions that follow glycolysis. Instead, fermentation reactions are used to regenerate NAD+ and to balance the redox reactions. ATP is made through substrate-level phosphorylation, so fermentation doesn't require the use of a membrane.

Because fermenting microorganisms lack an electron transport chain, they're unable to completely oxidize organic compounds and the final products still contain energy. This is the unavoidable result of having to balance redox reactions. The products of fermentation are different depending on both the enzymes present and the type of microorganism, but they include things like lactic acid, ethanol, CO_2, H_2, and many others, but in all cases they are considered waste by the microorganism and are excreted into the media. Because of their fermentative properties, some of these microbes are used to make wine, cheese, yogurt, cured meats, and in industrial processes to manufacture solvents and other useful things.

The types of fermentation are named mainly for the products that are made. *Embden-Meyerhof fermentations* (those beginning with classic glycolysis) all begin with glucose and lead to a myriad of fermentation products, including the following:

- **Homolactic fermentation** produces lactic acid only. Examples of homolactic fermentative bacteria include members of the genera *Streptococcus, Lactobacillus,* and *Bacillus*.
- **Mixed acid fermentation** produces lactic acid, acetic acid, formic acid, succinate, and ethanol and sometimes CO_2 and/or H_2. Examples of mixed acid fermenters include bacteria from the genera *Escherichia* and *Salmonella.*
- **Butanediol fermentation** can make the same products as mixed acid fermentation as well as 2,3 butanediol. Bacteria from the genus *Enterobacter* ferment in this way.
- **Butyric acid fermentation** is performed by *Clostridia* bacteria and produces butyric acid, CO_2, and H_2, as well as ethanol and isopropanol.
- **Butanol-acetone fermentation** is also done by a species of *Clostridium* and produces acetone, which was used to make gunpowder during World War I.
- **Propionic acid fermentation** produces acetic acid, CO_2, and propionic acid from lactic acid (this is called *secondary fermentation,* meaning the fermentation of a fermented product) and is especially useful for making Swiss cheese. The propionic acid gives the cheese its flavor, and the CO_2 makes the distinctive holes. Species of *Propionibacterium* are used for this type of fermentation process.

The *phosphoketolase pathway* is used during a type of fermentation that produces lactic acid along with ethanol and CO_2 called heterolactic fermentation or heterofermentation because more than one product is made. This pathway breaks down glucose in a different way than the classic glycolytic pathway and produces about half the amount of energy (see Figure 10-8).

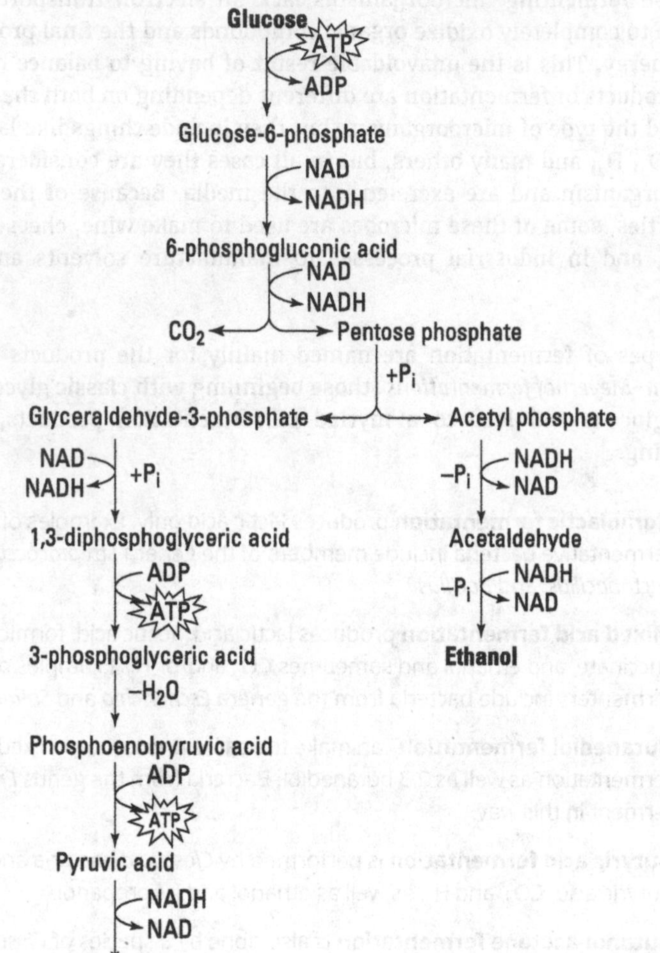

FIGURE 10-8: The phosphoketolase pathway of heterofermenters.

As in classic glycolysis, in the phosphoketolase pathway the six-carbon glucose molecule is split into a three-carbon molecule of pyruvic acid, which goes on to be reduced to lactate. But instead of a second molecule of pyruvic acid, the other three carbons are split between CO_2, which is released, and acetyl phosphate, which gets converted to ethanol. Microorganisms that use this pathway, like

Lactobacillus, are best known for being helpful in making sauerkraut and a type of yogurt called kefir.

The *Entner-Doudoroff pathway* is another type of glycolysis that produces less energy than classic glycolysis (see Figure 10-9).

The microorganisms that use this type of fermentation alone are few and come from a group of *Zymomonas* bacteria that are involved in the production of tequila and mescal. For many other microbes, however, this pathway is used to break down glucose that feeds into respiration.

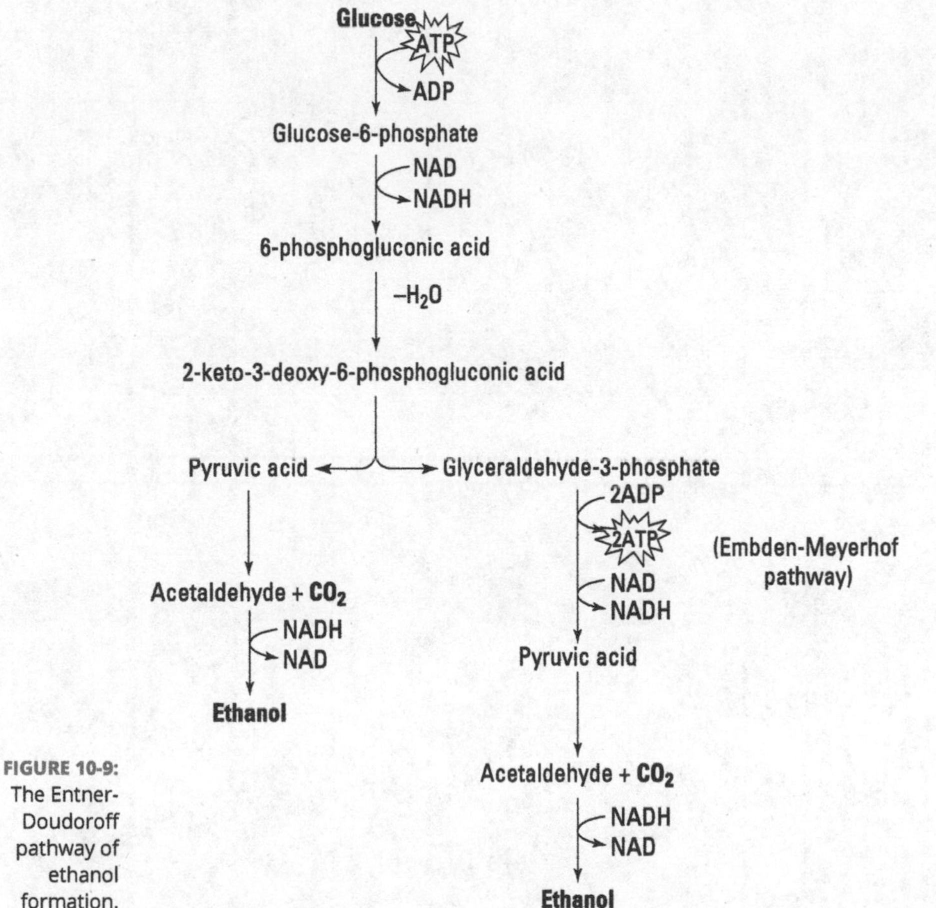

FIGURE 10-9: The Entner-Doudoroff pathway of ethanol formation.

IN THIS CHAPTER

» **Getting to know the earth's nutrient cycles**

» **Exploring the many places microbes live**

» **Seeing how microbes cooperate and communicate**

Chapter 11
Uncovering a Variety of Habitats

Living organisms don't exist in isolation, every one is surrounded by a living and nonliving environment with which it interacts.

Before talking about the many habitats where microorganisms are found, we want to fill you in on some terminology for describing microbes in an ecosystem, as well as the factors that influence the individuals within a habitat:

- **Colonization:** The growth of microorganisms either on a surface or within something.
- **Population:** An accumulation of the several members of the same species. For microorganisms, especially prokaryotes, a population often arises from one individual cell.
- **Community of microorganisms:** Formed by many populations existing together; can form on a very small physical scale.
- **Guilds:** Populations of microorganisms in a community that use similar kinds of metabolism to exploit the same resource. An example of this is the use of sunlight by populations of different phototrophic bacteria.

- **Niche:** The environment shared by a guild of microbes that provides what they need to grow.
- **Habitat:** A general environment where populations of microorganisms live. Habitats often have several niches occupied by different guilds using different resources. As an example, a pond is a habitat with more than one niche for microorganisms; the photic zone where light can penetrate and the sediment where it's dark.
- **Ecosystem:** All the interconnected parts of an environment, living and nonliving. This includes all the plants, animals, microbes, rocks, soil, and water that affect one another and cycle nutrients. Ecosystems are composed of many different habitats, some of which contain all kinds of life, and others of which, like deep inside rocks, contain only microbes.

In this chapter, we describe the many types of microbial habitats and explain how microorganisms get what they need to survive in each one. We also cover the lifestyle strategies used by microbes to survive and thrive in different habitats.

Defining a Habitat

A habitat has a set of chemical and biotic features that define it. These include water, oxygen, pH, and temperature (all of which are covered in Chapter 4), as well as the following:

- **Energy inputs:** Energy inputs include light, organic carbon, and reduced inorganic compounds.
- **Resources:** Resources include nutrients containing carbon, nitrogen, sulfur, phosphorous, iron, and many other micronutrients. Two important resources are electron donors and electron acceptors, which are essential for energy harvesting by all cells (see Chapter 5).
- **Activities:** Activities include primary production of organic compounds by autotrophs, the consumption of these organic products by heterotrophs, and the conversion of both inorganic and organic compounds in the environment by chemotrophs (see Chapters 9 and 10 that go over this in detail).

A habitat can be stratified in terms of temperature, oxygen, nutrients, and sunlight, and these stratifications make up different niches to which a specific microorganism, or group of microorganisms, can be uniquely suited (see Figure 11-1).

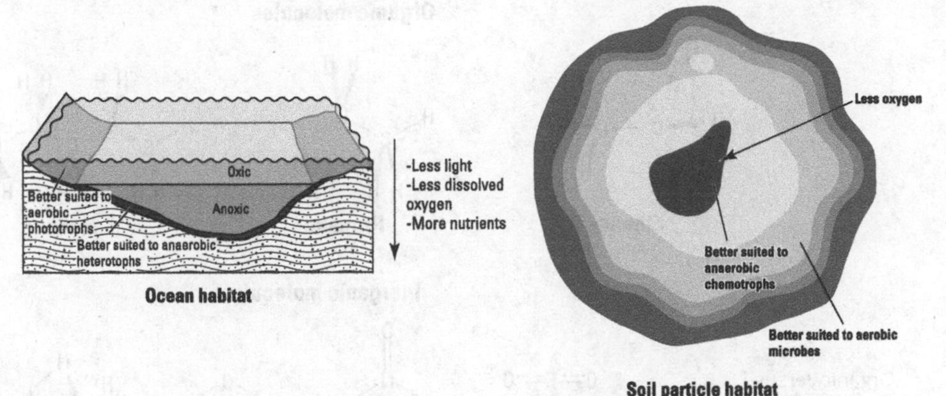

FIGURE 11-1: Stratification of a habitat.

For microorganisms, these habitats can be on very small size scales. For instance, a particle of soil (refer to Figure 11-1) can have several different gradients of chemical conditions that define different niches.

Understanding Nutrient Cycles

Because the earth is a closed system, nothing is ever lost — it only changes form. Nutrients cycle between environments changing from oxidized to reduced and from simple molecules to complex molecules. The main elements that make up all of the major nutrients are carbon, oxygen, nitrogen, phosphorous, and sulfur. These processes are called *biogeochemical processes* because cycles are affected not only by biological processes but also by geological and chemical ones.

Despite their small size, microorganisms have a large impact on biogeochemical processes. As long ago as the 19th century, bacteria were suspected of recycling elements between environments and today we know just how extensive their roles are.

For every biogeochemical cycle, there are reservoirs where compounds are stored and active processes that transform and move compounds around.

Carbon cycling

Organic compounds are carbon-containing compounds that make up all the cellular life on earth. These include carbohydrates, fats, and proteins. Carbon dioxide (CO_2) is not an organic carbon source because it doesn't have any carbon-carbon bonds (see Figure 11-2).

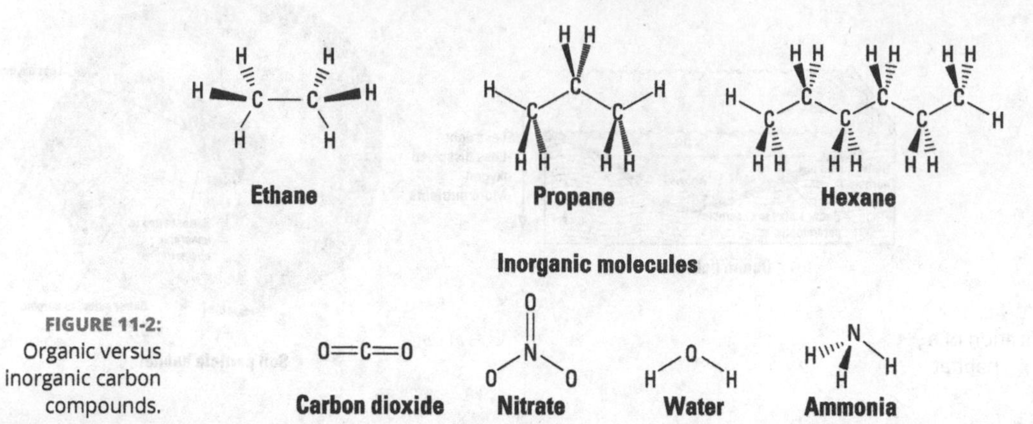

FIGURE 11-2: Organic versus inorganic carbon compounds.

Taking the inorganic carbon in the environment and making organic compounds is the job of the autotrophs that are also called *primary producers.*

REMEMBER

All cells need a carbon source. Autotrophs use inorganic carbon in the form of CO_2. Autotrophs incorporate the CO_2 into organic molecules; heterotrophs use this organic carbon directly. Also, remember that cells need a carbon source and an energy source. Heterotrophs use organic molecules for both — energy from the bonds between atoms and the carbon within the molecules — whereas autotrophs have a separate energy source (like light or oxidation of another molecule) and CO_2 for their carbon needs.

The main autotrophs responsible for carbon fixation on earth are plants and photosynthetic microorganisms like algae, cyanobacteria, and other types of phytoplankton. Other nonphotosynthetic microbes fix carbon and contribute a bit more to the global organic material.

Heterotrophs, including animals and microbes, consume the organic carbon and either release CO_2 as waste or hold onto the organic carbon until they die. Decomposition of dead organisms by microorganisms breaks down the organic matter into CO_2 and methane (CH_4), but complete breakdown can take decades for some of the more resistant compounds. CH_4 can be converted back to CO_2 by methanotrophs, and the cycle is complete. The complete carbon cycle is illustrated in Figure 11-3.

The main processes of carbon utilization on earth are from heterotrophic microbes that either consume it through respiration (both aerobically and anaerobically) or fermentation. Ocean and land animals also consume a small amount of the organic matter produced.

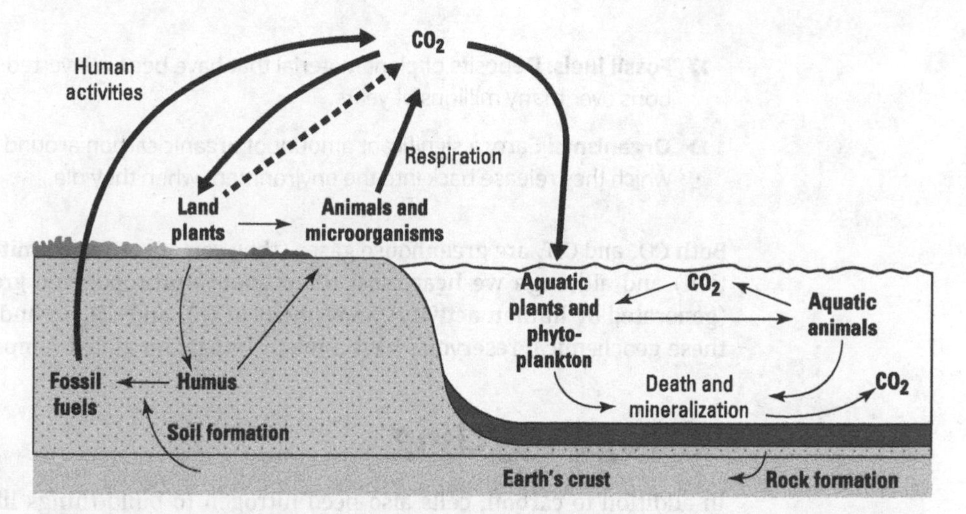

FIGURE 11-3: The carbon cycle.

Within the carbon cycle, there is also a side loop for CH_4, which is produced by *methanogenic* (methane-producing) microorganisms called *methanogens.* These microbes are found in the sediments of the ocean and belong exclusively to the Archaea. Methanogens can only use a small number of carbon compounds to make CH_4, CO_2, and acetate. To make CH_4 from organic matter, they need help from the syntrophs (see the sidebar on syntrophs in Chapter 5 for a description of these interesting bacteria).

Major processes include

- **Carbon fixation by autotrophs:** $CO_2 \rightarrow$ organic carbon
- **Respirationandfermentation:** Organic carbon $\rightarrow CO_2$
- **Methanogenesis:** Organic carbon or $CO_2 \rightarrow CH_4$
- **Methanotrophy:** $CH_4 \rightarrow CO_2$

Major reservoirs of carbon include, in order of highest percent to lowest percent of carbon on earth, the following:

- **Rocks:** Contain carbon in the form of limestone. Nearly 99.5 percent of the carbon on earth is contained in the rocks in the earth's crust.
- **Oceans:** Contain the highest amount of CO_2, stored as carbonic acid. Absorbing more CO_2 from the atmosphere as the levels of CO_2 rise.
- **Methane hydrates:** Trapped molecules of CH_4, made by microbes, frozen deep in ocean sediments. CH_4 leaks out at different rates, depending on the temperatures in the sea.

- **Fossil fuels:** Deposits of plant material that have been converted to hydrocarbons over many millions of years.
- **Organisms:** Carry a significant amount of organic carbon around with them, which they release back into the environment when they die.

Both CO_2 and CH_4 are greenhouse gases (they can adsorb and emit infrared radiation) and although we hear most often about anthropogenic greenhouse gases (generated by human activity), movement of CO_2 and CH_4 in and out of some of these geochemical reservoirs have a major impact on global temperatures.

Nitrogen cycling

In addition to carbon, cells also need nitrogen to build things like proteins and nucleic acids. But unlike carbon, where the different forms are more complex organic molecules, the nitrogen cycle involves only different oxidation states of nitrogen.

Nitrogen in the atmosphere is very plentiful, making up 79 percent of the air, but it's in a form that most cells can't use: nitrogen gas (N_2). Microbial processes are essential for converting N_2 into ammonia (NH_3), called *nitrogen fixation.* Aside from fixing it from the air, nitrogen is available from decomposing organic matter. When something dies, microorganisms break down the proteins into amino acids and then into ammonia. Together these processes make up the nitrogen cycle, an overview of which is shown in Figure 11-4. As with carbon, there are nitrogen sources and reservoirs between which different processes act to cycle around nitrogen-containing molecules:

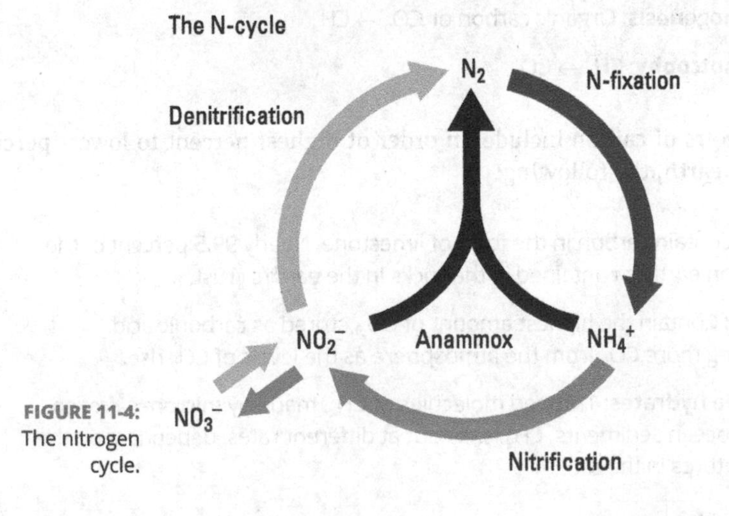

FIGURE 11-4: The nitrogen cycle.

» **Ammonification:** Organic N to NH_4^+. The breakdown of proteins and nucleic acids to ammonium. Performed by many microorganisms. When ammonia is used to make amino acids and nucleic acids, it's called *assimilation.*

» **Denitrification:** NO_3^- to N_2. Happens under anaerobic conditions. It's detrimental when denitrification happens in waterlogged fields because it causes fixed nitrogen to escape as N_2. It's good when denitrification happens in sewage treatment because it lowers the amount of fixed nitrogen in treated water that ends up in lakes and rivers.

» **Nitrification:** NH_4^+ to NO_2^- then NO_2^- to NO_3^-. Each reaction is catalyzed by different bacteria that live together in neutral soils that are well drained, because flooded soils become anoxic quickly. This happens extensively in soils, especially after fertilizer (natural or chemical) is added. Because NO_3^- is very soluble, it will wash away quickly or succumb to denitrification if the soil becomes flooded.

» **Anammox:** NO_2^- and NH_4^+ to N_2. Only a small number of strict anaerobes can do this. Happens extensively in marine sediments and sewage treatment.

» **Nitrogen fixation:** N_2 to NH_3. Very few microbes can fix nitrogen, so they don't contribute greatly to the total amount of fixed nitrogen in general, but because fixed nitrogen is often in limited supply when nitrogen fixers are present, they really give plants a boost. Because nitrogen fixation is so expensive metabolically, it can only happen when a lot of energy is available.

There are two categories of nitrogen-fixing bacteria — those that live inside of plant tissues (called *symbiotic*) and those that don't (called *free-living*):

- Free-living nitrogen-fixing bacteria include *Azotobacter, Cyanobacteria,* and *Clostridium.*
- Symbiotic nitrogen-fixing bacteria include *Rhizobia, Cyanobacteria,* and *Frankia,* all discussed next.

One enzyme is responsible for nitrogen fixation: *nitrogenase.* This bacterial enzyme is very sensitive to inactivation by oxygen, so you'd think that the bacteria making it would stick to anaerobic environments, but instead many have come up with elaborate strategies to keep oxygen away from nitrogenase:

- *Frankia,* a filamentous member of the Actinomycete bacteria, make swellings that have thick cell walls and reduce diffusion of oxygen (O_2) into the cell.
- *Cyanobacteria* use heterocysts.
- *Rhizobia* live in nodules and produce an oxygen-binding protein called leghemoglobin that keeps free O_2 levels low.
- Others have high rates of oxygen consumption so that there is never a large concentration within the cell.

HAPPY COUPLES

Microbial processes are important for completing most cycles and, in the process, coupling one nutrient's cycle with another. The calcium cycle in the ocean is tightly coupled with the carbon cycle. The levels of CO_2 in the air determine how much will be absorbed by the oceans. When absorbed, CO_2 becomes carbonic acid (H_2CO_3), decreasing the pH of seawater. Organisms that use Ca^{2+} in their exoskeletons, like *Foraminfera* and corals, need a more basic ocean pH; otherwise, the calcium carbonate deposits that they make get dissolved. These organisms are important food sources for other ocean life, and when they die, they're a major part of the cycling of organic matter and Ca^{2+} to deep in the ocean, where anoxic processes break down organic compounds very slowly, an important part of the carbon cycle.

From most oxidized and negatively charged to most reduced and positively charged, the nitrogen compounds discussed in this section are: nitrate (NO_3^-), nitrite (NO_2^-), nitrogen gas (N_2), ammonia (NH_3), and ammonium (NH_4^+). NH_3 is a gas and spontaneously converts to NH_4^+ in water when the pH is neutral where it's soluble.

Sulfur cycling

The sulfur cycle is about the different oxidation states of sulfur — there are many more states than there are for nitrogen. The ocean is a major reservoir of sulfate (SO_4^{2-}). The major volatile gas hydrogen sulfide (H_2S) is the most reduced form of sulfur and is used by many bacteria that oxidize it either to elemental sulfur (S_0) or to sulfate. Some bacteria even store S_0 in their cells as a source of electrons for later (see Chapter 10).

Sulfate-reducing bacteria are abundant and widespread. They reduce sulfate when organic carbon is present. The sulfide produced from sulfate reduction, combines with iron to form insoluble black deposits of iron sulfide minerals (FeS and FeS_2). S_0 can be oxidized by *Thiobacillus*, producing sulfuric acid (H_2SO_4) and causing acidification of the environment at the same time. S_0 can also be reduced to sulfide by hyperthermophilic archaea.

Phosphorous cycles in the ocean

Unlike nitrogen and sulfur cycles where the elements change oxidation states, the phosphorous cycle follows the change in solubility (in the nearby sidebar you'll see that the calcium cycle is affected by pH). More soluble phosphorous is more available to plants, and vice versa. These elements also cycle due to the activities of microorganisms, but unlike the other cycles, there are no volatile forms that can escape. As with the other nutrient cycles, keeping things in balance is very important to ocean and terrestrial life.

Microbes Socializing in Communities

Microbes living together in communities interact with one another in positive, negative, and neutral ways. They compete with the other members of their guild for resources, and they compete with everyone for space. Microbes also cooperate with each other to use resources most efficiently. They orchestrate these interactions by communicating with members of their own species and with other species through chemical signaling molecules. When coordinated, microbial cells can produce structures called *biofilms* within which they're protected from outside stresses.

Using quorum sensing to communicate

Quorum sensing is the process in which regulatory pathways within the cells of a population of bacteria are controlled by the density of cells of their own kind. As the name implies, if a sufficient numbers of cells (a *quorum*) are present, they can do something that requires more than one cell to accomplish. Quorum sensing controls biofilm formation and toxin production, among many other things, some of which are not fully understood.

Cells produce a signal molecule called an *autoinducer* that is sensed by other cells in the vicinity. When enough cells produce the autoinducer, the concentrations become high enough within cells to trigger gene expression. Some autoinducers are specific for the same species of bacteria, whereas others can signal across species. Gram-negative bacteria make acyl homoserine lactones (AHL) and autoinducer 2 (AI-2) molecules; Gram-positive bacteria use short peptides as signals.

Living in biofilms

Whenever you have a fluid washing over a surface, like the water over rocks in a stream or saliva over the teeth in your mouth, biofilms form. A *biofilm* is a collection of microbes, usually bacteria but also sometimes archaea, within a sticky matrix attached to a surface. The bacteria make the substances that bind the biofilm together; these substances include polysaccharides, proteins, and even DNA. There can be an impressive number of species of microorganisms inside of a biofilm, or it can contain only a small number of species. The bacteria aren't just stuck in the matrix, however; they're a living, growing community.

From afar, biofilm looks like a film on a surface, but up close a biofilm has spaces (as shown in Figure 11-5), which allows liquid to flow through so that gases and nutrients can be exchanged.

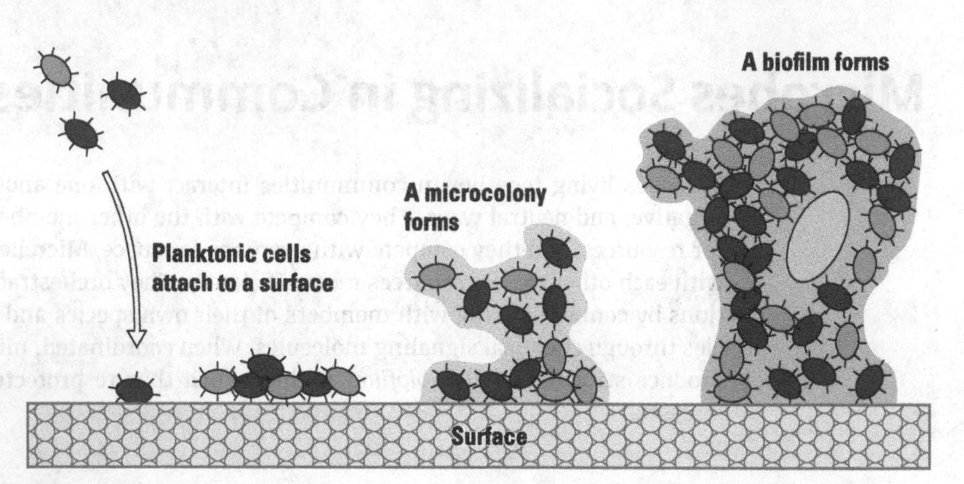

FIGURE 11-5: A biofilm.

Here are the main reasons bacteria form biofilms:

- **Forprotection:** Biofilms are thought to form in order to protect cells from predators, from environmental stresses, and from being mechanically removed from a surface. One important feature of a biofilm is that cells within them are resistant to antibiotics, mostly because the drugs can't penetrate the matrix.
- **To keep cells in place:** Biofilms are also thought to trap nutrients.

CONTROLLING BIOFILMS

Bacterial biofilms protect the inhabitants from a variety of things, including mechanical disturbance, exposure to toxic substances like antibiotics, and predators. They can become a problem for humans, however, because they tend to clog up pipes and filters, form on medical equipment and devices, and protect human pathogens during an infection.

Because biofilms are resistant to antibiotics and protected from phagocytosis by our immune cells, organisms in a biofilm are really hard to get rid of. Current research is aimed at preventing biofilms from forming, as well as biofilm-busting treatments with mechanical means and chemicals.

» **For proximity:** When cells are close enough together, they can communicate and exchange genetic material and other molecules.

» **Because it's the default:** Biofilms are likely very common in nature, where nutrients are available but sometimes hard to get.

Within solid parts of a biofilm, oxygen gradients can exist. This is where oxygen levels on the surface or near a hole in the biofilm are highest and decrease as you move toward the center of a solid area. These areas of low oxygen are a perfect place for anaerobic bacteria or archaea to colonize.

Exploring microbial mats

You can think of a microbial mat as an extreme example of a biofilm. They started forming 3.5 billion years ago and were the main type of ecosystem for a long time. When plants arrived and started competing with mats for light, and when predators arrived and started eating the bacteria, the number of mats declined. Mats are still found today, mainly in habitats with extreme temperatures or high levels of salt.

Microbial mats are many centimeters thick, made of several layers, each with different species of bacteria. Microbial processes, as well as physical factors, result in each layer having different oxygen concentrations, nutrients, and pH conditions. Filamentous cyanobacteria, as well as filamentous *chemolithotrophic* (sulfate-oxidizing) bacteria are common members of microbial mats. Because of the dynamics within mats and due to their age, they're among some of the most complex ecosystems, containing the highest number of different species known. They're ecosystems in the true sense of the word because they contain primary producers (either cyanobacteria or chemolithotrophs that produce organic carbon compounds from CO_2) and consumers (heterotrophs) that cycle all the key nutrients, like carbon and nitrogen.

Discovering Microbes in Aquatic and Terrestrial Habitats

Everything is not everywhere, so although microbes can be found in every habitat, species have their preferred habitats, along with some less-than-ideal ones that they occupy only occasionally. Most aquatic habitats (saltwater and freshwater) and terrestrial habitats contain plants, animals, invertebrates, and microbes. This is unlike subsurface habitats, which are entirely microbial; life has been found down to nearly 2 miles underground, possibly accounting for 40 percent of the earth's biomass.

Aside from the nutrient cycles, a few other things describe a microbial community: *membership* (who is there), *diversity* (how many different species are thriving there together), and *biomass* (how big the populations are). The factors that affect these things are nutrient concentrations, mixing, and oxygen concentrations.

Thriving in water

The photosynthetic microorganisms are commonly found in both freshwater and saltwater habitats. These include algae and cyanobacteria, which either float in the water column (planktonic) or attach to surfaces (benthic).

Oxygen levels in freshwater habitats influence the types of communities that can live there. The oxygen levels fluctuate depending on the amount of primary production. High rates of carbon fixation by primary producers is a problem for all aquatic habitats because it can lead to spikes in heterotrophic activity that consume all the oxygen. These bursts of oxygen consumption affect rivers and oceans as well, but freshwater lakes in summer are particularly vulnerable because they tend to be stratified without much mixing.

The organic material and dead autotrophs from the top layer sink down to the bottom of the lake. Because diffusion rates of oxygen into water are low, heterotrophs at the bottom quickly use up all the available oxygen when consuming the organic matter. This zone lacking oxygen is called the *anoxic zone* and is unsuitable for fish or invertebrate life, but it can be perfect for anaerobic microorganisms.

Coastal oceans and the deep sea are two other aquatic habitats for microorganisms. Figure 11-6 shows the different habitats that exist in the ocean.

OXYGENIC VERSUS ANOXYGENIC PHOTOTROPHS

Both are autotrophs using CO_2 as their carbon source, but each uses different electron donors. Oxygenic phototrophs use water (H_2O) as an electron donor, whereas anoxygenic phototrophs use other reduced molecules like H_2S and H_2). So, although we sometimes think of algae and cyanobacteria as using CO_2 and producing O_2, which they do, we should think of them as using CO_2 and H_2O (for different reasons of course) and producing O_2.

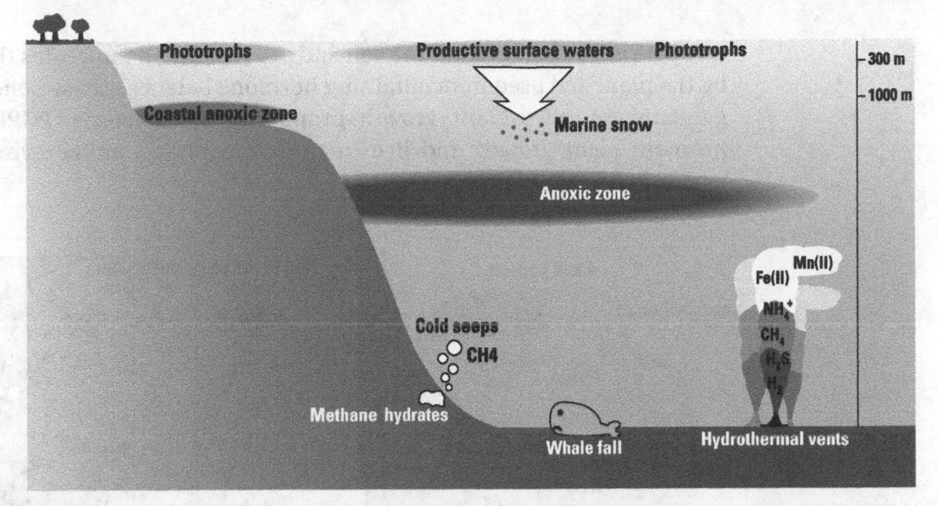

FIGURE 11-6: Ocean habitats.

The photic zone extends down to 300 meters, where light penetrates the water, providing an energy source for a large variety of phototrophic marine microorganisms like algae, phytoplankton, and bacteria. Immediately below the photic zone, down to 1,000 meters, there is no light but there is still biological activity. The levels of nutrients here are very low, but there are bacteria and archaea that are adapted to low nutrients and they're abundant here. Below 1,000 meters is the deep sea, where there is far less biological activity because of the low temperature, low levels of nutrients, and high pressure, which increases with depth. There are some unique microbial habitats in the deep sea for bacteria and archaea able to withstand the great pressure, such as those near hydrothermal vents and in the sediments.

Swarming soils

Another important microbial habitat is soil. Soil is made of minerals, organic matter, water, and microorganisms, with pockets of air mixed in. Soil composition varies greatly. For instance, dry soil has very little water, compacted soil has less air, mineral soil has little organic matter, and organic soil has a lot. The smallest unit of soil is the particle, which can have many habitats within it (see Figure 11-7). Different parts of a soil particle contain different micro-colonies of bacteria or archaea or have fungal hypha growing through them. Some parts are aerobic, but many are anoxic and home to anaerobic microorganisms.

One very important soil habitat for microorganisms is around the roots of plants, called the *rhizosphere* (refer to Figure 11-7). Plant roots excrete many compounds into the soil such as organic acids, amino acids, and sugars. These compounds attract and support the growth of many kinds of microbes, including bacteria and fungi. Some of these rhizosphere microbes are beneficial to plants because they compete with pathogens in the soil and produce small molecules that are taken up by the plant and used in maintaining hormone balance. These beneficial microorganisms are called plant-growth-promoting rhizobacteria (PGPR) because they promote plant growth and live in the rhizosphere, unlike symbiotic bacteria, which live inside plant tissues.

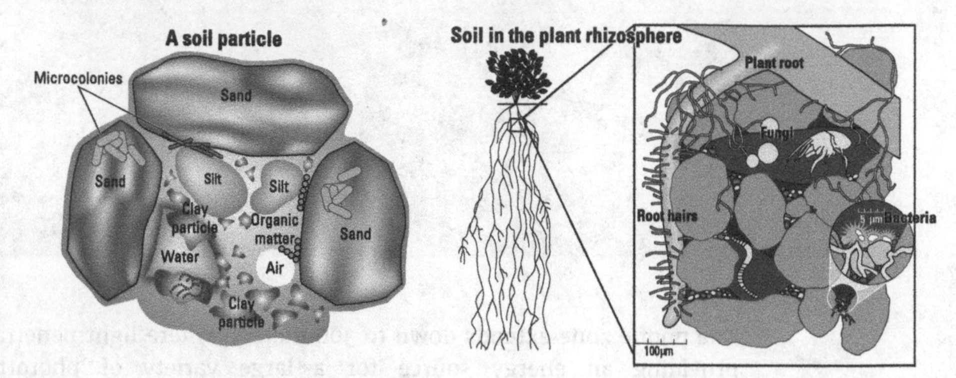

FIGURE 11-7: Soil habitats.

Getting Along with Plants and Animals

Microorganisms don't just inhabit the nonliving parts of ecosystems, there are many microbial habitats on and in other organisms, including plants, animals, and even humans. The intensity of the interaction between microbes and their living habitat can be anywhere from very high (where both need each other to live) to low enough that neither really notices the other. Very intimate relationships between organisms are called *symbiosis*, with the microorganism called the *symbiont* and the other called the *host*. The nature of the relationship between organisms can be positive, negative, or neutral. For the most part, only positive and neutral relationships are covered here because negative relationships, such as infections, are discussed in Chapters 15 and 17.

Microorganisms can form a symbiotic relationship with one another, as well as with other organisms. Lichen are an example of this where algae or cyanobacteria live in very close association with fungal hypha to the mutual benefit of both of them. The algae or cyanobacteria make organic compounds through photosynthesis, and the fungus provides support and protection. When a lichen reproduces, spores are made of an algal cell wrapped in a small fungal hyphae (see Figure 11-8).

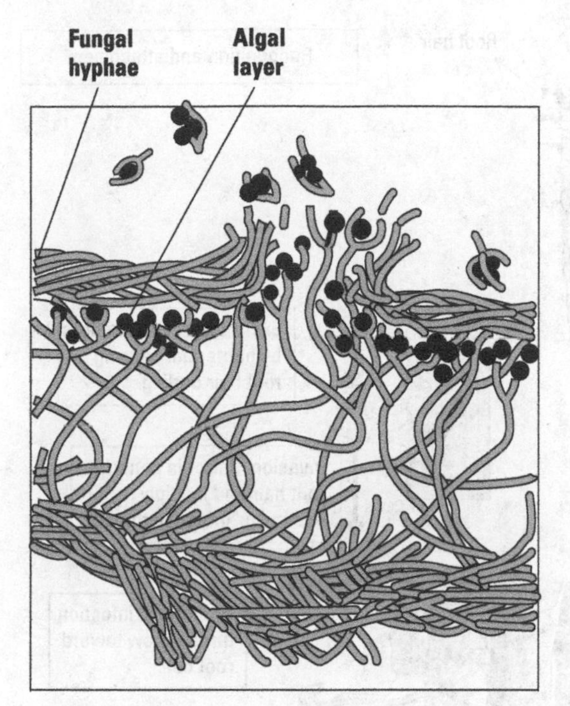

FIGURE 11-8: Lichen.

Living with plants

Plants are teeming with microorganisms. Every surface of a plant — both above ground and below ground — is colonized with microorganisms. Plants form intimate relationships with some microorganisms, like bacteria and fungi, that provide them with fixed nitrogen, small molecules, and protection from pathogens in exchange for sugars. There are many different types of plant-microbe interactions, a few of which are covered here.

As mentioned in the earlier section on the nitrogen cycle, plants are limited by a lack of fixed nitrogen and often live in association with nitrogen-fixing bacteria, which occur either in the soil as free-living microbes or within plant root tissues.

Legumes are plants that form a root nodule within which rhizobia live. The term *rhizobia* refers to the group of nodule-forming rhizobacteria that include species of *Rhizobium* and *Bradyrhizobium*, among others. Figure 11-9 gives an overview of how bacteria infect the root hair and then form the nodule within which they fix nitrogen that benefits the plant. Legumes are extremely important to agriculture because they return some fixed nitrogen to the soil when planted in rotation with other crops. Rhizobia have a preference for which legumes they'll colonize so there is a different species for each type of plant, including alfalfa, clover, soybean, and peas.

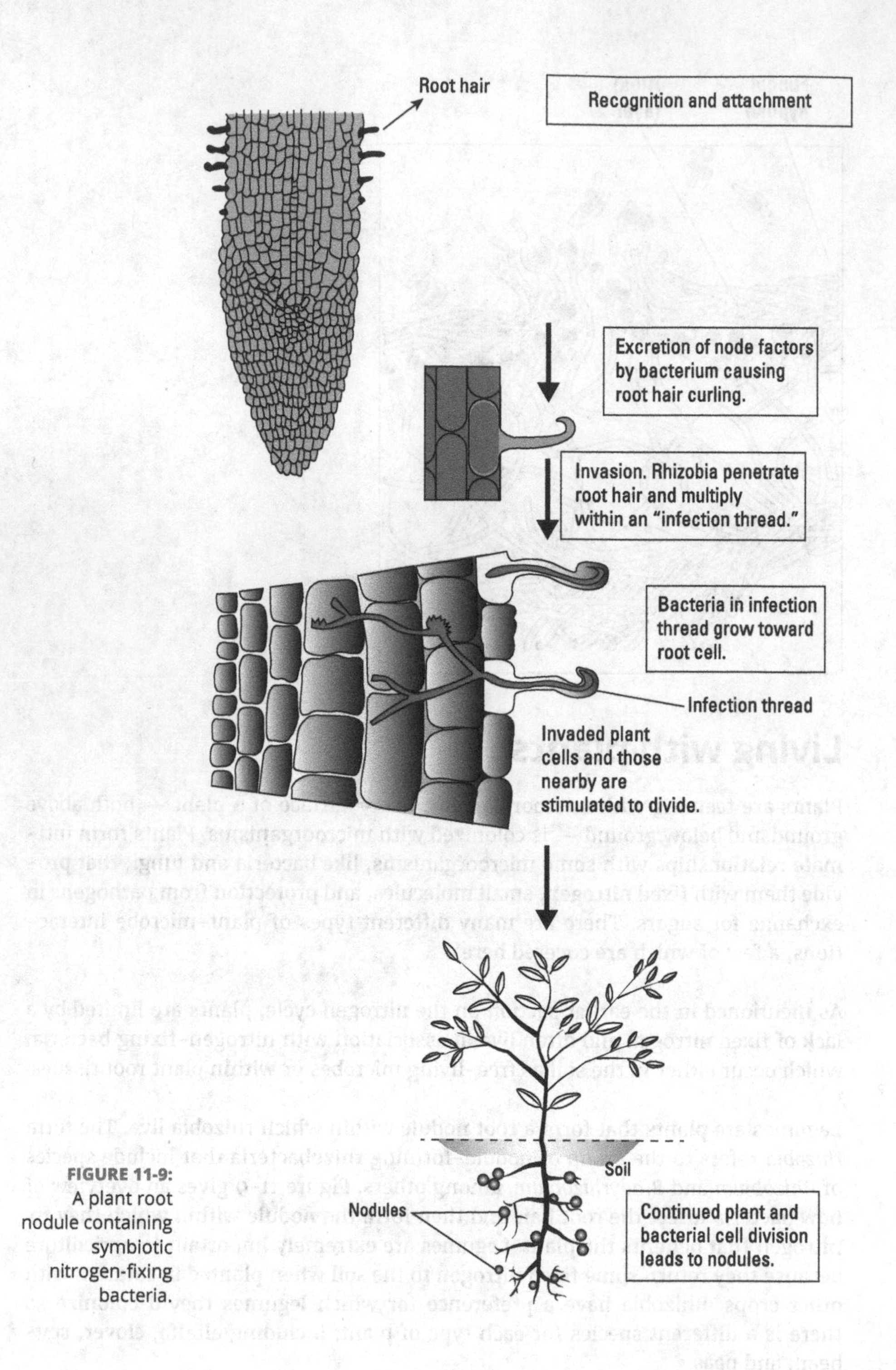
Root hair
Recognition and attachment
Excretion of node factors by bacterium causing root hair curling.
Invasion. Rhizobia penetrate root hair and multiply within an "infection thread."
Bacteria in infection thread grow toward root cell.
Infection thread
Invaded plant cells and those nearby are stimulated to divide.
Soil
Nodules
Continued plant and bacterial cell division leads to nodules.

FIGURE 11-9: A plant root nodule containing symbiotic nitrogen-fixing bacteria.

Nonlegume plants also form an interaction with nitrogen-fixing bacteria. One example is the alder tree, which can associate with the filamentous bacteria *Frankia.* The partnership generates enough fixed nitrogen that alder trees can grow in nitrogen-poor soils. *Frankia* is not as picky as the rhizobia and will also colonize other woody plants.

Mycorrhizal fungi form important symbiotic relationships with many different plants, some of which are dependent on the fungi for survival. (See Chapter 13 for a complete discussion of mycorrhizal fungi.) Mycorrhiza is the name for the fungal growth, which extends out to cover an area that is much greater than that of the plant's own root system. Although the mycorrhiza doesn't perform nitrogen fixation, it does provide the plant with access to more water and nutrients than it would otherwise get.

The bacterial genus *Agrobacterium* forms a parasitic association with plants, called crown gall disease, causing large tumor-like growths on plant tissues. *Agrobacterium* species carry a large plasmid, called the tumor-inducing (Ti) plasmid, which they transfer into the DNA of the host plant. Once integrated into the plant's DNA, the genes from the Ti plasmid direct the plant cells to make modified amino acids called *opines* that can only be metabolized by the bacteria and are used as a source of carbon and nitrogen. *Agrobacterium* is considered a plant pest, but research on the Ti plasmid has been useful in biotechnology strategies that aim to genetically alter plant cells (see Chapter 16).

Living with animals

Although omnivores get some nutrients from fermentation of plant material in the gut, herbivores are completely reliant on it for survival. Cellulose is the structural material in plant tissues and is made up of glucose molecules. Because only microbes have the enzyme to digest cellulose into glucose subunits, they're essential to the survival of ruminant animals (like cows). Ruminants have a specialized stomach, rumen, where plant material is fermented by bacteria.

Most of these bacteria are *cellulolytic,* and they require the proper pH, temperature, and anoxic environment within the rumen to happily go about fermenting cellulose. After the microbes break down the cellulose into glucose, the glucose is fermented into small fatty acids, which are absorbed by the animal for nutrients. Then the bacteria themselves are digested, providing protein and vitamins. CO_2 and CH_4 gas are produced as waste products and are belched out. Besides cellulose-degrading and glucose-fermenting bacteria, the rumen also contains methanogenic archaea, which use the H_2 produced by fermenters to reduce CO_2 to CH_4.

All animals are likely colonized by microbes, and the human body is no exception. It has been estimated the there are ten times more microbial cells than human cells in a single person. Considerable effort has been put into cataloguing microbes that inhabit the human body using 16S rRNA gene sequencing. Numerous studies using this method have found evidence of bacteria at sites including the entire length of the digestive tract, the respiratory tract, the urogenital tract, as well as the skin. The role of these microorganisms in human health and disease is still unclear, but a concerted effort is being made to see if there is a pattern of microbial colonization associated with human health. It's not yet clear, but it has been suggested for some time that even sites traditionally considered sterile are colonized with bacteria, such as the brain and the fetus *in utero*.

Living with insects

Like all other animals, insects are colonized by a diverse set of microbial species. Unlike other animals, as far as we know, insects can have symbionts that affect their reproduction. It's estimated that 60 percent of insect species are infected with microbes that are passed from parents to offspring and affect both the sex and survival of the larvae. An example of a reproductive manipulator bacteria is *Wolbachia*, which is discussed in the sidebar in Chapter 12.

The termite hindgut also acts like a rumen, digesting cellulose and hemicellulose with the help of microbial symbionts. Some termites have mainly bacteria in their hindguts, whereas others have both bacteria and anaerobic protists. The protists within the termite gut also have symbionts that make either CH_4 or acetate as a waste product.

Living with ocean creatures

Some of the most interesting microbial symbionts are those of ocean invertebrates.

For example, a small species of squid has a light organ colonized by a bacterial species called *Aliivibrio fischeri*. The light organ emits light at night, which protects the squid from predators swimming below it that mistake the light for moonlight. The squid selectively chooses its symbiont from among the many other microbes in seawater and then lets the chosen bacteria grow to high cell densities within its body; then it expels the whole population once a day.

Around hydrothermal vents, giant tube worms take up reduced inorganic compounds like H_2, H_2S, or NH_4^+, which are delivered to bacterial symbionts that use them as electron donors for metabolism. These bacterial autotrophs make organic material that supplies the tube worms with all the nutrients they need.

Corals form some of the most ecologically important structures in the ocean, providing habitats for countless fish. Although not technically an obligate symbiosis between an algae and a marine invertebrate, corals are much less healthy without the algae.

Tolerating Extreme Locations

Eukaryotes don't tolerate extreme conditions well. The most resilient cells are those of bacteria and, especially, archaea.

Thermophiles are organisms that thrive at high temperatures, in particular enjoying growth above 45°C. Hyperthermophiles require even hotter temps, with optimum growth happening above 80°C, and the most heat-tolerant archaea has been found to grow at above 122°C. Thermophiles are found in hot springs and other geothermal locations, but only environments under pressure, like hydrothermal vents deep within the ocean, allow water to heat up to temperatures above 100°C.

Another essential characteristic of microbes found deep in the ocean is the ability to tolerate high pressure. *Piezophiles* can not only withstand the high pressures of deep sea environments but thrive there. They grow optimally at pressures between 300 and 400 atm (a unit of pressure: the standard atmosphere) but can also grow at 1 atm (the atmospheric pressure at sea level, which would be considered normal for us). Extreme piezophiles from the Mariana trench require more than 400 atm and grow optimally at 700 to 800 atm.

Acidophiles are bacteria and archaea that require low acidity to survive. Acidic environments include soils near volcanic activity, in the stomach, and in acid mine drainage. Although the term *acidophile* is defined as something with an ability to grow at a pH below 4, the environments mentioned all have a pH of less than 2 and are known to harbor bacteria and archaea.

Psychrophiles do well in cold temperatures like polar seawater, arctic snow, and the depths of the ocean. These bacteria and archaea and algae function best below 15°C and many can be killed by 20°C temperature.

KEEPING IT TOGETHER

Archaeal cell membranes are perfectly suited to hot temperatures because they don't melt like bacterial cell membranes. This is because the cell membranes of archaea are a single layer of phytanyl subunits that are actually bonded to one another. In bacteria, the cell membrane is made up of two layers of phospholipids that associate with each other because of hydrophobic interactions (that is, because they're not soluble in water). This means that where the bacterial cell membrane is flexible and will start to disintegrate at high temperatures, the archaeal cell membrane is more rigid and will stay intact even when temperatures rise.

Detecting Microbes in Unexpected Places

It seems like everywhere we look for microorganisms, we find them. What is still uncertain is how they impact both their own environments and our lives. There is a lot of interest in finding microbes where we haven't looked for them before and measuring their activities.

Finding microbes in our built environment has really taken off in recent years, with studies of everything from showerheads to airplanes to concrete. Locating microbes in hospitals and on hospital equipment is increasingly important as the numbers of hospital-borne infections and antibiotic-resistant pathogens keep growing.

Are there microbes in space? There is some controversial evidence of something on two meteorites, but it's not clear if they represent fossilized microbial remains or something else.

MICROBES IN SPACE

If we do find microorganisms that we think originated in space, we really want to make sure that we didn't put them there. NASA spends a great deal of energy making sure that the equipment that it sends out into space isn't carrying some tiny earthly passengers. Anything that will be launched into space, especially for missions to Mars or farther, is built in a clean room that has special air handling and is cleaned and sterilized every day. Clean rooms are not unique to the space industry — they're used by the pharmaceutical and the medical equipment manufacturing industries.

To make sure that their clean rooms are free of life, NASA microbiologists look carefully for any viable organisms. A new species of bacteria named *Tersicoccus phoenicis* was found in two different NASA clean rooms. Ironically, it's only because the areas are frequently cleaned with chemicals, heat, and drying that microorganisms capable of surviving these harsh conditions are selected for. It's also likely that these organisms live happily somewhere in the environment, but because no one is looking for them, they've never been seen before.

4
Meeting the Microbes

IN THIS PART . . .

Get acquainted with microorganisms from the three domains of life — from those we know a lot about (like bacteria, viruses, fungi, and protists) to those we know much less about (like the archaea and sub-viral particles).

Get friendly with the many kinds of bacteria, whether they're important for geochemical cycles or human health.

Get an overview of eukaryotic microorganisms including the yeasts, fungi, and the great diversity of protists that include the algae, the phytoplankton, and the amoeba, among others.

Discover the structures and behaviors of the viruses, including those that infect plants, animals, and bacteria.

IN THIS CHAPTER

» Becoming familiar with the Bacteria

» Introducing the Archaea

Chapter 12
Meet the Prokaryotes

Along with viruses, the prokaryotes make up most of the evolutionary diversity on the planet. A rough estimate puts the number of bacterial and archaeal cells on earth at around 2.5×10^{30}. The number of species is harder to pin down. Some scientists think that there are far more prokaryotic species than all eukaryotic organisms combined, whereas others think that it's the reverse. Either way, more prokaryotic species are being discovered every year, and it's likely that we've just hit the tip of the diversity iceberg!

WARNING

Prokaryote is sort of a misnomer because it's used to talk about all non-nucleated cells, as opposed to eukaryotes, which have a nucleus and organelles, among other things. Both the Bacteria and the Archaea fall into this category, but they're more distantly related to one another than are the Archaea and the Eukaryota (the third major domain of life) and so they technically shouldn't be grouped together. Because the Bacteria and the Archaea have many other similarities, it's simply more convenient to consider them at the same time in this book. However, archaea and bacteria are fundamentally different from one another in terms of cellular structures and genes, including those used to determine ancestry.

Making sense of the vast numbers of different species and lifestyles is no easy task. In truth, scientists will be working for many years and there still won't be a tidy sorted list. With this in mind, we've put together a chapter describing the major differences between the different prokaryotes based roughly on how they're related to one another and how they live.

Another term for how things are related to one another in the evolutionary sense is *phylogeny*. Phylogeny is measured by comparing the genetic code in each organism. There are several ways to do this, which are summarized in Chapter 11.

TIP

There are three domains of life: Bacteria, Archaea, and Eukarya, and within each are several *phyla*. A phylum is a major evolutionary division that is then divided again as *class*, then *order*, then *family*, then *genus*, then *species*. This type of organization is called *taxonomic classification* and each of these divisions is called a *taxonomic rank*.

TECHNICAL STUFF

Kingdom used to be the highest taxonomic rank until recently when the higher rank of domain was added. Kingdom is still an important rank when describing major groups within the domain Eukarya, but it's less useful for describing the Bacteria and the Archaea domains. For this reason, kingdom isn't used in this chapter.

Getting to Know the Bacteria

Of the two domains of prokaryotes, the Bacteria are the best studied and contain all known prokaryotic pathogens. In reality, only about 1 percent of all bacteria have been studied in any detail and of these only a small proportion cause disease. Some, like *Pseudomonas*, take the opportunity to colonize humans when their immune system is down, but they aren't primarily human pathogens thriving mainly as free-living bacteria in soils. Others, like *Wolbachia* and *Mycoplasma*, lack a cell wall and cannot live outside a host cell. Figure 12-1 shows a general view of the known phyla in the domain Bacteria.

The Gram-negative bacteria: Proteobacteria

This phylum contains all kinds of interesting metabolic diversity that doesn't match the evolutionary paths of diversity. This might be because members have been swapping DNA and have taken on traits that other bacteria had to evolve. This type of genetic transfer is called lateral gene transfer (LGT, or sometimes horizontal gene transfer, HGT) and makes deciphering bacterial evolution a bit tricky. The Proteobacteria can be divided genetically into five major classes named for letters of the Greek alphabet: alpha (α), beta (β), delta (δ), gamma (γ), and epsilon (ε).

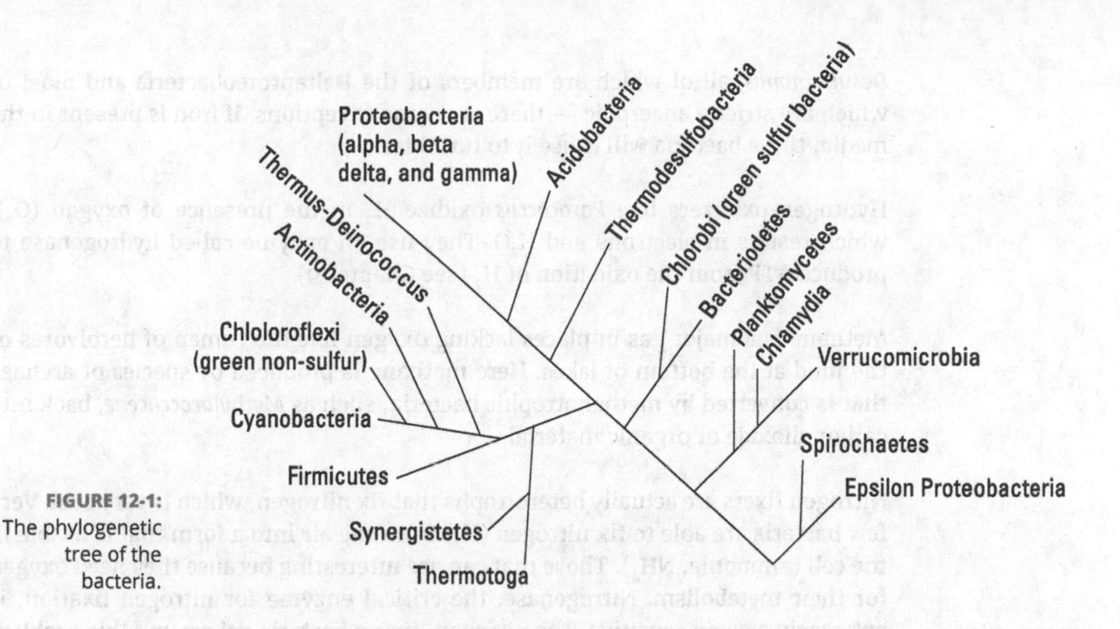

FIGURE 12-1: The phylogenetic tree of the bacteria.

This group seems to have the largest number of species, and many of them have been isolated in laboratory culture. Many members of the Proteobacteria are models for the study of microbial systems like genetics *(E. coli)* and anoxic photosynthesis (purple sulfur bacteria).

Autotrophic lifestyles

Nitrifiers oxidize inorganic nitrogen compounds like ammonia and nitrate for energy. All are environmental, found in sewage treatment plants as well as soil and water. They're different in that they have internal membranes that help with compartmentalizing toxic compounds made as a part of the oxidation process.

TIP

Ammonia oxidizers have names that start with *Nitroso*– (for example, *Nitrosomonas*), and nitrate oxidizers have names that start with *Nitro*– (for example, *Nitrobacter*).

Sulfur oxidizers live either in acidic or neutral environments rich in sulfur compounds. The acid-tolerant sulfur oxidizers (like *Thiobacillus*) acidify their environment by making sulfuric acid as a waste product during metabolism, and many can also use iron as an energy source. Neutral sulfur environments like sulfur springs and decomposing matter in lake sediments are home to sulfur oxidizers like *Beggiatoa* that grow in long chains and often have sulfur granules deposited within their cells.

On the other side of the coin, sulfate and sulfur can be used by sulfate and sulfur-reducing bacteria. These include members like *Desulfobacter*, *Desulfovibrio*, and

Desulfomonas, all of which are members of the Deltaproteobacteria and most of which are strictly anaerobic — there are some exceptions. If iron is present in the media, these bacteria will cause it to turn black.

Hydrogen oxidizers like *Paracoccus* oxidize H_2 in the presence of oxygen (O_2), which results in electrons and H_2O. They use an enzyme called hydrogenase to produce ATP from the oxidation of H_2 (see Chapter 9).

Methane is a major gas in places lacking oxygen like the rumen of herbivores or the mud at the bottom of lakes. Here methane is produced by species of archaea that is converted by methanotrophic bacteria, such as *Methylococcaceae*, back into carbon dioxide or organic material.

Nitrogen fixers are actually heterotrophs that fix nitrogen, which is very cool. Very few bacteria are able to fix nitrogen (N_2) from the air into a form that is usable in the cell (ammonia, NH_4). Those that can are interesting because they need oxygen for their metabolism. Nitrogenase, the critical enzyme for nitrogen fixation, is extremely oxygen sensitive. The nitrogen-fixing bacteria get around this problem in two ways. Free-living nitrogen fixers form a thick slime around their cells that lets them have just the right amount of oxygen but not too much. Others, like *Rhizobium*, live in an intimate association with the roots of plants (such as soybean) inside which they aren't exposed to too much oxygen.

Heterotrophic lifestyles

The pseudomonads are ecologically important in soil and water and can break down things like pesticides. They can only metabolize compounds through respiration (they can't use fermentation), but most of the group can do this both aerobically and anaerobically. They can metabolize many organic compounds (more than 100) but don't make hydrolytic enzymes, which means that they can't break down complex food sources like starch. Members of the group include *Burkholderia, Ralstonia,* and *Pseudomonas.* Several pseudomonad species are opportunistic human pathogens and specific plant pathogens.

The genera *Neisseria, Moraxella, Kingella,* and *Acinetobacter* are all aerobic, non-swimming Proteobacteria with a similar shape, so they're often grouped together. The interesting thing about their cell shapes is that many (all except *Neisseria,* which has a round shape called *coccoid* all the time) are rod shaped during log growth and then switch to a coccoid shape in stationary phase. *Moraxella* and *Acinetobacter* use twitching motion (see Chapter 4) to get around. Most are found as commensals associated with moist surfaces in animals (such as mucous membranes), but some species of each are human pathogens and *Acinetobacter* in particular is more common in soil and water.

The enteric bacteria are facultative aerobes (not inhibited by oxygen) that ferment sugars with many different waste products. The bacteria in this group are all closely related within the Gammaproteobacteria and so are sometimes difficult to tell apart. Many are of medical and industrial importance. Most are rod shaped, and some have flagella, but for the most part they're distinguished from the pseudomonads based on the fact that they produce gas from glucose and don't have specific proteins needed to make the electron transport chain (cytochrome c) needed for respiration. This group includes the following genera: *Salmonella*, *Shigella*, *Proteus*, *Enterobacter*, *Klebsiella*, *Serratia*, *Yersinia*, and *Escherichia*. Of note is the genera *Escherichia* that includes the best-studied species of bacteria, *E. coli*, which has been used in countless research and industrial applications. The genus *Yersinia* contains the species *Y. pestis* that was responsible for the plague of the Middle Ages.

A group of Proteobacteria similar to the enteric bacteria are the Vibrio bacteria. Members of this group do have a cytochrome c gene, but otherwise they're pretty similar in other respects to the enterics. The group is named for the genus *Vibrio*, which contains not only the pathogen *V. cholera* but many other aquatic bacteria that produce fluorescent light in a process called bioluminescence. Other members of this group include the genera *Legionella* and *Coxiella*.

The Epsilonproteobacteria include bacteria found as commensals and pathogens of animals like *Campylobacter* and *Helicobacter* that are also common in environmental samples from sulfur-rich hydrothermal vents.

Interesting shapes and lifecycles

The *Spirillia* are spiral-shaped cells with flagella for moving around. They're different from the *Spirochaetes*, which are distantly related and have different cellular structures. Two interesting examples of spiral-shaped Proteobacteria include *Magnetospirillum*, which have a magnet inside each cell (see the example in Figure 12-2) that helps them point north or south, and *Bdellovibrio*, which attacks and divides inside another bacterial cell.

A sheath is like a tube inside which many bacterial cells divide and grow protected from the outside environment. Sheathed bacteria are often found in aquatic environments rich in organic matter like polluted streams or sewage treatment plants. When food gets scarce, the bacteria all swim out to look for a better place to live, leaving behind the empty sheath. Some bacteria, such as *Caulobacter*, form stalks that they use to attach themselves to surfaces in flowing water. Budding bacteria, such as *Hyphomicrobium*, reproduce by first forming a long hyphae at the end of which forms a new cell in a process called *budding*.

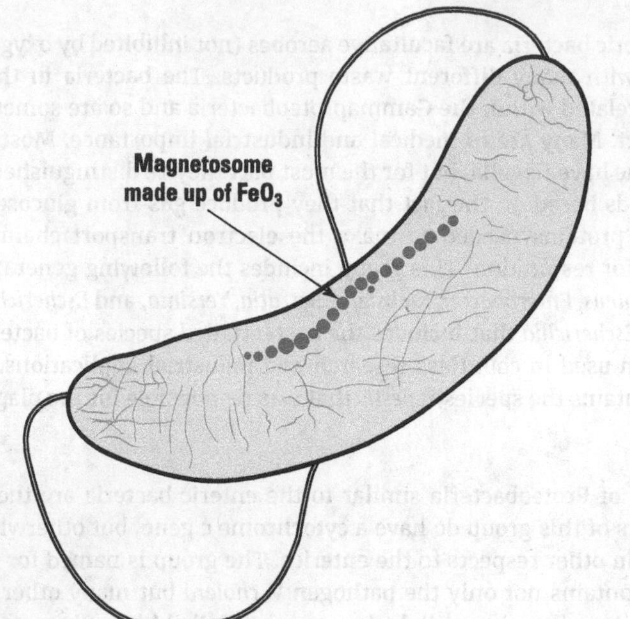

FIGURE 12-2: Magnetic bacteria.

REMEMBER

Budding is different from *binary fission* (where the cell divides into two equal parts) because the cell doesn't have to make all the cell structure before it starts to divide. Budding is often used by bacteria with extensive internal structures that would be difficult to double inside of one cell.

The *Rickettsias* are obligate intracellular parasites of many different eukaryotic organisms, including animals and insects.

The *Myxobacteria* have the most complex lifestyle of all bacteria that involves bacterial communication, gliding movement, and a multicellular life stage called a *fruiting body*. When the food sources are exhausted in one site, myxobacterial cells swarm toward a central point where they come together and form a complex structure called a *fruiting body* that produces mixospores. These mixospores can then disperse to a new location where a new food source can be found.

More Gram-negative bacteria

Many of the known Gram-negative bacteria are from the phylum Proteobacteria, but there are several other phyla that are also Gram-negative. Each is unique and an important part of the microbial world:

WHO'S YOUR DADDY? *WOLBACHIA!*

Species of *Wolbachia* live inside the cells of their host and infect countless species of beetle, fly, mosquito, moth, and worm (among many others) — more than 1 million species in all. In some cases, it's a parasite, causing its host harm; in other cases, it forms a mutualistic relationship with its insect host, a situation that is beneficial for both parties.

Some species of insect actually need to be infected with *Wolbachia* in order to reproduce successfully. In many cases, infection alters how or if the embryos develop. Here's an example: The *Wolbachia* bacteria can infect female eggs but not the male sperm. Infected females then produce female offspring without being fertilized. Infection makes the male sterile so that he can't fertilize an uninfected female.

Other strategies to increase the number of infected female offspring include killing male embryos and changing males into females after they've developed. Some of the insects that these bacteria infect are themselves parasites of animals. For example, heartworm that infects dogs requires a *Wolbachia* infection to reproduce; if the worm is treated with antibiotics, it dies.

As we talk about in Chapter 15, however, using antibiotics this way eventually leads to antibiotic resistance in bacteria, so ideally it won't catch on as a treatment. We still don't understand a lot about this phenomenon, but research into how it works and how it affects insect, animal, and plant populations is ongoing.

- **Cyanobacteria:** The phylum Cyanobacteria were likely the first oxygen-making organisms (through photosynthesis) on earth and were critical for converting the earth's atmosphere into the pleasantly aerobic one it is today. They come in all shapes and sizes, as shown in Figure 12-3, from single cells to colonies and chains with specialized structures where nitrogen fixation occurs (called *heterocysts*).
- **Purple sulfur bacteria:** The purple sulfur bacteria use hydrogen sulfide (H_2S) as an electron donor to reduce carbon dioxide (CO_2) and are found in anoxic (oxygen-free) waters that are well lit by sunlight and in sulfur springs. This group contains more than 40 genera with examples such as *Lamprocystis roseopersicina* and *Amoebobacter purpureus,* as well many species of *Chromatium.*
- **Purple nonsulfur bacteria:** The purple nonsulfur bacteria can live in the presence and absence of oxygen in places with lower concentrations of hydrogen sulfide. They're photoheterotrophs, meaning that they can use photosynthesis for energy but use organic compounds as carbon sources. Many have *Rhodo–* in their names like *Rhodospirillium, Rhodovibrio,* and *Rhodoferax,* among others.

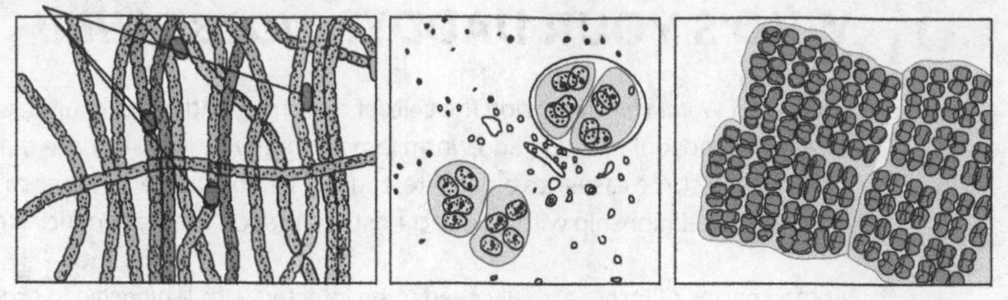

FIGURE 12-3: Cyanobacteria.

- **Chlorobi:** The phylum Chlorobi are called the green sulfur bacteria and are also *phototropic* (gathering energy from light), but they're very different from the green Cyanobacteria. For one thing, they live deep in lakes where they use hydrogen sulfide (H_2S) as an electron donor and make sulfur (S^0) that they deposit outside their cells. For another, they don't produce oxygen during photosynthesis, so they didn't contribute to the oxygenation of the earth's atmosphere like the Cyanobacteria did.
- **Chloroflexi:** The phylum Chloroflexi is also known as the green nonsulfur bacteria. These bacteria are found near hot springs in huge communities of different bacteria called *microbial mats* (see Chapter 11), where they use photosynthesis to gather energy without producing oxygen.
- **Chlamydia:** The phylum Chlamydia is made up entirely of obligate intracellular pathogens. These bacteria can't live outside a host cell, so they must continuously infect a host. Members of this group cause a myriad of human and other animal diseases and are transmitted both sexually and through the air where they invade the respiratory system.
- **Bacteroidetes:** The phylum Bacteroidetes contains bacteria common in many environments, including soil, water, and animal tissues. The genus *Bacteroides* can be dominant members of the large intestine of humans and other animals and are characterized by being anaerobic and producing a type of membrane made of sphingolipids that are common in animal cells but rare in bacterial cells. Other important genera include *Prevotella,* which are found in the human mouth, and *Cytophaga* and *Flavobacterium,* found in soils around plant roots.
- **Planktomycetes:** Members of the phylum Planktomycetes stretch the concept of prokaryote because they have extensive cell compartmentalization, (see Figure 12-4), usually only seen in eukaryotic cells. These compartments are especially useful to keep by-products like hydrazine (a component of jet fuel) contained (see Chapter 9).

  These bacteria live mainly in aquatic environments like rivers, streams, and lakes where some attach to surfaces by a stalk so that they can take up more nutrients from the surrounding water. These stalked bacteria divide by budding to produce a swimmer cell that takes off to find a new place to attach.

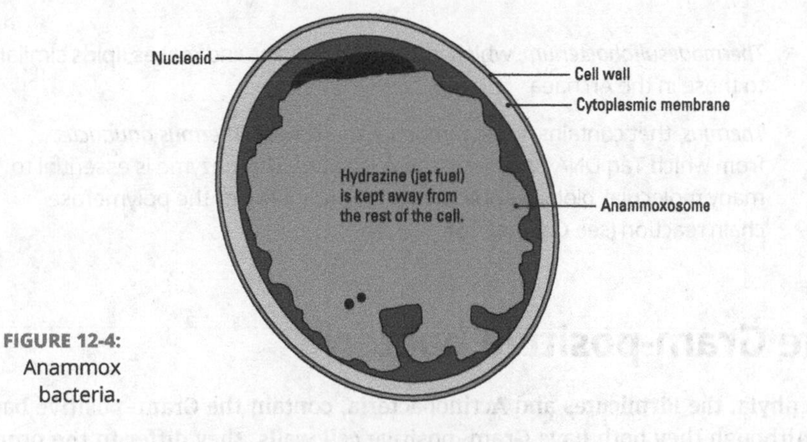

FIGURE 12-4: Anammox bacteria.

- **Fusobacteria:** The phylum Fusobacteria contains bacteria with cells that are long and slender with pointed ends. Some of the species of this group are found in the plaque of teeth as well as in the gastrointestinal tract of animals. They are anaerobic and members include *Fusobacteria* and *Leptotrichia.*
- **Verrucomicrobia:** The phylum Verrucomicrobia are named the warty (from the Greek *verru*) cells not because they cause warts but because some members look warty. The group is widespread in water and soil, but one Genus in particular is associated with the mucosal membranes of humans. *Akkermansia mucilagina* is more often associated with the guts of lean people.
- **Spirochaetes:** The Spirochaetes are highly coiled bacteria common in aquatic environments and associated with hosts. The latter group includes human pathogens such as *Treponema pallidum* that cause syphilis, species of *Borellia* that cause Lyme disease, as well those that help to break down wood in the guts of termites.
- **Deinococci:** The Deinococci share many structures with the Gram-negative bacteria, but because they have a very thick cell wall they stain Gram-positively. Members of this group are so tough that they can withstand levels of radiation 1,500 times higher than would kill a person. Not only do they have a tough cell wall, but they have many different DNA repair enzymes that can take a complete *Deinociccus radiodurans* chromosome that has been shattered into hundreds of pieces by radiation, and put it all back together in the right order.
- **Thermotolerant bacteria:** Several bacterial groups spanning many different phyla are thermotolerant. Some examples include
  - *Aquifex,* which are the most thermotolerant bacteria known.
  - *Thermotoga,* which makes a sheath (hence, *toga* in the name) and contains genes similar to those in the Archaea.

- *Thermodesulfobacterium,* which is a sulfate reducer and makes lipids similar to those in the Archaea.
- *Thermus,* that contains, most famously, the species *Thermus aquaticus,* from which Taq DNA polymerase was isolated. This enzyme is essential to many molecular biology applications because it drives the polymerase chain reaction (see Chapter 16).

The Gram-positive bacteria

Two phyla, the Firmicutes and Actinobacteria, contain the Gram-positive bacteria. Although they both have Gram-positive cell walls, they differ in the proportion of Gs (for guanine) and Cs (for cytosine) in their DNA. The Firmicutes are also known as the low G + C Gram-positive bacteria (with between 25 percent and 50 percent G + C), and Actinobacteria are also known as the high G + C Gram-positive bacteria (with between 50 percent and 70 percent G + C).

Low G + C: Firmicutes

The Firmicutes can be split roughly based on their ability or lack of the ability to form endospores. Dividing the group this way is mainly for convenience because it's easy to tell endospore formers from nonendospore formers by heating a culture up to kill everything but the spores. Within the two groups, there is quite a bit of phylogenetic and metabolic diversity.

Endospore formers, including species of *Clostridium* and *Bacillus,* live mostly in soil where endospore formation comes in handy when it's dry. Some infect animals and cause nasty diseases, but for the most part this is accidental. One important member of this group is *Bacillus thuringiensis* (Bt), which makes an endospore that contains a crystalline toxin called the Bt toxin (see Figure 12-5), which is particularly effective against many species of insect. Bt toxin is used extensively as an insecticide in agriculture (see Chapter 16).

The bacterial genera that don't form endospores can be grouped further into the *Staphylococci* and the *Lactococci.* Both groups contain commensal and pathogenic bacteria of animals and are distinguished by where they're found and their metabolism. For instance, the *Staphylococci* are tolerant of salt and are found on the skin, whereas the *Lactococci* are fermentative bacteria *(Peptostreptococcus* and *Streptococcus),* found in the guts of animals *(Enterococcus)* and in milk *(Lactococcus).*

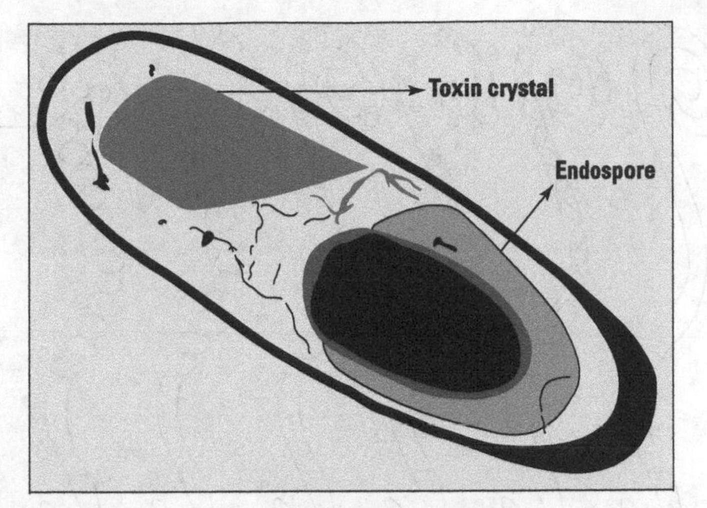

FIGURE 12-5: *Bacillus thuringiensis* endospore with toxin crystal.

High G + C: Actinobacteria

The phylum Actinobacteria contains many very common soil bacteria and several bacteria that are commensal of the human body, as well as a few notable human pathogens such as *Mycobacterium tuberculosis* and *Corynebacterium diphtheria.* Here are three important genera represented in this phylum:

- Members of the genus *Proprionibacterium* ferment sugars into propionic acid and CO_2 gas and are the main bacteria used to make Swiss cheese. The gas makes the holes in the cheese, and the acid gives it a nutty flavor.
- Colonies of *Mycobacteria* have a waxy surface because of special acids in their cell walls called *mycolic acids* that make them difficult to stain in the regular way. Instead, heat and acid are used to stain cells red so that they can be visualized under a microscope. This group has many non-pathogenic members as well as *M. tuberculosis.*
- The *Streptomyces* were thought for a long time to be a type of fungus because they make big filamentous clusters. They are, in fact, bacteria that, instead of dividing by binary fission into individual cells, form mycelia that make spores, which then pop off to populate new areas (see Figure 12-6). More than 500 different antibiotics have been isolated from this group, many of which are used in medicine today.

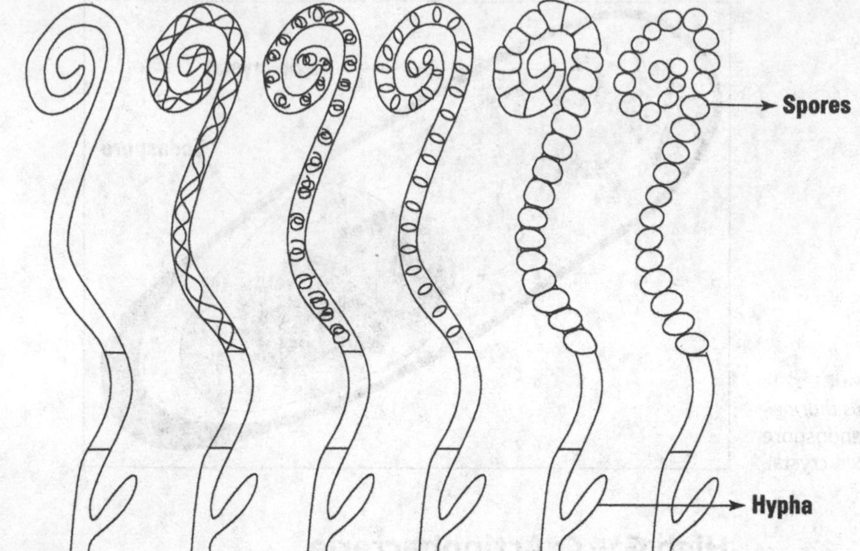

FIGURE 12-6: *Streptomyces* spore formation.

Acquainting Yourself with the Archaea

Also known as archaebacteria (*archaea*, from the Greek, means "ancient"), the archaea are thought to be the oldest forms of cellular life on earth. They differ from the bacteria in a few fundamental ways but until recently were thought to be part of the domain Bacteria. When sequencing genes to test the evolutionary relationship between microorganisms became popular, it became clear that the Archaea weren't part of the Bacteria at all but made up a division of their own.

Since their discovery in the late 1970s, there has been a steady increase in the number of described members. Each time a new group is found, information is added to what is known about the evolution of the entire group, because new members help to resolve the branching in the phylogenetic tree, shown in Figure 12-7. It's likely that many more archaea will be discovered and that the current tree will change quite a bit.

TECHNICAL STUFF

Currently, there are two main phyla in the domain Archaea: the Euryarchaeota and the Crenarchaeota. However, within the Crenarchaeota, there may soon be a few new phyla, including the Thaumarchaeota, the Korarchaeota, and the Aigarchaeota.

REMEMBER

As new archaeal strains are discovered, the gaps in what we know about how all archaea are related get filled in.

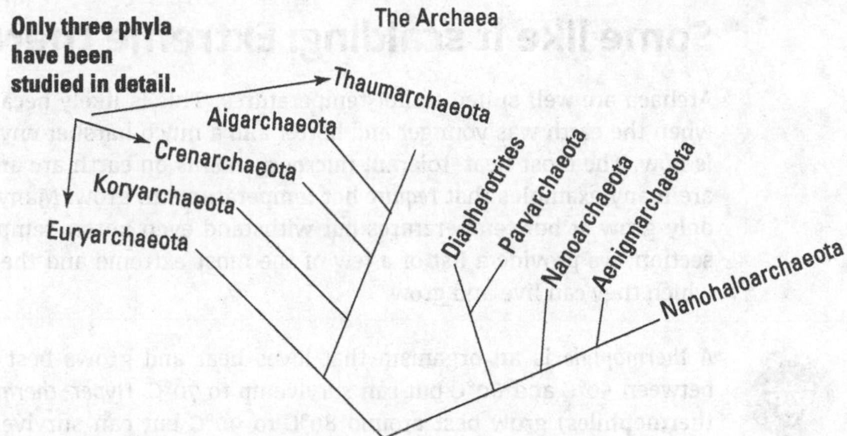

FIGURE 12-7: The phylogenetic tree of Archaea.

As with the Bacteria, there are far too many archaeal species to describe them all here but you can go to `www.ncbi.nlm.nih.gov/Taxonomy/Browser/wwwtax.cgi?id=2157` for a complete list. In this section, we discuss representatives of the different forms of archaeal life, filling you in on their ability to tolerate extremes of temperature, acidity, and salinity. It's likely that the most extreme of the Archaea were some of the first life forms on earth, evolving during a time when the earth was hotter and harsher than it is now. How they're able to thrive in extreme conditions is covered in Chapter 11.

WHERE DO MY GENES COME FROM?

The Archaea are interesting because they have many genes that resemble those in bacteria and others that resemble the genes in eukaryotes. This is part of the reason why they confounded microbiologists for years — they couldn't squarely be placed within the domain of Bacteria or Eukarya.

A great example of this is an archaeon (singular for archaea) called *Methanocaldococcus jannaschii,* which has core metabolic genes that bear some resemblance to those in bacteria, but most of the genes for molecular processes (things like RNA transcription and protein translation) have similarities to those in eukaryotes. More than a third of its genome (40 percent) contains genes that don't resemble those in either bacteria or eukaryotes.

Archaea likely evolved around the same time as the earliest bacteria. It's even possible that eukaryotes came from an early archaeal ancestor. It's mysteries like this that make the microbiology of the archaea so fascinating.

Some like it scalding: Extreme thermophiles

Archaea are well suited to hot temperatures. This is likely because they evolved when the earth was younger and hotter and a much harsher environment than it is now. The most heat-tolerant microorganisms on earth are archaea, and there are many examples that *require* hot temperatures to grow. Many archaea can not only grow at hot temperatures but withstand even hotter temperatures. In this section, we provide a list of a few of the most extreme and the temperatures at which they can live and grow.

TECHNICAL STUFF

A *thermophile* is an organism that loves heat and grows best at temperatures between 50°C and 60°C but can survive up to 70°C. *Hyper-thermophiles* (extreme thermophiles) grow best around 80°C to 90°C but can survive in much higher temperatures. Some hyper-thermophiles have been found to survive above 120°C in the high-pressure environment of the deep sea near hydrothermal vents.

The following archaea are thermophiles and extreme thermophiles:

- ***Thermococcus*** and ***Pyrococcus*** are strict anaerobes that get energy from metabolizing organic matter in many different thermal environments. *Thermococcus* grows fine in a range of temperatures between 55°C and 95°C, and *Pyrococcus* grows best at 100°C.
- ***Methanopyrus*** is a hyperthermophilic methanogen (it produces methane). This group contains a unique kind of cellular membrane not found in any other organism. One species of this group, *M. kandleri,* is the current record holder for growth at the hottest temperature, at 122°C. Water can attain temperatures this high only in deep ocean environments where great pressure stops water coming out of hydrothermal vents from boiling.
- ***Nanoarchaeum*** are very small in size and, as shown in Figure 12-8, live as parasites on another hypothermophilic archaea, *Ignicoccus*. These two archaea can be found together in hydrothermal vents and hot springs at temperatures between 70°C and 98°C.
- ***Ferroglobus*** can oxidize iron anaerobically. It's likely that *Ferroglobus* and others like it were oxidizing iron before the earth's atmosphere contained oxygen, creating blankets of iron deposits on the ocean floor. As time went on, this layer of iron got trapped and is now seen as banding patterns in ancient rocks.
- ***Sulfolobus*** lives in sulfur-rich, acidic environments like those around hot springs where it attaches to sulfur crystals oxidizing the elemental sulfur for energy (see Figure 12-9).

» ***Desulfurococcus*** and ***Pyrodictium*** are strictly anaerobic sulfur-reducing archaea that thrive around marine hydrothermal vents. *Desulfurococcus* grows best at 85°C, whereas *Pyrodictium* grows best at 105°C.

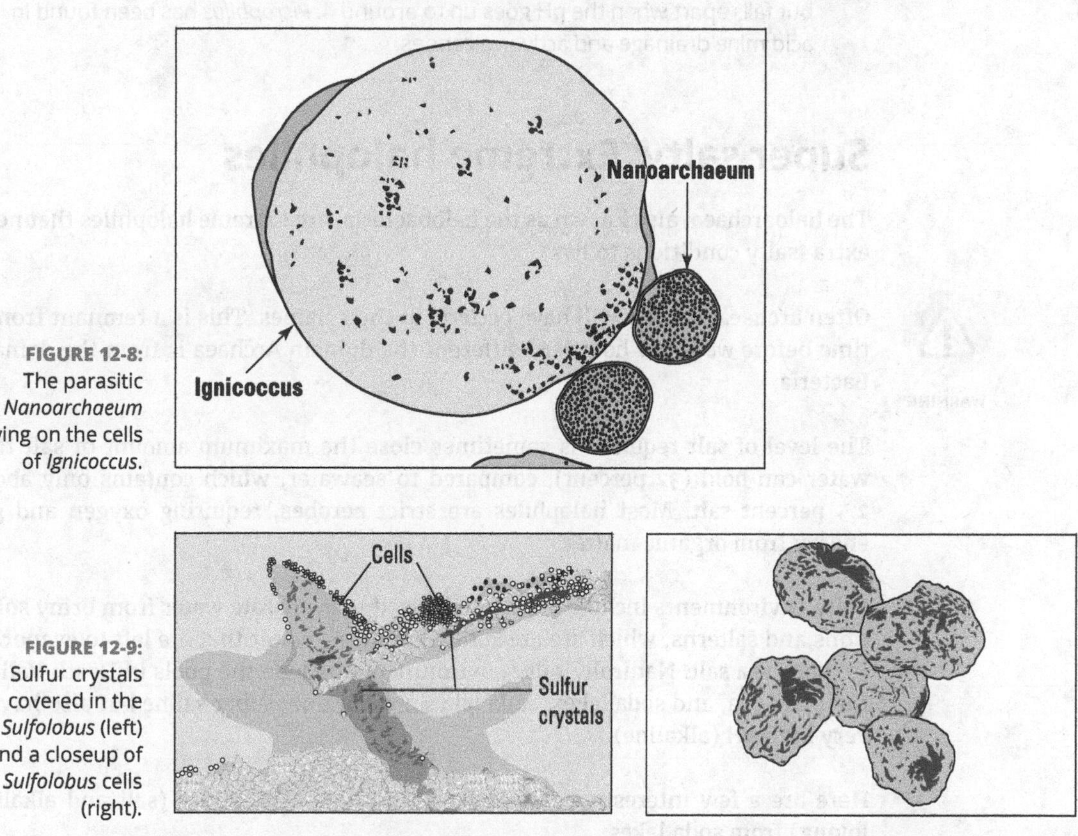

FIGURE 12-8: The parasitic *Nanoarchaeum* living on the cells of *Ignicoccus*.

FIGURE 12-9: Sulfur crystals covered in the *Sulfolobus* (left) and a closeup of *Sulfolobus* cells (right).

Going beyond acidic: Extreme acidophiles

Some of the most acid-tolerant microorganisms known are archaea, many of which are also thermophilic. Extremely hot and acidic environments are some of the most difficult to get to and sample from, which explains why so few microorganisms from these environments have been isolated. Here are some examples of extreme acidophiles:

» ***Thermoplasma*** lacks a cell wall and can live by sulfur respiration in coal refuse piles at temperatures around 55°C and hot springs.

- **Ferroplasma** also has no cell wall but lives in very acidic mine drainage at medium temperatures. It breaks down the pyrite in the mine waste, which acidifies its environment down to a pH of 0.
- **Picrophilus** is so well adapted to acidity that it can live at a pH of 0 and lower but falls apart when the pH goes up to around 4. *Picrophilus* has been found in acid mine drainage and active volcanoes.

Super salty: Extreme halophiles

The haloarchaea, also known as the halobacteria, are extreme halophiles that need extra-salty conditions to live.

Often archaeal species will have *bacteria* in their names. This is a remnant from a time before we knew how very different the domain Archaea is from the domain Bacteria.

The level of salt required is sometimes close the maximum amount of salt that water can hold (32 percent), compared to seawater, which contains only about 2.5 percent salt. Most halophiles are strict aerobes, requiring oxygen and get energy from organic matter.

Salty environments include brine ponds used to evaporate water from briny solutions and salterns, which are areas filled with sea water that are left to evaporate to make sea salt. Naturally salty environments include the pools in Death Valley, the Dead Sea, and soda lakes. Soda lakes are not only super saline but also have a very high pH (alkaline).

Here are a few interesting Haloarchaea and halo alkaliphiles (salt and alkaline loving) from soda lakes:

- **Halobacteria** was the first salt-loving archaeon studied and is the poster child for the group. It was used to learn most of what we know about the cellular structure and adaptations of highly salt-tolerant archaea. *Halobacteria* have a cell wall made of glycoprotein that is stabilized by the sodium ions (Na^+) in the environment.
- **Haloquadratum** lives in salterns and was named for its unusually shaped square cells, which are thin and filled with gas pockets that let it float to the surface where the oxygen is.
- **Natronococcus** is a halo alkaliphile found in soda lakes with a pH of between 10 and 12.

Some have regular shapes like rods and cocci, whereas others can have very unexpected shapes like squares or cup-shaped disks.

Because water has a tendency to move from an area of low solute concentration to an area of high solute concentration (which is the concept of osmosis), cells have to maintain a higher ion concentration inside than the environmental ion concentration. This accumulation of *compatible solutes* inside the cell is the only thing that stops it from losing water to the hypersaline environment. *Halobacterium* accumulates massive amounts of potassium (K^+) inside its cytoplasm to counteract the ultra-high concentration of Na^+ outside the cell.

These microorganisms are so well adapted to their super-salty environments that they can't live without very high levels of sodium in the environment. Sodium stabilizes the outside of the cells. In addition, they need a large supply of potassium, which is required for the proteins and other components inside the cell.

Not terribly extreme Archaea

Despite making up much less of the known microbial world, archaea have a big impact on the earth's geochemical cycles. For instance, many primary producers in aquatic and terrestrial habitats are archaea that contribute the carbon cycling in these places. The ammonia oxidizing archaea are another example that are important players in the nitrogen cycling in the oceans because they're part of the nitrification process. Methanogenic archaea are those that produce methane and live in environments lacking oxygen, such as the digestive tracts of animals (and humans), aquatic sediments, and sewage sludge digesters. They're important members of carbon cycling, catalyzing the final step in the breakdown of organic matter. Examples include

- *Methanobacterium,* the cell wall of which contains chondroitin-type material. Chondroitin is a major component of cartilage.
- *Methanobrevibacter*
- *Methanosarcina*
- *Nitrosopumilus,* the ammonia-oxidizing ocean archaea

There are archaea living in nonextreme environments, both aquatic and terrestrial, including under polar ice in the Arctic Ocean. Scientists have evidence that they're there, but none have either been grown in laboratory culture or been fully described.

IN THIS CHAPTER

» Finding out about microorganisms that are not prokaryotes

» Figuring out how fungi reproduce

» Sorting through the many different protists

Chapter 13
Say Hello to the Eukaryotes

In Chapter 8, we discuss the relatedness of all organisms and how the tree of life has three main branches, or domains, consisting of the Bacteria, the Archaea, and the Eukarya. This third branch gave rise to, and contains all of, the multicellular organisms, as well as many microorganisms, which we cover in this chapter.

Until the advent of DNA sequencing, classification of eukaryotic microorganisms was done by comparing the physiology of eukaryotic groups to determine how they related to one another. Although many organisms can be classified in this way, many of the evolutionary relationships between groups were fuzzy. Now, thanks to modern techniques, we know more about evolution within the domain Eukarya with some interesting changes to how we think about this group. For instance, the fungi, thought to be closely related to plants, are in fact closely related to animals.

There is a lot of diversity within eukaryotic microbes but they can be divided roughly into fungi and protists, with the latter containing much of the diversity within the entire domain.

Fun with Fungi

For years, scientists thought fungi were closely related to plants, but it turns out that they're more closely related to animals. Fungi take many different forms, from single-celled yeast to some of the largest and oldest organisms on the planet. This diverse group can be split into mushrooms, molds, and yeasts, all of which have important roles in nature. They are helpful in that they break down decaying plant and animal material in the environment, and are used extensively in the food and drug industries. Some of them, though, can be harmful as they are responsible for many economically important plant diseases. Some fungi cause disease in humans and animals, but for the most part they're benign and even delicious.

Figuring out fungal physiology

When growing vegetatively (not reproducing), fungi can grow as single cells or as filaments. Some grow in both ways, but most fungi use only one form of growth. *Unicellular* (single-cell) fungi include the yeasts that divide either by budding or by fission (see Figure 13-1).

WHAT IS A EUKARYOTE?

The term *eukaryote* is used as a general term for all organisms within the domain Eukarya and was originally coined from the Greek words *eu* (for true) and *karyo* (for kernel), referring to the fact that eukaryotic cells have a nucleus within them. Following this naming strategy, the Bacteria and Archaea are called prokaryotes (meaning prenucleus), which although convenient is a poor term for the other two major domains of life (see Chapter 8 for reasons why).

The nucleus is not the only difference between eukaryotic and prokaryotic organisms. Eukaryotic cells are usually much larger and contain membrane-bound organelles. Next to the nucleus, the other organelle considered to be a hallmark of eukaryotic cells is the mitochondrion. These are present at many copies within cells and function to provide the adenosine triphosphate (ATP) necessary to power cellular processes. Another essential organelle is the chloroplast, which contains the structures necessary for photosynthesis. Both the mitochondria and the chloroplasts are thought to be the remnants of prokaryotic cells that were engulfed by an ancestor of current eukaryotes in a process called *endosymbiosis*, theories of which are described in Chapter 8.

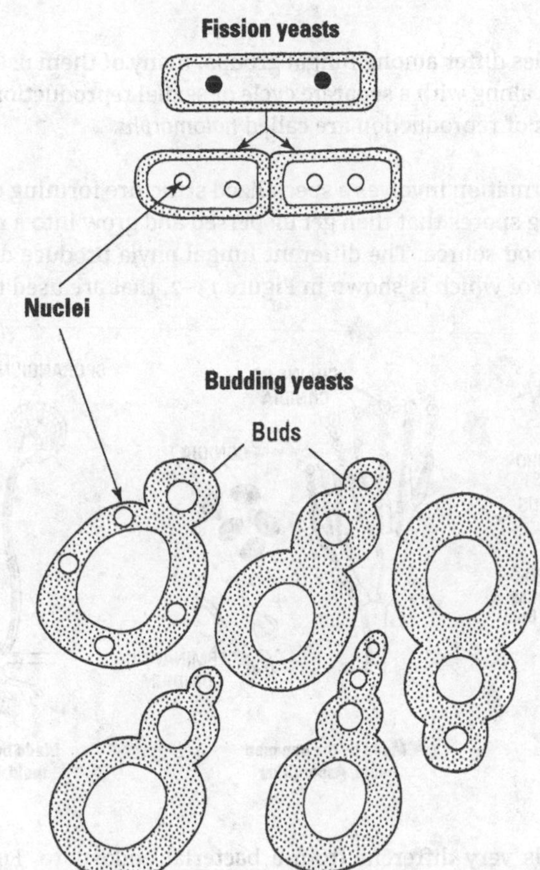

FIGURE 13-1:
Unicellular fungi.

Most fungi fall into the filamentous category and are multicellular organisms made up of *hypha,* which are long filaments of interconnected cells. The walls between cells are called *septa;* septa don't always completely separate one cell from another so cell contents can move from one compartment to another. Other fungal hypha — called *coenocytic hypha* — are not separated at all; instead, many nuclei exist within the cytoplasm, which is continuous throughout. The dense cluster of fungal hypha that form as the fungi grow is called *mycelia.*

TIP

Fungal cell walls are made of *chitin,* a polymer made from glucose. Chitin is a lot like cellulose in plants or keratin in animals — it gives cell walls their rigidity.

The main way that fungi get nutrients is by secreting hydrolytic enzymes that break down complex organic matter into simple subunits like amino acids, nucleic acids, sugars, and fatty acids. Some of the toughest polysaccharides in wood are digested only by fungi, making them important decomposers in an ecosystem. Fungi are found everywhere. They often contaminate food and culture media because they're versatile, and their spores can be spread very easily.

Although lifecycles differ among fungal groups, many of them use a cycle of asexual reproduction along with a separate cycle of sexual reproduction. Fungi that use these two means of reproduction are called *holomorphs.*

Asexual spore formation involves a specialized structure forming on the end of the hypha, producing spores that then get dispersed and grow into a new fungus after they land on a food source. The different fungal phyla produce different types of spores, a sample of which is shown in Figure 13-2, that are used to identify them.

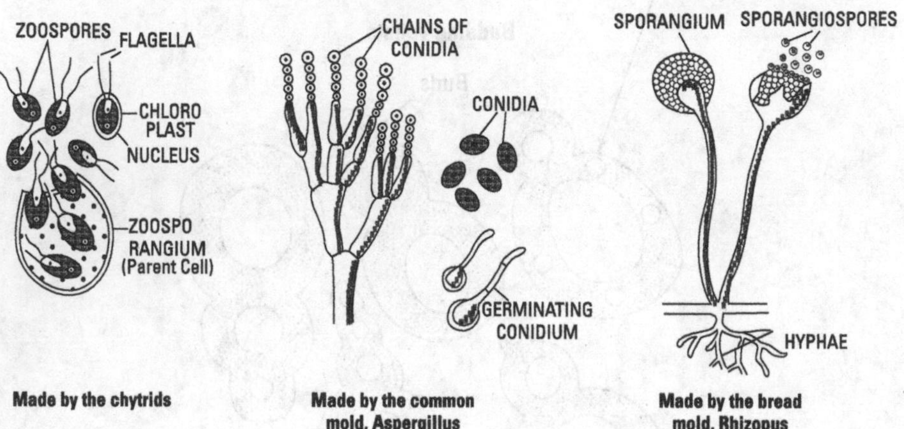

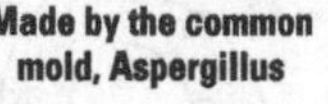

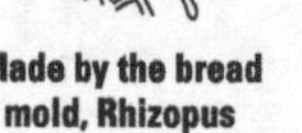

FIGURE 13-2: Types of asexual spores.

WARNING

A fungal spore is very different from a bacterial endospore. Fungal spores are reproductive. After they're dispersed, they give rise to new and separate fungi. They aren't overly heat tolerant, but they can survive drying rather well. A bacterial endospore is not produced for reproductive reasons but as a survival mechanism when conditions are unfavorable. An endospore is formed within one bacterial cell, is highly resistant to heat and other stresses, and will germinate as the original bacterial cell when conditions improve.

Sexual reproduction is a way of increasing the genetic diversity of individuals and involves two different fungal hypha coming together to form a structure containing spores produced through *meiosis.* Meiosis is the process of making cells that contain half of the genetic information of the parent cell, called *haploid cells,* necessary in preparation for mating.

TIP

In order to reproduce sexually, two compatible fungal cells have to come together. The two different yet compatible cell types are analogous to male and female if instead of two genders there were many different ones. The result is that fungi encounter compatible mating types more often than if they could exist only in two types.

The new fungus, the product of the previous two compatible fungal cells, then undergoes meiosis at some point to produce haploid fungi again, allowing another chance of meeting a different compatible fungus, thereby increasing genetic diversity. In the phylum Ascomycetes many species have lost the ability to reproduce sexually, and are referred to as *anamorphs*. Sexual reproduction in the Basidiomycetes involves producing a large structure to disperse spores called a *fruiting body*, which is commonly recognized as a mushroom.

TECHNICAL STUFF

Fungi are generally *haploid*, meaning that nuclei contain one copy of their genome. When two haploid nuclei fuse they become *diploid* because the result is two copies of the genome. Meiosis is splitting of diploid nuclei into two haploid copies, along with a bit of mixing so that the resulting cells don't have an exact copy of either parent but a combination of the two.

TECHNICAL STUFF

Plasmogamy and karyogamy are separate but related events in sexual reproduction. When two cells fuse and their cytoplasmic contents mix but the nuclei don't fuse it's called *plasmogamy* and the resulting cell is called *dikaryotic*. When the nuclei in a dikaryotic cell fuse, it's called *karyogamy* and the result is a *zygote*. Plasmogamy is more common in the fungi, whereas karyogamy is widespread in nature; one great example of karyogamy is the fusing of animal egg and sperm cells during fertilization.

Itemizing fungal diversity

One of the unique things about fungi is that many of them change dramatically throughout their lifecycles — so much so that the different stages have often been described as a separate species. Over time *mycologists* (microbiologists who study fungi) have begun to clean up our understanding of many fungal groups, relabeling those that were originally thought to be different species but are, in fact, two different life stages of the same species.

Because there are so many different forms of fungi it can be hard to keep them all straight. But they can be organized into several different groups based on *phylogeny* (how they are related) and lifestyle — the five major phyla for fungi are the Chytridiomycetes, Zygomycetes, Glomeromycetes, Ascomycetes, and Basidiomycetes. There is still a lot we don't know about fungal evolution, and many species of fungi have yet to be discovered that will change how we organize this list.

Most fungal groups are benign to animals and humans, but some do cause animal diseases called *mycoses*. Mycoses are difficult to treat because drugs aimed at fungal cells are also highly toxic to animal cells. Here are some examples of fungi-causing mycoses:

- Opportunistic infections from *Microsporidia, Pneumocystis,* and *Cryptococcus* cause life-threatening disease in immunocompromised people. *Microsporidia* is in its own phylum, whereas *Pneumocystis* is an Ascomycete and *Cryptococcus* is a Basidiomycete.
- Some members of the Chytridiomycete phylum, called *chytrids,* cause a serious disease in frogs by infecting the skin and reducing respiration leading to death. A large number of frog species have suffered tremendous losses in their populations recently due to this pathogen.
- Less serious, yet inconvenient, infections include thrush, caused by the yeast *Candida albicans,* and athlete's foot, caused by the fungi *Trichophyton* (both Ascomycetes).

There are a number of fungal plant pathogens, many of which cause large economic losses of crops and the heartbreaking loss of mature trees. Here are some examples:

- Apple scab starts as a brown discoloration on the fruit and leaves of apple and pear trees and eventually turns into dark dry scabs that crack, causing a lot of fruit loss. It's caused by an Ascomycete fungi called *Venturia* and can only be eradicated by removing all diseased plant material from near healthy plants.
- Dutch elm disease and chestnut blight are both caused by different Ascomycete fungi. They're native to China and Japan where trees that have evolved along with the fungi have some immunity to disease. In North America and Europe, however, *Ophiostoma* species decimate elm tree numbers and *Cryphonectria parasitica* has almost completely wiped out the American chestnut.
- Wheat rust and corn smut are two economically important plant pathogens from the Basidiomycete group. Rusts have in the past devastated nearly half of North America's wheat crops and new, more virulent strains are currently threatening much of the wheat varieties grown in Africa. They need at least two hosts to complete their lifecycles because they spend the winter on one and then infect the other during the summer months. Smuts, on the other hand, need only a single host, but they often get into the reproductive structures of flowering plants and infect the seeds before they even have a chance to be dispersed.

In addition to pathogenic and benign fungi, there are beneficial fungi that associate with plant roots, called *mycorrhizae* (literally "fungus roots" in Greek). These provide an essential symbiosis — many plants, like pine trees, can't grow without their mycorrhizal partners.

Interacting with plant roots

Several different types of fungi form intimate associations with plant roots in a beneficial symbiotic relationship and up to 90 percent of plants have a mycorrhizal component underground. Most forest soils are rich in mycorrhizal fungi that will associate with new seedlings, helping them gain nutrients and moisture, producing enzymes, and offering protection from pests. In nutrient-poor soils, mycorrhizal fungi may tip the balance in favor of plant survival, and some plants absolutely need them to grow. Pines, for instance, would not survive in their preferred sandy soils if it weren't for these associations.

These fungi form extensive structures with plant root tissue that function to transfer nutrients between themselves and the plant cells. We'll talk about two different kinds here: the *endomycorrhizal* ("endo" meaning inside) and the *ectomycorrhizal* ("ecto" meaning outside) fungi.

As their name suggests the endomycorrizal fungi form extensive structure within plant root tissues that, in addition to hypha, include fingerlike projections called *arbuscules,* important for nutrient exchange, and balloonlike structures called *vesicles,* used for fungal storage of plant carbon (see Figure 13-3). For this reason, these fungi are called *arbuscular mycorrhizal fungi* (AMF) and they belong to the phylum Glomeromycete. The fungi supply the plant with much higher levels of phosphorous that it would absorb on its own. In exchange, the plant provides all the mycorrhiza's carbon needs.

Instead of penetrating extensively into plant tissues, ectomycorrhizal fungi form a dense layer of mycelia around plant roots and extend only slightly into plant roots (refer to Figure 13-3). At least three different phyla of fungi have members that form ectomycorrhizal relationships with plant roots, many of which form aboveground structures that are easily recognized as mushrooms. The sheath of mycelia protects plant roots from pathogens and allows increased uptake of water and nutrients. To interact with plant roots, fungi produce a hyphal network, called a *Hartig net,* that extends a few cell layers into root tissue and acts as the site of nutrient exchange.

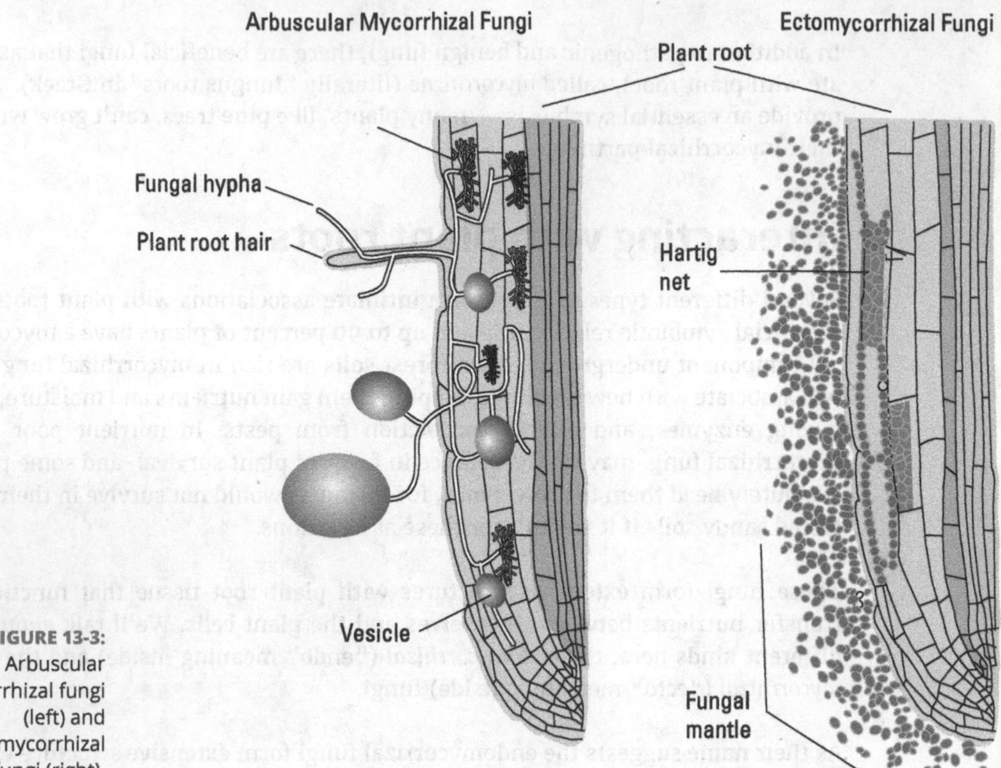

FIGURE 13-3: Arbuscular mycorrhizal fungi (left) and ectomycorrhizal fungi (right).

Ask us about the Ascomycetes

Members of this group of fungi produce spores inside an ascus (Greek for sac). Figure 13-4 shows a section through the fruiting body of a conspicuous type of cup fungi from this group, often found on the forest floor. The formation of this structure starts with the meeting of two individual fungal hypha that interact to form hypha containing two nuclei, a process called *plasmogamy*. These special hypha with a double nuclei extend upward and become diploid briefly as the nuclei fuse and then undergo meiosis to produce haploid ascospores that get dispersed when the asci (plural of *ascus*) rupture.

Some Ascomycetes have a very different lifestyle from the filamentous fungi just mentioned. The most well known of these is the brewer's yeast *Saccharomyces* that lives mainly as a single cell and divides by asexual budding. Sexual reproduction begins with the fusing of two haploid cells that can then remain diploid and undergo asexual budding for a long time before meiotically dividing to form haploid ascospores that then germinate.

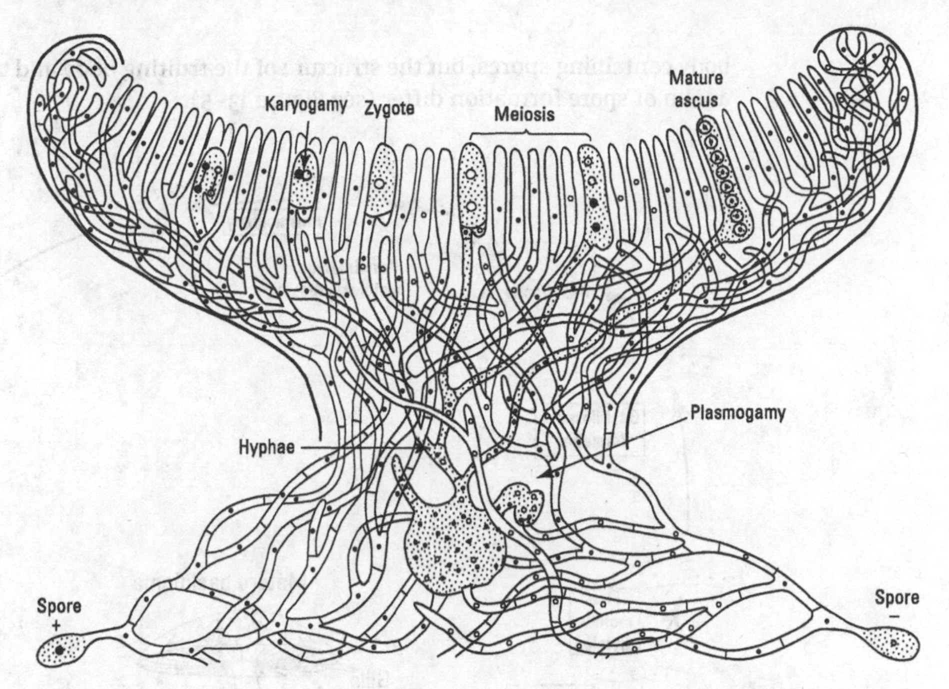

FIGURE 13-4: The ascocarp of a cup fungus.

Unlike the teleomorphs that reproduce sexually, anamorphs can only reproduce asexually, forming *conidia* (asexually produced spores) that are dispersed to new areas in the search for food. There are several Ascomycetes that have done away with sexual reproduction; an example is *Penecillium*, which grows along happily until it runs out of food and then produces conidia at the terminal ends of a specialized hyphal structure (refer to Figure 13-2).

Mushrooms: Basidiomycetes

Along with the Ascomycetes, the Basidiomycetes group makes up a large part of the diversity of the fungi, with a variety of shapes and spore dispersal strategies. Some of the reproductive methods can be quite complicated, involving several cycles of asexual and sexual reproduction in association with a number of different hosts. Some members of this group are known only by their anamorph (or asexual) stage, and it's still unclear whether these have completely lost the ability to reproduce sexually or if evidence of these forms will turn up as scientists describe more species of fungi.

Several Basidiomycetes are well known, but none more so than the mushroom, whose clublike shape was the inspiration for naming the group (*basidio* means club). The mushroom lifecycle bears similarity to the Ascomycete group in that two individual fungi come together, combining their hypha to produce a fruiting body containing spores, but the structure of the fruiting body and the exact mechanism of spore formation differ (see Figure 13-5).

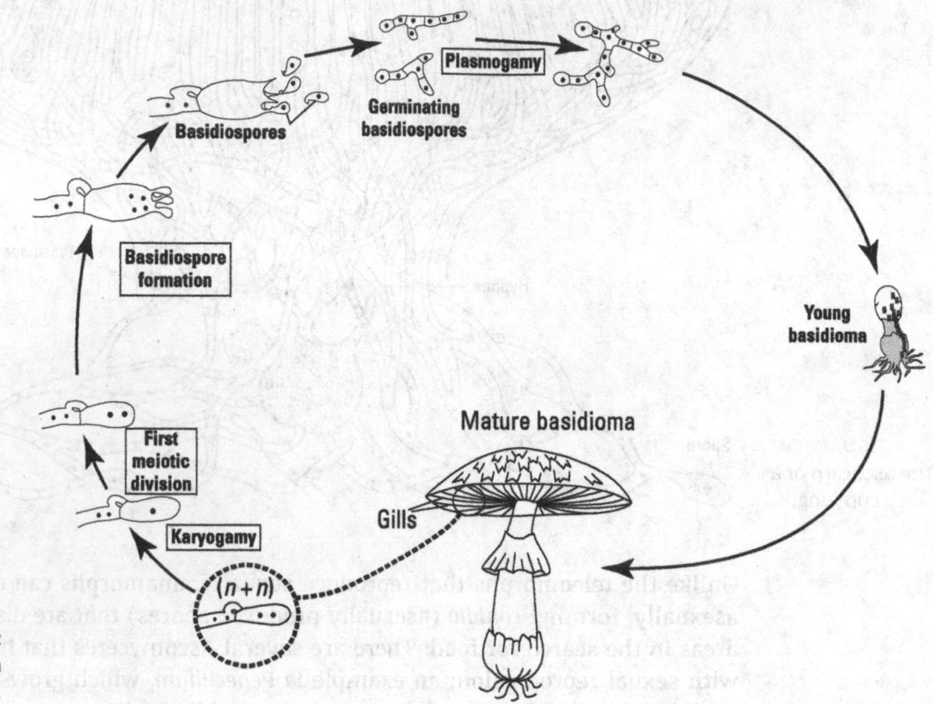

FIGURE 13-5: The mushroom lifecycle.

Here dikaryotic mycelia grow to make the fruiting body, which is called a *basidioma*, inside which the basidiospores are formed. It's under the cap of the mushroom, between the gills, that spore formation happens. There, the two nuclei inside cells at the end of fungal hypha fuse and then undergo meiosis to produce haploid basidiospores. These can then disperse and find a place to germinate. Some mushrooms are very long-lived; the oldest one to date, *Armillaria ostoyae*, which is approximately 2,400 years old, is also the biggest at 2,200 acres, located in Oregon.

Perusing the Protists

As a catchall for Eukaryotic microorganisms that are not fungi, the Protist group contains many different phyla making up most of the diversity within the Domain Eukarya. Many shapes, sizes, and lifestyles are represented in this group, making it difficult to keep them all straight.

Most are single-celled organisms; some like the algae form multicellular structures and others like some slime molds live as single cells that congregate to form multicellular structures. There is no single formula for reproduction either, which ranges from simple division to complicated cycles with many different structures.

Instead of discussing each case in great detail, we offer a few examples here that show some of the diversity of this group. We present the protists by their major habitat, starting with human parasites, plant pathogens, free-living amoeba and ciliates, and finally the algae and other photosynthetic eukaryotes.

Making us sick: Apicoplexans

Although some members of the other groups are known to cause human disease, these are noteworthy because the diseases they cause are particularly unpleasant and/or deadly. Although some can live in the environment before infecting animals, others have adapted completely to a parasitic lifestyle, having lost many of the genes necessary for a free-living lifestyle.

Plasmodium, *Toxoplasma*, and *Cryptosporidium* are part of the same group, called Apicoplexans, which are obligate parasites meaning that they can't live outside a host. Malaria, a disease that affects 10 percent of people worldwide is caused by species of *Plasmodium* that has an insect host and a human host. This parasite reproduces sexually within the mosquito, producing motile sporozoites that are transmitted to a human host by the insect's bite. In the human, cycles of asexual reproduction take place in the liver and blood cells, leading to the characteristic fevers and chills associated with the disease. Mosquitoes that feed on infected humans then become infected themselves and start this complex cycle over again (see Figure 13-6).

One strategy used by species of *Cryptosporidium* and *Toxoplasma* to move between hosts is a process called *encystment*. This involves making a cyst that is excreted in the host's waste. The cyst allows the organism to survive long enough to be picked up by another animal, where it can divide asexually and set up a new infection.

Species of *Trypanosoma* (see Figure 13-7) cause nasty diseases, including African sleeping sickness, where the parasite invades the spinal cord and brain of those infected. Although not a member of the Apicoplexans group, it's passed to humans from a biting insect — the tsetse fly. Species of *Trypanosoma* have a long, slender cell shape that turns in a corkscrew when swimming, thanks to a long flagellum that undulates under the cytoplasmic membrane along the length of the cell. This swimming motion makes it possible for the organisms to move in viscous liquids like blood.

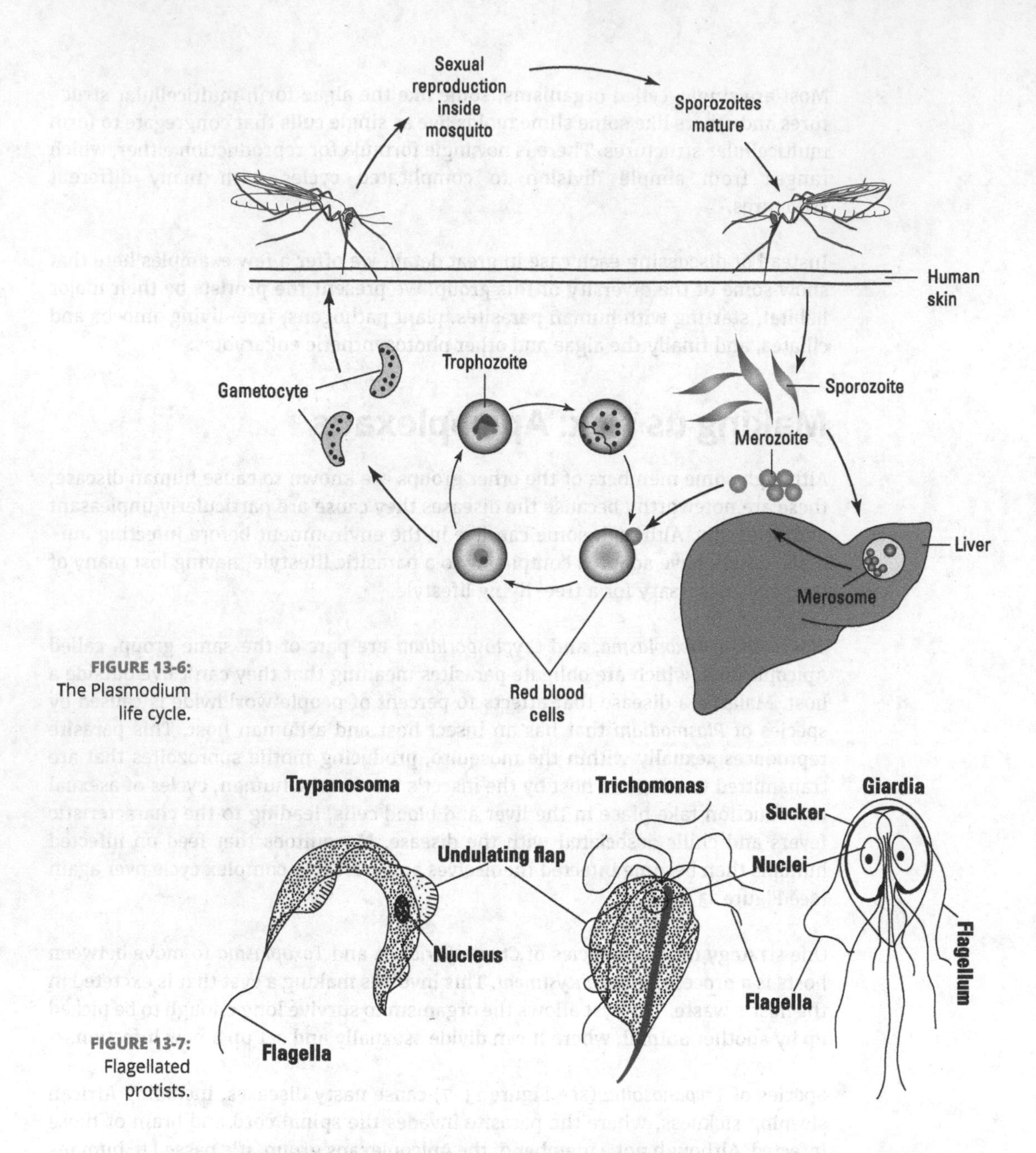

FIGURE 13-6: The Plasmodium life cycle.

FIGURE 13-7: Flagellated protists.

Another flagellated parasite, *Giardia lamblia* (refer to Figure 13-7) can survive in rivers and streams and cause a nasty diarrheal disease called giardiasis in humans and other animals. Another, *Trichomonas vaginalis* (refer to Figure 13-7), is transmitted sexually but can survive outside the body for a limited time. Both of these pathogens lack mitochondria, but they do have a *mitosome*, which is a remnant of a mitochondrion that has lost many of the mitochondrial genes. This means that these microbes likely once had mitochondria, but due to their strictly parasitic lifestyle, they've lost the need to generate much of their own ATP.

Making plants sick: Oomycetes

Once thought to be fungi because of their filamentous growth, the Oomycetes are responsible for many plant diseases as well as some animal ones. They have cell walls and are responsible for breaking down decaying organic matter on the forest floor, but as it turns out, they're more closely related to diatoms (see the "Encountering the algae" section, later in this chapter) than to fungi.

Downey and powdery mildews are plant pathogens from this group, but the most notorious one, late blight of potato, caused widespread crop losses in the 19th century. This pathogen hit Europe, North America, and South America hard, but nowhere was as badly hit as Ireland, where it wiped out potato crops, all of which were of the same vulnerable variety, causing widespread famine.

Chasing amoeba and ciliates

Ciliates and amoeba are not part of closely related groups. In fact, they aren't very similar at all. But they're often grouped together based on the fact that they move around chasing their food and ingesting it by phagocytosis.

TECHNICAL STUFF

Phagocytosis is the process where the cell membrane moves outward to surround a particle of food on all sides forming a pocket for it called a *vacuole*. The vacuole contents are then completely enclosed inside the cell, separate from the cytoplasm, where digestive enzymes can be transferred from cytoplasm to vacuole. Once the food is digested, the vacuole breaks open to release nutrients into the cellular cytoplasm.

The ciliate *Paramecium* is a widespread example of this group of microorganisms. It's covered with cilia that are used for moving around and directing food into the equivalent of a ciliate mouth, the *oral groove* (see Figure 13-8), where food is ingested by phagocytosis.

REMEMBER

Cilia are shorter and finer than flagella and beat in unison to create movement.

Most ciliates are abundant in aquatic environments where some swim freely and others attach to surfaces by a stalk using their cilia for feeding. Very few ciliates are pathogenic to animals, but some do exist.

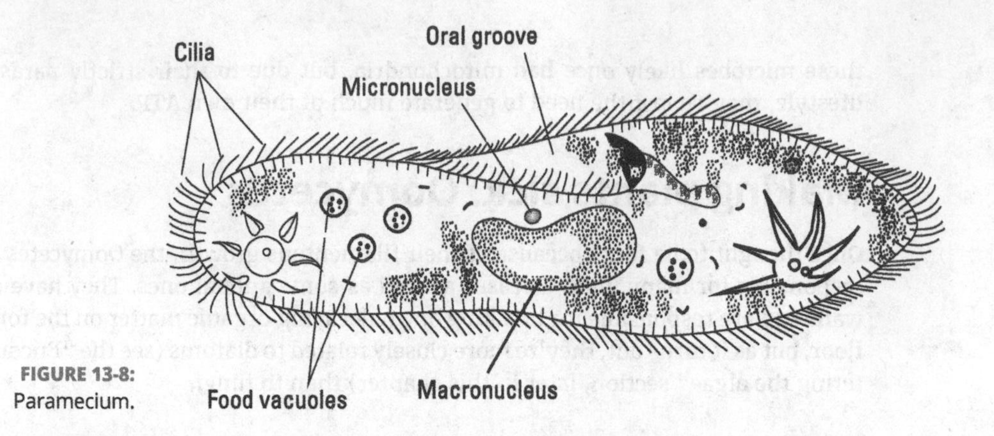

FIGURE 13-8: Paramecium.

Amoeba, unlike ciliates, move around using a process that's named after them, *amoeboid movement*. This type of movement involves extending part of the cell outward to form a *pseudopodia*, within which the cytoplasm streams more freely than in the rest of the cell. When the pseudopodia has reached forward, the rest of the cell is pulled forward by contraction of the microfilaments inside the cell (see Figure 13-9).

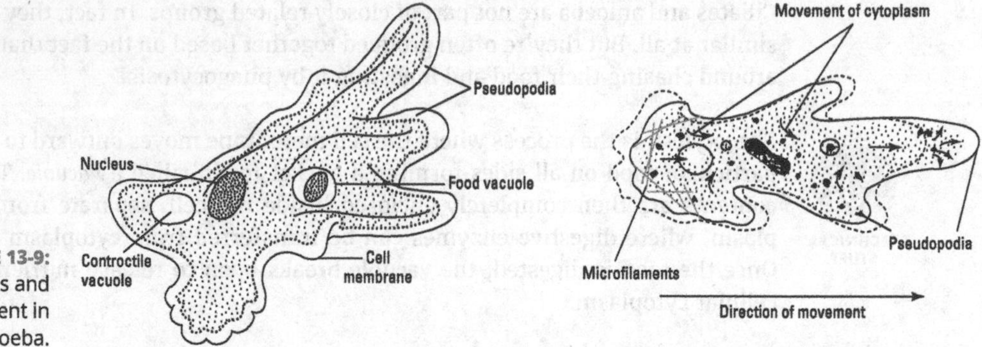

FIGURE 13-9: Structures and movement in amoeba.

Amoeba also have a specialized structure called a *contractile vacuole* that is involved in getting rid of waste.

TIP

Some amoeba live happily in aquatic and soil environments without ever causing problems for humans, whereas others are responsible for a deadly form of amoebic dysentery.

Slime molds live in environmental habitats and have a very interesting life cycle. Until recently, they were thought to be fungi because they produce fruiting bodies during reproduction, but now they're known to be closely related to amoeba. There are two types of slime molds, one that spends most of its time as a single cell (cellular) and another that spends its life as a huge mass of protoplasm containing many nuclei but without individual cells (plasmodial). Plasmodial, or acellular, slime molds move around with amoeboid movement looking for food; then when resources are gone, they produce haploid flagellated cells that swim off, and eventually two of them fuse to form a new diploid plasmodium.

Cellular slime molds live as individual haploid cells, moving around and consuming food. Then, when the food runs out, many of them come together to form a slug that eventually stops moving and forms a fruiting body in which spores are formed. Each spore is released and becomes a new single-celled individual (see Figure 13-10).

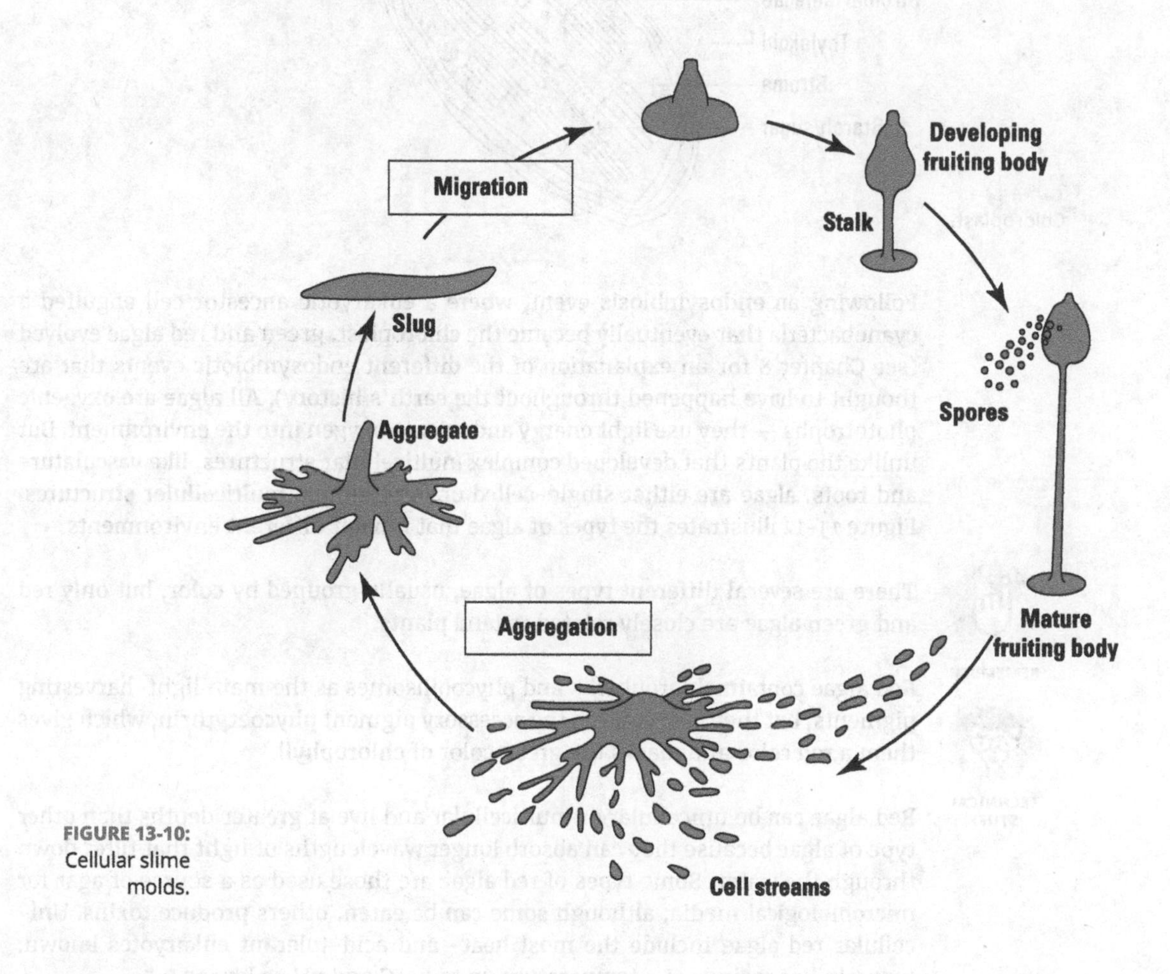

FIGURE 13-10: Cellular slime molds.

Encountering the algae

In this section, we cover both algae and eukaryotic microorganisms that make up plankton. The term *algae* is not a taxonomic classification — it's used to describe eukaryotic microorganisms that live a photosynthetic lifestyle thanks to chloroplasts (see Figure 13-11) inside the cytoplasm of the cell.

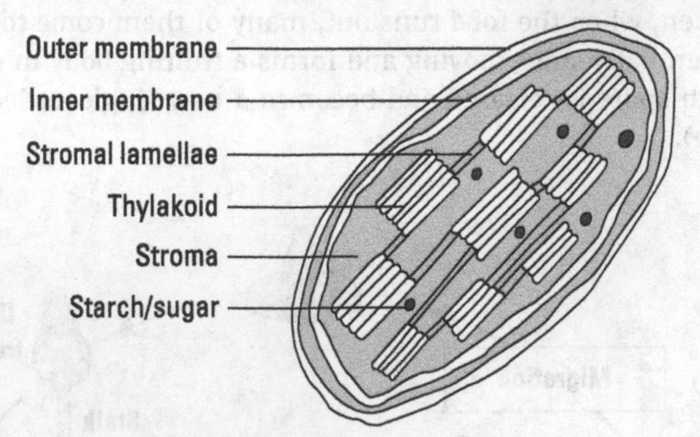

FIGURE 13-11: Chloroplast.

Following an endosymbiosis event, where a eukaryotic ancestor cell engulfed a cyanobacteria that eventually became the chloroplast, green and red algae evolved (see Chapter 8 for an explanation of the different endosymbiotic events that are thought to have happened throughout the earth's history). All algae are oxygenic phototrophs — they use light energy and release oxygen into the environment. But unlike the plants that developed complex multicellular structures, like vasculature and roots, algae are either single-celled or form simple multicellular structures. Figure 13-12 illustrates the types of algae that inhabit different environments.

REMEMBER

There are several different types of algae, usually grouped by color, but only red and green algae are closely related to land plants.

TECHNICAL STUFF

Red algae contain chlorophyll a and phycobilisomes as the main light-harvesting pigments, but they also contain the accessory pigment phycoerythrin, which gives them a red color and masks the green color of chlorophyll.

Red algae can be unicellular or multicellular and live at greater depths than other type of algae because they can absorb longer wavelengths of light that filter down through the water. Some types of red algae are those used as a source of agar for microbiological media; although some can be eaten, others produce toxins. Unicellular red algae include the most heat- and acid-tolerant eukaryotes known, living in hot springs at a temperature up to 60°C and pH as low as 0.5.

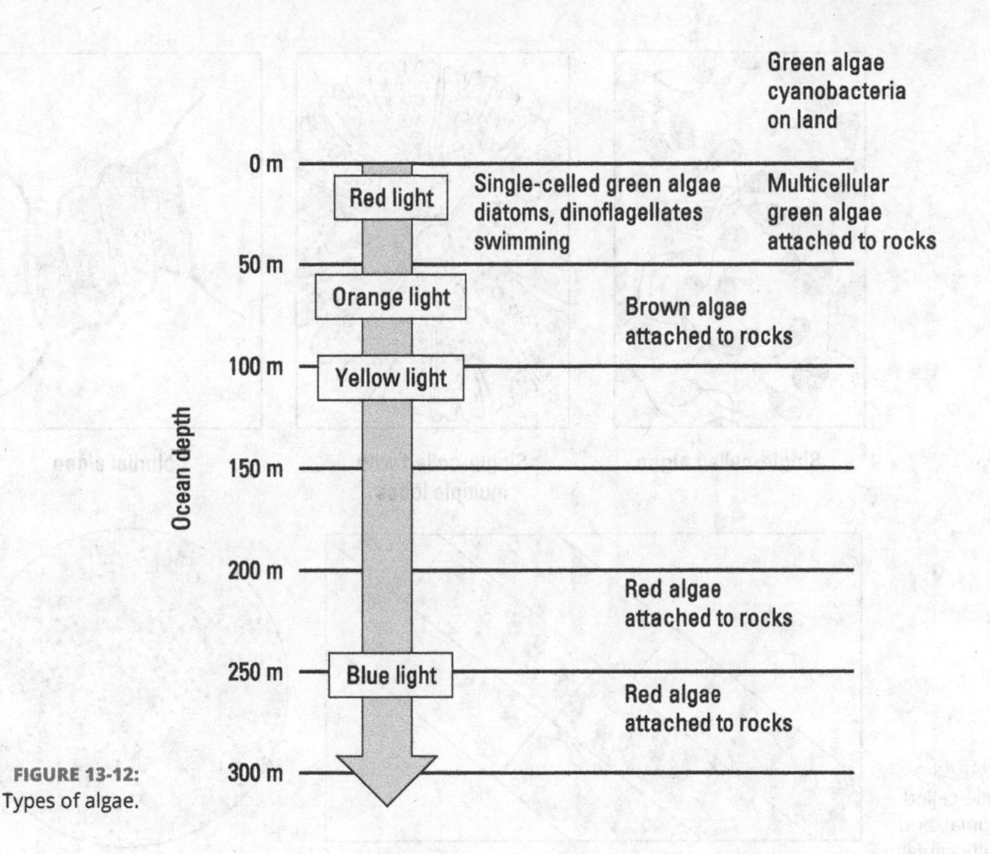

FIGURE 13-12: Types of algae.

Brown algae are called kelp. They're large multicellular organisms that can grow rapidly in their ocean habitat. They produce alginate, which is used as a food thickener.

Green algae are most like plants. They have cellulose in their cell walls, contain the same chlorophylls as plants, and store starch. Most green algae are unicellular; however, others are either colonial (growing together in a colony), filamentous, or able to form multicellular structures (see Figure 13-13). Some green algae live in soil; others live inside rocks, using the light that filters through their semi-transparent home.

Lichen are a symbiotic partnership between a single-celled green algae and a filamentous fungi.

Diatoms are a major component of photoplankton. They use photosynthesis for energy, but instead of storing it in starch like the green algae do, they store it as an oil, which can be lethal if ingested in a high enough concentration. They make a cell wall of silica, the outermost part of which is called the *frustule*; the frustule remains long after the cell dies. The shapes of diatom frustules are often very ornate and beautiful and are either pinnate (elongated) or centric (round). (see Figure 13-14).

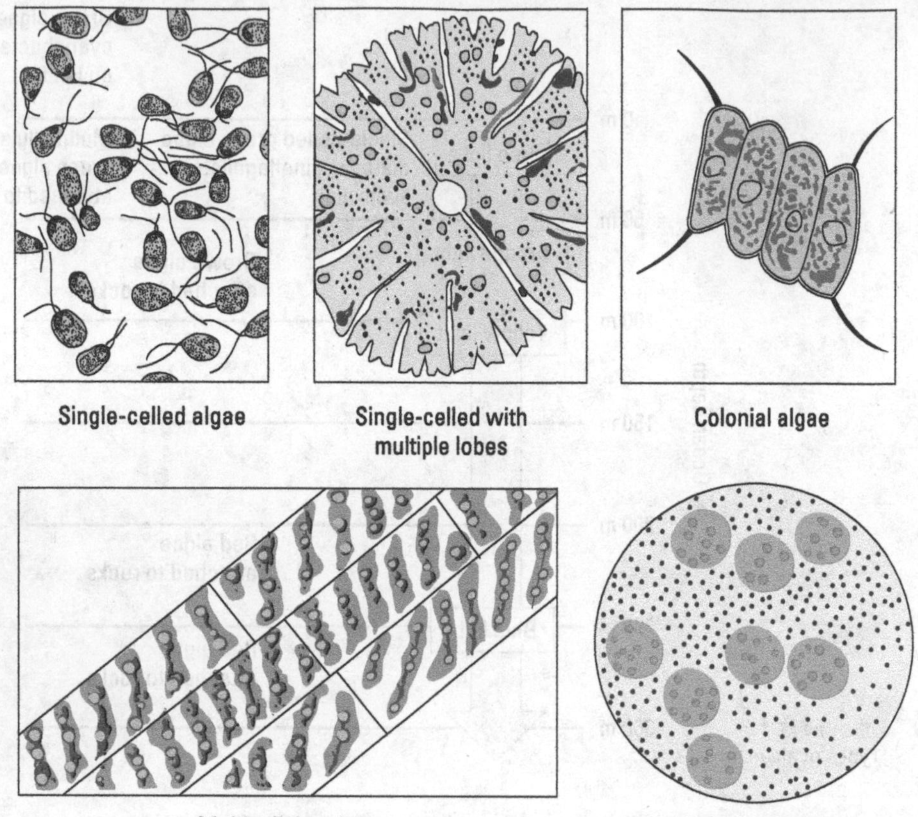

FIGURE 13-13: Single-celled, colonial, and multicellular green algae.

Radiolarians and cercozoans are amoebalike microbes that live inside a structure called a *test*, which is made either of silica (radiolarians) or of organic material strengthened by calcium carbonate (cercozoans). Unlike diatoms, they aren't photosynthetic; instead, they feed on bacteria or other particulate matter in the sediments of aquatic environments. They extend part of their cells out as needle-thin pseudopodia to gather food and move around.

Dinoflagellates (refer to Figure 13-14) also make up a large part of oceanic plankton. They're photosynthetic, and they swim in a spinning motion using two flagella. They have cellulose within the plasma membrane, giving their cells a distinct shape. An overgrowth of members of this group can be deadly for fish because they produce neurotoxins. The famous red tide is due to overgrowth of a red-colored dinoflagellate named *Alexandrium* that turns the water a deep red and causes massive fish kills.

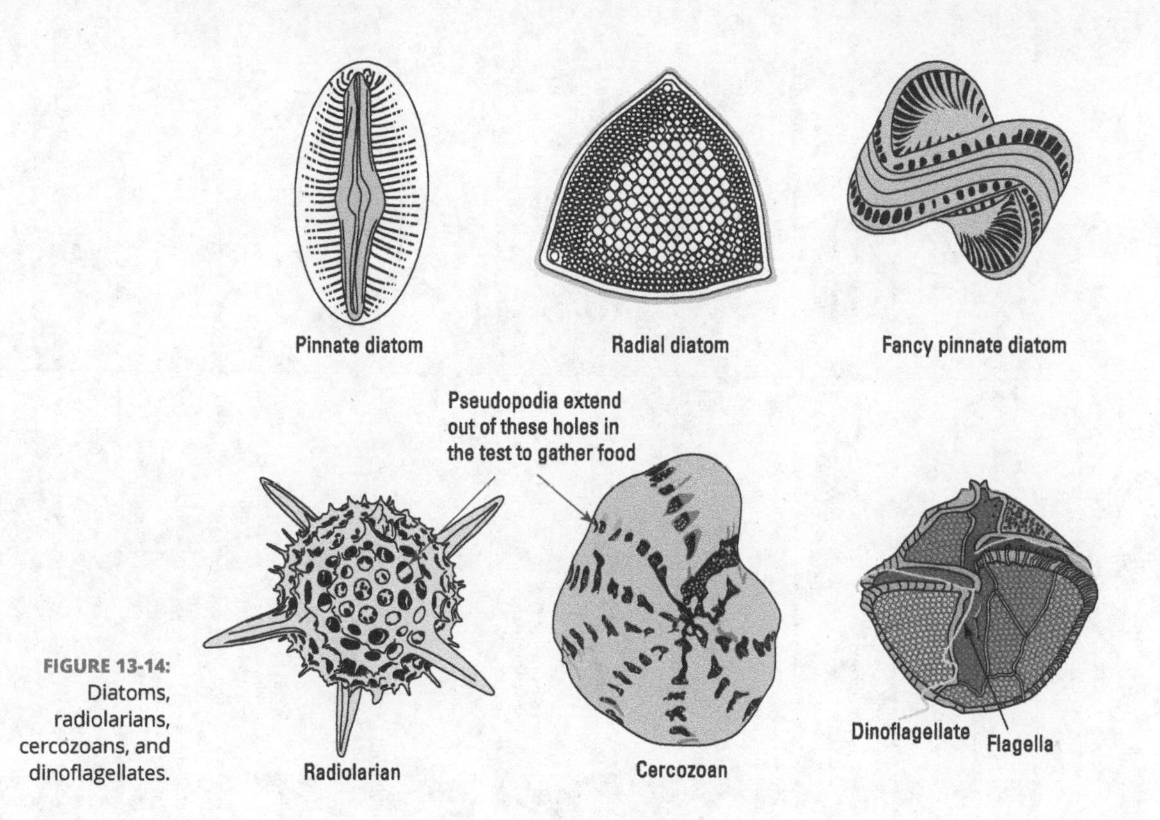

FIGURE 13-14: Diatoms, radiolarians, cercozoans, and dinoflagellates.

IN THIS CHAPTER

» Going over viral structure and function

» Catching up with bacteriophage

» Contrasting plant and animal viruses

» Watching host cells defend themselves

Chapter 14 Examining the Vastness of Viruses

There are many more viral particles on earth than there are bacterial cells — by some estimates ten times more. There is likely a virus for every organism, and our guess is that there's more than one. Viral diversity is so great that scientists have only scratched the surface when it comes to describing it. As with bacterial diversity, scientists are much more interested in viruses for which there is a known disease or impact on human lives.

In this chapter, we explore the shape and function of viruses, along with the differences in the types of viruses that infect the different types of cells — from bacteria to people to tobacco plants. In addition to viruses, there are also virus-like particles that don't completely fit the definition of a virus. These include sub-viral particles and prions, which we cover in this chapter.

Hijacking Cells

Viruses are obligate parasites, meaning they must have a host. Because they aren't cells and don't have the machinery necessary to make proteins, viruses take over the machinery of a host and use it for their own replication.

Viruses range in size from 28 to 200 nanometers (nm). In contrast, bacteria are in the 1,000 to 10,000 nm range and animal cells are in the 10,000 to 100,000 nm range.

Frugal viral structure

There are many different viral forms with different structural components, but all viruses are much less complex than living cells. Viruses lack the machinery to perform biochemical reactions. All viruses contain nucleic acid, which encodes the *viral genome* (the nucleic acid that codes for the viral genes) and the *viral capsid* (the protein shell that encases the viral genome), which may or may not contain viral enzymes. Together, the viral genome and viral capsid are called the *nucleocapsid.* Other viral structures can include the envelope and the tail.

The basic structure of some viruses is shown in Figure 14-1.

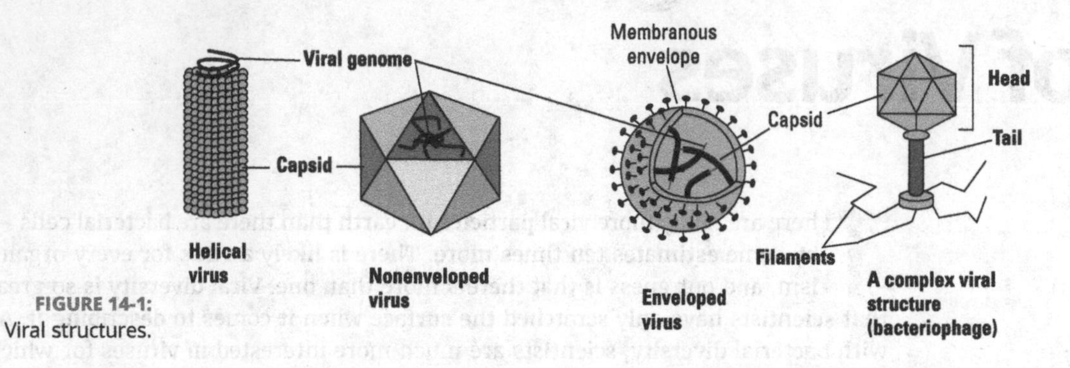

FIGURE 14-1: Viral structures.

TIP

Another name for a single virus particle is a *virion.* This term is used when talking specifically about the physical attributes of a virus particle. In comparison, the term *virus* is used when talking about the behavior of the organism.

Both the size and type of nucleic acid found within the viral particle can differ between different groups of viruses. Here are a few of the main types of nucleic acids, along with how each affects viral structure and replication:

- **DNA viruses:** Can have either single-stranded (ss) or double-stranded (ds) DNA molecules within the nucleocapsid. If it's ds, the DNA can be directly used for transcription and replication; if it's ss, the DNA needs to form a ds DNA intermediate before it can be used.
- **RNA viruses:** Can either have ss or ds RNA molecules within their nucleocapsid. These viruses need to encode an RNA-dependent RNA polymerase

for replication of the genetic material because host cells don't have an enzyme that can make RNA from an RNA template.

- **Retroviruses:** Exist as ss RNA that has to first be made into DNA, which can then be used for transcription and as a template to make the genomic copy. These viruses carry the reverse transcriptase enzyme with them in their capsid. The DNA copy of the retrovirus genome then integrates into the host genome. Reverse transcriptase is an important enzyme for molecular biology and biotechnology applications because it allows the formation of DNA from an RNA template.

The viral capsid acts to package the nucleic acids so that they can be delivered to a new host cell. Within the capsid, the genetic material can come as one or several molecules, but in all cases the capsid proteins surround the genetic material.

Repeated units of one or a small number of proteins are arranged in a symmetrical pattern and give viral capsids a geometric appearance when observed by electron microscopy. Helical viruses coat their RNA molecule with a protein, and the entire capsid forms a helix that is very uniform. In other viruses, an icosahedral (20-sided) structure is used for the capsid. The use of an icosahedral is no accident — it's the most efficient way to form a spherical shape using the smallest number of proteins. Examples of the different capsid arrangements are shown in Figure 14-1.

Some viruses, mostly those that infect animal cells, have a viral envelope surrounding the viral nucleocapsid. The envelope can be studded with projections that help the virus interact with its target cell. An example of this is the influenza virus, whose viral envelope contains hemagglutinin (a glycoprotein) and neuraminidase (a protein) spikes that are important to its attachment and release from its host cell. The viral envelope is made of lipids that come from the host cell's lipid membrane during the last stage of viral replication. When a virus lacks an envelope, it's called a *naked virus.*

The most complex viral structures include those of the bacteriophage. One example, shown in Figure 14-1, includes an icosahedral head structure attached to a tail with filaments. The complicated tail structure is used to create a hole in the bacterial cell wall and then inject the viral genetic material into the cell.

Simplifying viral function

Viral replication is the entire process of making more viral particles, or virions. It's driven by viral genes that have taken control of the host cell's machinery for this purpose.

TIP

Some viral replication causes all host processes to stop. This interferes with the rhythm of biochemical processes normally carried out, a fact that is often irrelevant because the cell will die at the end of viral replication. Other viruses only take over a small number of processes, leaving the host cell to carry on as usual.

Here is a description of the steps involved in this process, the details of which differ between viral groups:

1. A viral particle attaches to the host cell surface.

   The surface of the viral particle has proteins that interact with receptors on the surface of the host cell. The virus may use one or more host receptors, but they have to be present for attachment to occur. Different types of cells express different receptors, making viral attachment specific for a particular cell type.

2. The entire virus or just the viral nucleic acids enter the host cell.

   The viral genome has to enter the cell so that transcription of viral genes can occur. If other enzymes are needed — for example, a reverse transcriptase — then they must also enter the host cell.

   The process of releasing the viral contents into the host cell is called *uncoating,* which can happen in a few different ways (see Figure 14-2). For example, the viral envelope is necessary to gain entry into animal cells, whereas the bacteriophage inject their DNA through a hole that they make in the bacterial cell wall.

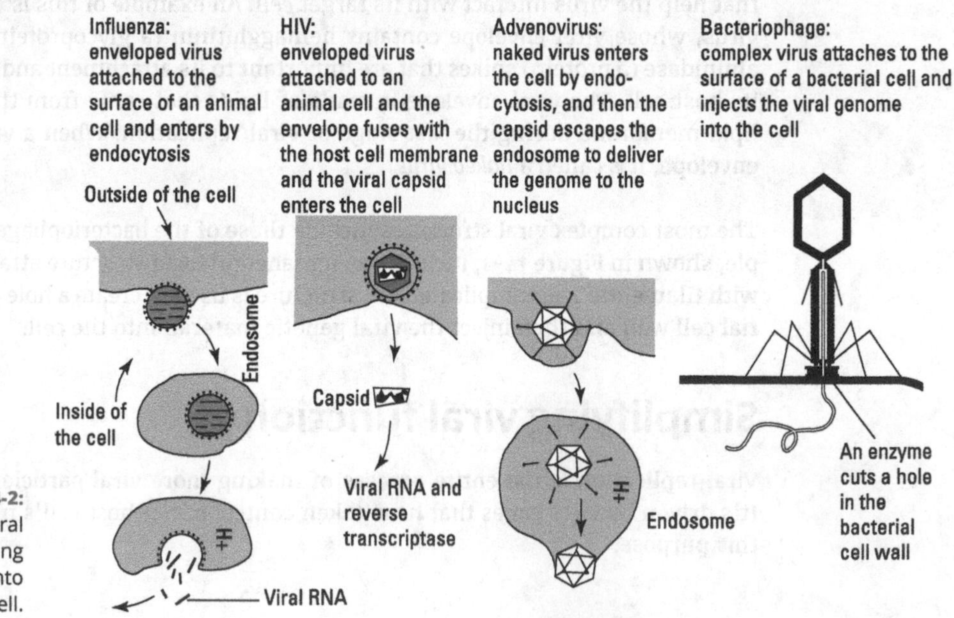
Influenza: enveloped virus attached to the surface of an animal cell and enters by endocytosis
Outside of the cell
Endosome
Inside of the cell
Viral RNA
HIV: enveloped virus attached to an animal cell and the envelope fuses with the host cell membrane and the viral capsid enters the cell
Capsid
Viral RNA and reverse transcriptase
Advenovirus: naked virus enters the cell by endocytosis, and then the capsid escapes the endosome to deliver the genome to the nucleus
Endosome
Bacteriophage: complex virus attaches to the surface of a bacterial cell and injects the viral genome into the cell
An enzyme cuts a hole in the bacterial cell wall

FIGURE 14-2: Uncoating of viral particles during penetration into the host cell.

3. The host cell transcribes and then translates the viral genes.

   The early viral proteins made are those involved in copying the viral genome and transcribing viral genes. Mid and late viral genes are for capsid and other viral proteins needed to make parts of the viral structure and are produced in large quantities.

4. Viral proteins are assembled and the genetic material is loaded into the capsid.

5. Complete viral particles, *virions,* are released from the host cell. Some bud off of the host cell, taking a bit of the cell membrane with them, which becomes the viral envelope. Some cause the host cell to *lyse* (the cell membrane, and cell wall if present, are disrupted releasing the contents of the cell) and are released that way.

Because viral genes have different promoters than host genes, often the host machinery for things like mRNA synthesis or DNA replication won't work on the viral genome. There are three main strategies for dealing with this:

- Viruses use host promoters and signals so that viral DNA and RNA look like host nucleotides.
- Viruses modify the host enzymes to recognize viral promoters and DNA.
- Viruses bring their own enzymes to get the ball rolling and then encode the enzymes needed in their genome.

Making Heads or Tails of Bacteriophage

There is likely a phage for every bacterial species, but only a few have been studied. In this section, we fill you in on three main viral lifestyles: lytic, lysogenic, and transposable.

Lytic phage

Also called *virulent phage*, lytic phage cause the lysis and almost complete destruction of the host bacterial cell. An example is T4 phage that has been studied extensively because it infects one of the most studied bacterial species, *E. coli* (see Figure 14-3).

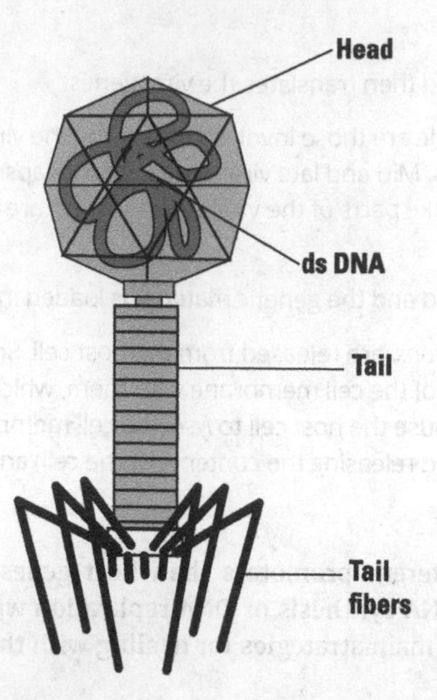

FIGURE 14-3: Bacteriophage T4, an example of a lytic phage.

Early in infection, T4 makes a lot of virally encoded proteins for DNA replication that are much quicker than the host proteins and have a preference for replicating viral DNA. This means that many copies of the genome are present in the host cell from an early stage. Viral capsid, the head, is made and is packaged full of DNA. Then the tail and fibers are added afterward. The viral genome is added to the head by threading a string of DNA into the opening at one end; when the pressure inside is sufficient, the DNA strand is cut and the head is sealed. The pressure inside of the bacteriophage T4 head is ten times the pressure in a bottle of champagne.

Because the phage copies its genome as a long string without stopping between copies, and because the head can fit more DNA than there is in the genome, there is always a bit more than one copy of the genome in each virion. Some genes are present in more than one copy, and the beginning of the genome isn't the same each time.

After the viral particles have been assembled, T4 lysozyme is made. T4 lysozyme degrades the bacterial cell wall and the viral particles escape, leaving the host cell extremely damaged.

Temperate phage

Lysogeny is a survival strategy for viruses, allowing them to lay dormant for a period instead of continuing the frenzied cycle of always looking for a new host.

When a temperate phage infects a host bacterial cell, it can either be lytic or lysogenic.

REMEMBER

The lytic pathway proceeds similarly to that of a lytic phage ending in lysis of the host cell to release the viral particles. Sometime before the production of structural viral proteins, the virus can switch to the lysogenic state where the viral genes are integrated into the host DNA and no viral particles are made.

See Figure 14-4 for an overview of the cycle for a lysogenic virus.

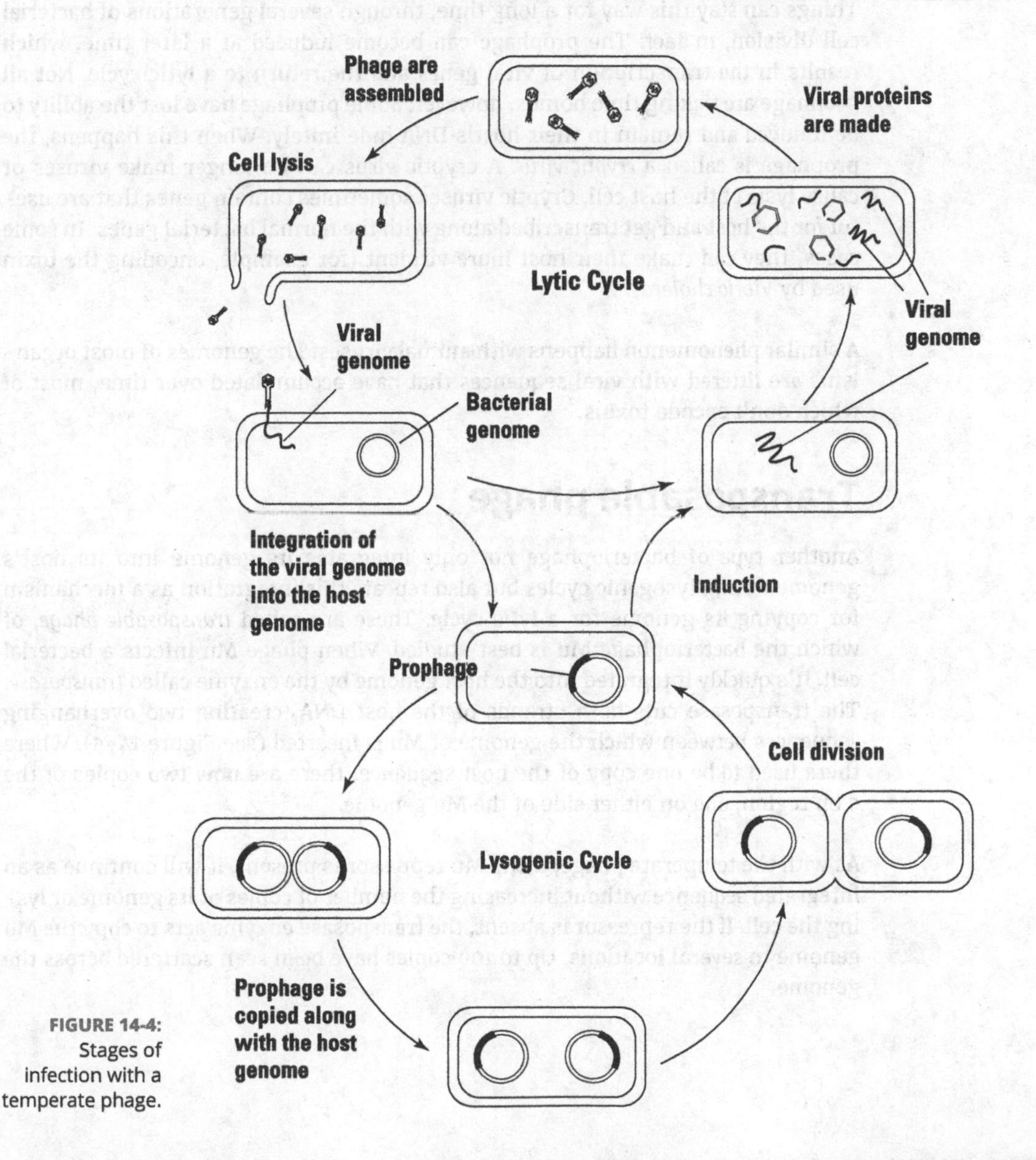

FIGURE 14-4: Stages of infection with a temperate phage.

Once lysogeny has been induced, the viral genome is integrated into the host genome. At this point, it's called a *prophage*, and the production of viral replication and capsid proteins stops. One viral protein is still expressed in the host cell; this is the one that represses the expression of all other viral proteins, which would induce the lytic cycle if induced.

REMEMBER

A bacterial host containing a prophage is called a *lysogen*. Some prophages are integrated into the bacterial chromosome, and others exist as a plasmid.

Things can stay this way for a long time, through several generations of bacterial cell division, in fact. The prophage can become induced at a later time, which results in the transcription of viral genes and the return to a lytic cycle. Not all prophage are ticking time bombs, however. Some prophage have lost the ability to be induced and remain in their host's DNA indefinitely. When this happens, the prophage is called a *cryptic virus*. A cryptic virus can no longer make viruses or cause lysis of the host cell. Cryptic viruses sometimes contain genes that are useful for the host and get transcribed along with the normal bacterial genes. In some cases, they can make their host more virulent (for example, encoding the toxin used by *Vibrio cholera*).

A similar phenomenon happens with animal viruses. The genomes of most organisms are littered with viral sequences that have accumulated over time, most of which don't encode toxins.

Transposable phage

Another type of bacteriophage not only integrates its genome into its host's genome during lysogenic cycles but also repeats this integration as a mechanism for copying its genome for a lytic cycle. These are called *transposable phage*, of which the bacteriophage Mu is best studied. When phage Mu infects a bacterial cell, it's quickly integrated into the host genome by the enzyme called *transposase*. The transposase cuts both strands of the host DNA, creating two overhanging sequences between which the genome of Mu is inserted (see Figure 14-5). Where there used to be one copy of the host sequence, there are now two copies of the 5 bp region, one on either side of the Mu genome.

As with the temperate phage, if the Mu repressor is present, it will continue as an integrated sequence without increasing the number of copies of its genome or lysing the cell. If the repressor is absent, the transposase enzyme acts to copy the Mu genome to several locations. Up to 100 copies have been seen scattered across the genome.

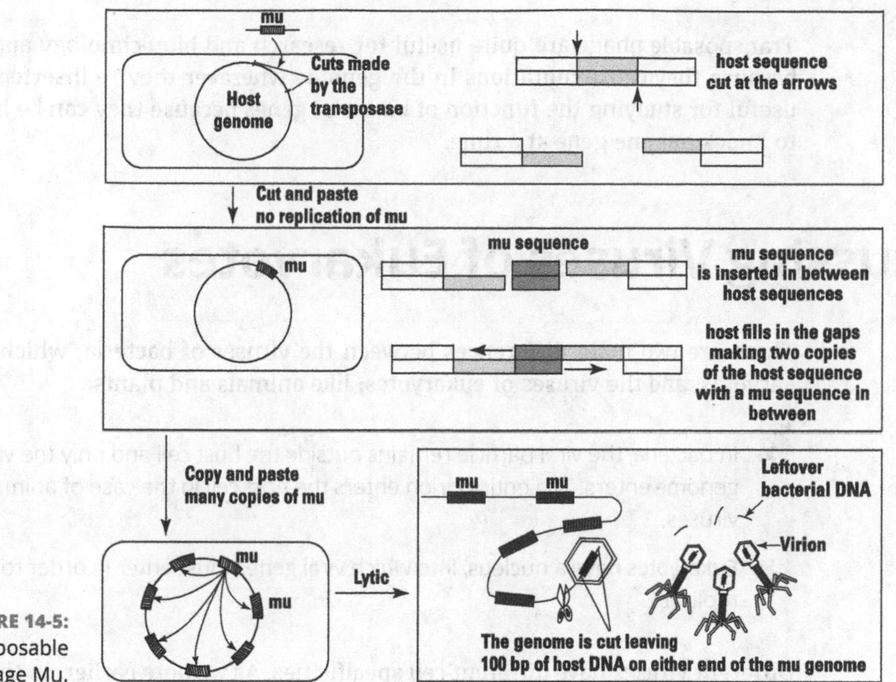

FIGURE 14-5: The transposable phage Mu.

REMEMBER

A sequence of DNA that can move from one place to another on the same genome is called a *transposable element*. These transposable elements may make a copy of themselves in the process (replicative transposition) or they may move as a complete unit to a new position (conservative transposition).

When the viral capsid is ready, the circular bacterial genome containing the many copies of the viral genome is cut around 100 base pairs (bp) before the beginning of the Mu genome, and then loaded into each viral head until full. The result is that, along with the viral genome, each virion contains around 1.8 kilobases (kb) of host sequence as well.

INFECTING THE ARCHAEA

Viruses of archaea are known to exist, and many have at least been imaged if not studied fully. What is known about archaeal viruses is that they must be tolerant of extreme environments because their hosts are often extremophiles. So far, all have shown to be double-stranded DNA genomes and many have heads and tails like bacteriophage, but none has been fully studied and not much is known about their method of infection and replication.

Transposable phage are quite useful for research and biotechnology applications because they cause mutations in the genome wherever they're inserted. They're useful for studying the function of bacterial genes because they can be harnessed to knock out one gene at a time.

Discussing Viruses of Eukaryotes

There are two main differences between the viruses of bacteria, which are prokaryotes, and the viruses of eukaryotes, like animals and plants:

- In bacteria, the viral particle remains outside the host cell and only the viral genome enters. The entire virion enters the host cell in the case of animal viruses.
- Eukaryotes have a nucleus, into which viral genes must enter in order to replicate.

Different viruses have different cell specificities. As we note earlier, in the "Frugal Viral Structures" section, viral particles have proteins on their surface that interact with host cell receptors, allowing attachment of the viral particle to the host cell surface. Which cell type a virus will infect is directly determined by which cell types it can enter. Because not all cells in an animal express the same receptors, they're not all susceptible to the same viruses.

Infecting animal cells

Of the grand diversity of viruses on earth, we probably know more about viruses that infect animal cells because they're the types that make people sick. There are animal viruses with each of the different types of genomes — for instance, double- and single-stranded DNA and RNA. Although the majority of animal viruses are enveloped, naked viruses also infect animal cells.

Here's a list of the different properties of animal viruses:

- **Virulent:** viruses that lyse their host cell.
- **Persistent:** viruses that are continuously shed from a host cell indefinitely.
- **Latent:** Some viruses can have latent infections (not causing an symptoms of disease) that emerge as virulent from time to time, but unlike with temperate bacteriophage, the viral genome usually doesn't integrate into the host DNA.
- **Fusogenic:** Some virus can promote the fusion of several cells.

» **Oncogenic:** Some viruses can cause cancers either by causing mutation in the host cell they infect by inserting into the genome, or by altering the normal regulation of cell growth control.

Instead of describing all the many types of viral structures and functions, we describe two interesting examples in detail here: retroviruses and prions.

Retroviruses

Retroviruses are unique animal viruses that carry with them a special enzyme, called *reverse transcriptase,* which copies the RNA viral genome into double-stranded DNA. Known retroviruses include the human immunodeficiency virus (HIV), which causes AIDS in people, and feline leukemia virus (FeLV), which affects domestic cats.

We'll focus here on the structure and activity of HIV as an example of a retroviral replication cycle (see Figure 14-6). Although other retroviruses infect different target cells, many of the steps in infection are similar.

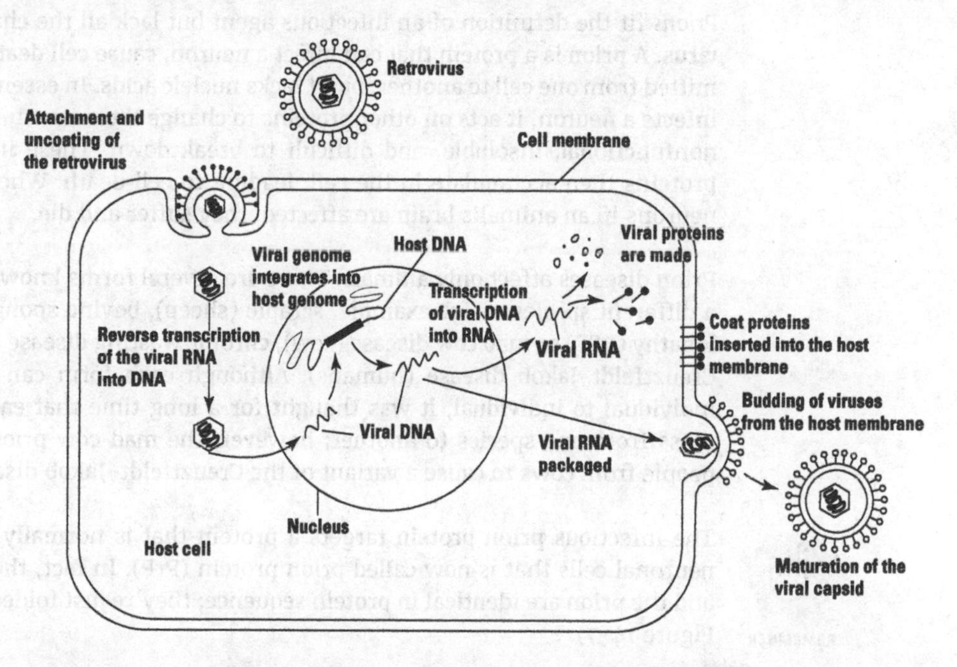

FIGURE 14-6: The replication cycle of a retrovirus.

Retroviruses are enveloped with a protein capsid containing two copies of their viral RNA genome, as well as three different viral enzymes needed for infection. The envelope is studded with proteins used for attachment to cells of the human immune system, T-lymphocytes. After it attaches to the cell surface, the virus is uncoated when the viral envelope fuses with the cell membrane, releasing the nucleocapsid into the cytoplasm. Within the nucleocapsid, the reverse transcriptase enzyme copies the RNA genome into double-stranded DNA, which then enters the host cell nucleus and is integrated into the host genome.

REMEMBER

After the HIV genome is integrated into the host cell's genome (and called a *provirus*), it can't be removed because it doesn't make an enzyme for excision.

RNA copies of the provirus are transcribed by the host cell machinery and are translated into viral proteins, including structural parts like the capsid and enzymes, like reverse transcriptase to be packaged into new virions. Other copies of the viral RNA are made that will be packaged into these new virions and after the nucleocapsid is ready, the virions bud out of the host cell membrane, taking a part of it along with them as their viral envelope. Viral proteins needed for interacting with new host cells are present on the surface of the host cell membrane before budding happens.

Prions

Prions fit the definition of an infectious agent but lack all the characteristics of a virus. A prion is a protein that can infect a neuron, cause cell death, and be transmitted from one cell to another, but it lacks nucleic acids. In essence, when a prion infects a neuron, it acts on other proteins to change their structure, making them nonfunctional, insoluble, and difficult to break down. These insoluble, useless proteins then accumulate in the cell, leading to cell death. When enough of the neurons in an animal's brain are affected, they suffer and die.

Prion diseases affect only animals. There are several forms known, each affecting a different species — for example, scrapie (sheep), bovine spongiform encephalopathy (BSE) or mad cow disease (cow), chronic wasting disease (deer), kuru, and Creuztfeldt-Jakob disease (humans). Although each form can be passed from individual to individual, it was thought for a long time that each type couldn't cross from one species to another; however, the mad cow prion has jumped to people from cows to cause a variant of the Creuztfeldt-Jakob disease.

REMEMBER

The infectious prion protein targets a protein that is normally expressed in all neuronal cells that is now called prion protein (PrP). In fact, the normal protein and the prion are identical in protein sequence; they're just folded differently (see Figure 14-7).

The normal protein in the correct conformation is labeled PrP^{C} (for cellular), whereas the prion protein is labeled PrP^{Sc} (for the first prion disease described, scrapie). When PrP^{Sc} enters the neuron, it acts to change PrP^{C} to PrP^{Sc}. As more PrP^{Sc} are made, they act to convert other PrP^{C} to the disease kind, and the misfolded proteins combine into tightly packed clumps called *amyloids.* One thing that distinguishes prions from viruses is that they can't induce the expression of proteins they need for infection; they simply act on the protein if it's present. The PrP^{C} protein has a normal function in neurons that isn't completely understood, but it isn't always expressed. If PrP^{Sc} arrives and no PrP^{C} is present to be acted on, then the infection can't happen.

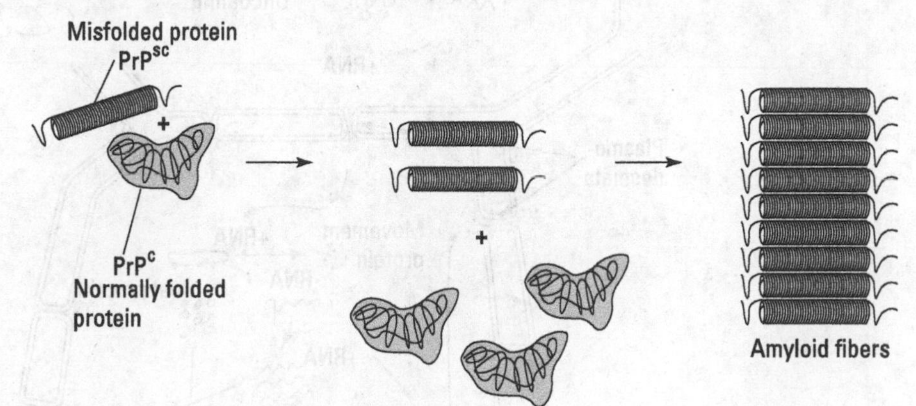

FIGURE 14-7: Prion disease.

Following plant viruses

Just like animals (and the prokaryotes), plants are also infected by viruses. These include important pathogens of agricultural crop plants. Moreover plants are also infected by viroids, which can be thought of as stripped down viruses.

Tobacco mosaic virus

Most plant viruses are single-stranded RNA viruses. Tobacco mosaic virus (TMV) was the first virus ever discovered and is the most well-studied plant virus. It's a helical RNA virus that is made entirely of a single strand of RNA surrounded by many coat proteins (see Figure 14-8).

The viral RNA genome is positive, meaning that it's the same as what the host cell would use as a template for translation. For instance, host mRNA is positive and the sequence of the nucleic acid sequence is complementary to the DNA sequence. The first step is the uncoating of the viral genome. Because the genome resembles a host mRNA, it's used directly for translation of proteins. One of the first enzymes made is the all-important RNA-dependent RNA polymerase used to make a negative RNA copy (–RNA) of the genome from which more positive RNA genomes (+RNA) can be made.

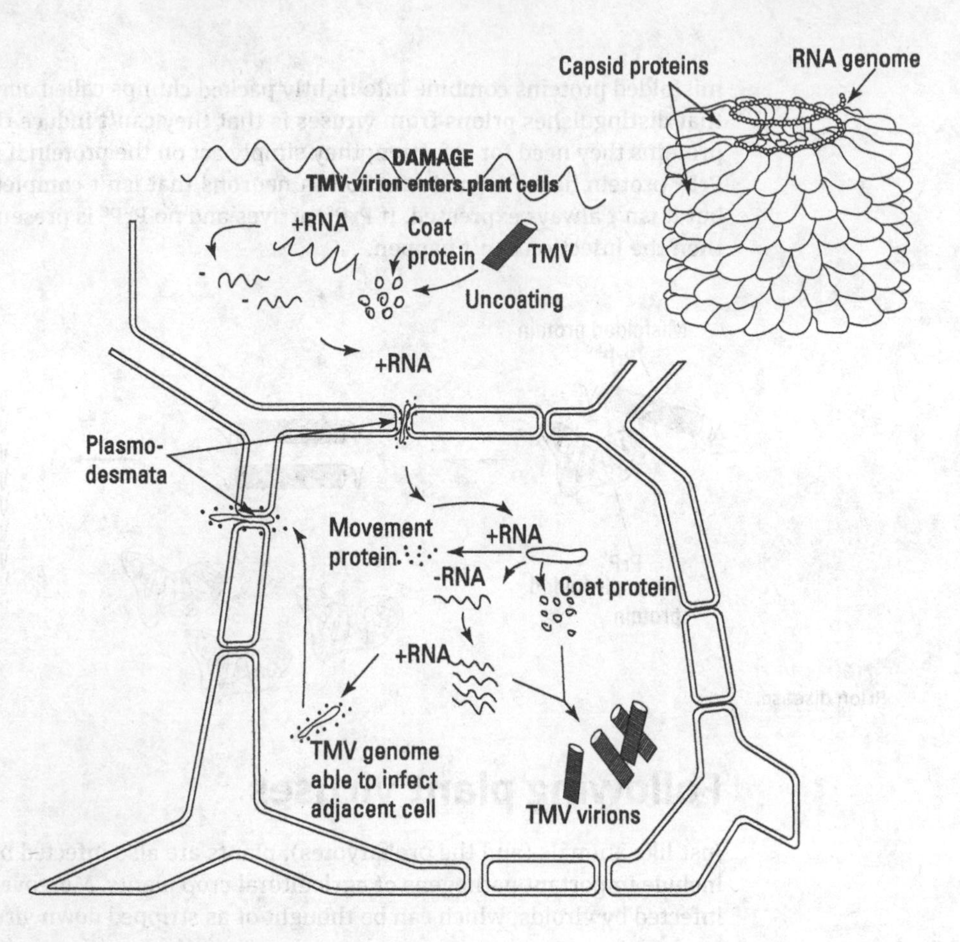

FIGURE 14-8: Tobacco mosaic virus.

Next, movement proteins and the coat proteins are made that enable the virus to infect other plants through cell damage from an insect or other herbivore. Another method that TMV uses to increase viral replication is through infection of cells via the cell-to-cell connections called *plasmodesmata,* which connect the cells throughout a plant but are too small for even bacteria or other viruses to pass through. A viral protein, called a *movement protein,* can bind to the viral RNA and help it to enter adjacent cells where it can begin a new infection.

Viroids

Viroids are short RNA sequences <400 bases that lack capsid proteins. No viroids are known to infect animals or prokaryotes, but there are several in plants. They're small and circular with secondary structure. Viroids enter plant cells after damage to the cell has compromised the integrity of the cell wall. They also move from cell to cell by the plasmodesmata.

How exactly the viroids cause disease is unknown. It's suspected that they create small interfering RNA molecules (siRNA) to be made that interfere with the translation of normal plant mRNAs. The way this works is that the small RNAs bind to the plant mRNAs, creating a double-stranded RNA complex that is flagged for degradation by the plant cell machinery. Plants use this type of regulation, called *small regulatory RNA* (srRNA), on their genes under normal conditions. So, it may not be a coincidence that plants are the only ones to be infected by these types of pathogens. Specifically, the viroids could have evolved accidentally from plant srRNAs.

How Host Cells Fight Back

Both viruses and their hosts exert evolutionary pressure on one another, meaning that each of them is going to take advantage of mutations that will give it an edge over the other. Because of their small genomes, the mutation rate in viruses is much higher than it is in host cells, but the host has sophisticated mechanisms to suppress viral replication. It's always a struggle to stay ahead of viruses.

Restriction enzymes

Bacteria and archaea have a way to protect themselves from viral attack. They use enzymes to cut up any foreign double-stranded DNA that is found in the cell. This method is aimed at stopping bacteriophage genomes, which have been injected into the host cell, from being transcribed, thus beginning the infection cycle. These enzymes are called *restriction enzymes*, and they act to recognize a short sequence, called a *recognition sequence*, and then cut double-stranded DNA. The system is called the *restriction system* because it restricts viral replication.

These enzymes are actually called restriction *endonucleases* because the enzyme cuts within a DNA molecule, as opposed to removing bases from the ends of a linear DNA molecule the way an *exonuclease* would.

Recognition sequences, also called *restriction sites*, have a few important features:

- They're short and range in size from around 4 to 8 bases. The shorter the sequence, the more likely it is to occur in a DNA molecule.
- They're usually palindromes, meaning that they have the same sequence in the forward and the reverse direction.
- They're present in the DNA of all organisms, not just viruses and bacteria.

» Modification of restriction sites in the bacterial DNA is necessary at the restriction sites to prevent cleavage of the host chromosome. By tagging host DNA with methyl molecules that stop the restriction enzyme from binding there — a process called *methylation* — host DNA is protected from cleavage.

A host organism sometimes has more than one restriction system, and there are thousands of different restriction enzymes, each named for the bacterial species that it was first discovered in. After binding to their recognition site, restriction enzymes then cut both strands of the DNA in a predictable way (see Figure 14-9). This predictable cleavage of DNA makes restriction enzymes useful for molecular biology and biotechnology, a topic covered in Chapter 16.

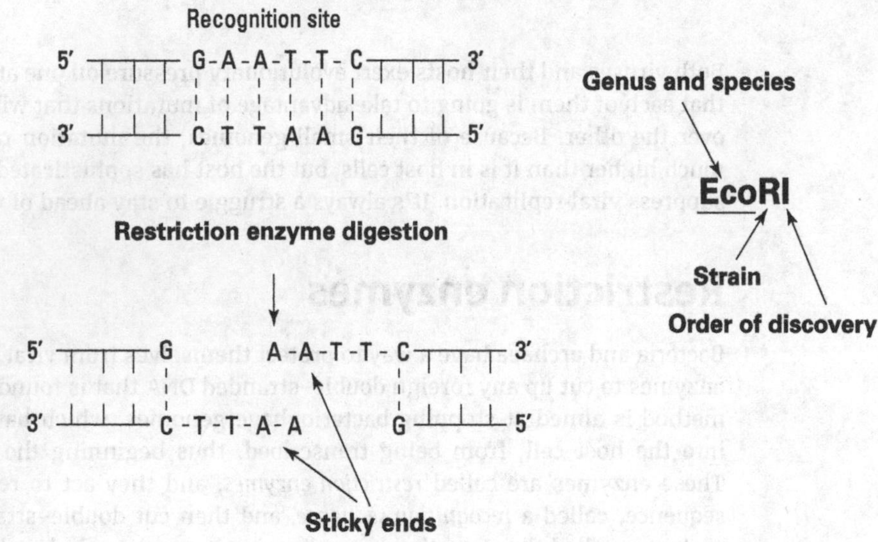

FIGURE 14-9: Restriction enzyme cleavage of a phage genome.

Viruses fight back by modifying their DNA by methylation and by glucosylation to protect it from bacterial restriction enzymes. Survival is a dirty war when it comes to bacteriophages. Many viruses encode their own restriction enzymes aimed at the host DNA, effectively removing competition for viral gene transcription and translation and stemming any inducible defense that the host could turn on.

CRISPR

Bacteria and archaea have another way to resist viruses that includes protection from not only double-stranded DNA viruses but also single-stranded DNA and RNA viruses. This method involves having short viral sequences in the bacterial genome that can be used to recognize and drive the destruction of matching viral sequences if they appear in the cell. This method, called the *CRISPR system* (pronounced "crisper"), is widespread in bacteria and archaea but is not yet completely understood.

What is known about CRISPR is that several viral sequences are present in an organism's genome separated by short sequences that are identical to each other (so they're called repeated sequences, or repeats) and are palindromic.

The nature of these *intergenic* (between genes) sequences gave the system its name — clustered regularly interspaced short palindromic repeats (CRISPR) — mostly because the short repeats were found in many bacterial genomes long before the sequences in between were recognized as viral.

The way that these viral genes protect bacterial and archaeal cells from viral attack is explained in Figure 14-10.

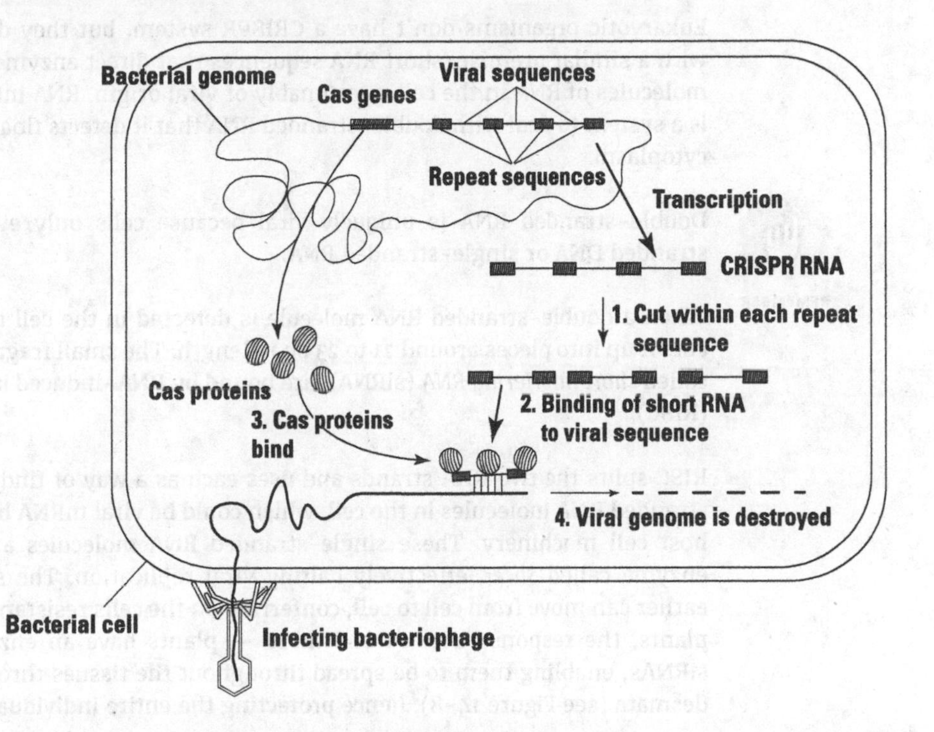

FIGURE 14-10: CRISPR antiviral protection.

First, the entire CRISPR region of the bacterial genome is transcribed. Then the repeat regions are cut so that a piece of RNA is present from each individual viral gene. These pieces are able to bind to viral DNA or RNA, which acts as the signal for the cleavage of the viral genome to begin. Cleavage of both the CRISPR RNA and eventually the offending viral genome is done by a group of proteins called *Cas proteins.*

Viruses are essentially transmissible genetic elements — the viral genes (genetic elements) move from cell to cell (transmissible). Another type of transmissible genetic element in bacteria is plasmid DNA, which is another target of the CRISPR system. Short sequences matching plasmids are often found in the recognition region of the CRISPR system, making them targets for cleavage by the CASCADE system.

Viruses fight back either by altering the sequence that corresponds to the CRISPR encoded version or by deleting the genes all together. This way, transcribed sequences for the CRISPR system either can't bind their target properly or have nothing to bind to.

Interfering with RNA viruses: RNAi

Eukaryotic organisms don't have a CRISPR system, but they do use something with a similar premise: short RNA sequences that direct enzymes to cut up other molecules of RNA in the cell, presumably of viral origin. RNA interference (RNAi) is a system to deal with double-stranded RNA that it detects floating around in its cytoplasm.

Double-stranded RNA is uniquely viral because cells only ever have double-stranded DNA or single-stranded RNA.

When a double-stranded RNA molecule is detected in the cell the enzyme Dicer cuts it up into pieces around 21 to 23 bp in length. The small fragments that result, called *short interfering RNA* (siRNA), are bound by RNA-induced silencing complex (RISC).

RISC splits the two RNA strands and uses each as a way of finding other single-stranded RNA molecules in the cell, which could be viral mRNA being made by the host cell machinery. These single-stranded RNA molecules are cleaved by an enzyme called *Slicer*, effectively halting viral replication. The siRNA mentioned earlier can move from cell to cell, conferring on the cells resistance to the virus. In plants, the response doesn't end there — plants have an enzyme to copy the siRNAs, enabling them to be spread throughout the tissues through the plasmodesmata (see Figure 14-8), hence protecting the entire individual.

Viruses can fight back in a number of ways:

- By making decoy RNAs that bind up all the Dicer-RISC complexes
- By replicating inside of compartments in the cell that aren't accessible by the RNAi machinery
- By making proteins that inhibit Dicer function

5 Seeing the Impact of Microbes

IN THIS PART . . .

See how the immune system interacts with microorganisms — both the nice and the not-so-nice ones — and how it deals with and remembers a bacterial infection.

Understand the ways that antibiotics work against bacteria and how, when they find ways to become resistant, they can be a real threat to human health.

See how microbial processes have been used for industrial and pharmaceutical purposes, among other things.

Get an overview of the strategies used to protect people from infectious diseases, such as public health policies, microbial identification, and vaccines.

IN THIS CHAPTER

» **Identifying immune defenses that protect against infection**

» **Considering antimicrobial therapies that target bacterial infections**

» **Understanding the mechanisms of antibiotic resistance**

Chapter 15 Understanding Microbes in Human Health and Disease

The intersection of microorganisms and the human body usually only interests people when they get sick, but the processes that go on between the body and microbes is pretty fascinating. In this chapter, we describe the ways that the body recognizes which microbes to keep out and, failing that, how it deals with an infection.

In Chapter 17, we go on to talk about the ways that we as a society try to protect ourselves from microbial diseases, but here we talk, for the most part, about what processes are going on in the body and at the level of the cell. Antibiotics and the repercussions of their use are an important part of the human health saga.

We end this chapter with a section on the ways that probiotics and prebiotics encourage good gastrointestinal bacteria and a brief discussion of antiviral therapies. For a more complete discussion of how cells deal with viral infections, see Chapter 14.

Clarifying the Host Immune Response

Pathogens are microorganisms that are harmful to our health. They invade the body and cause infections that can result in various symptoms, from a fever and cough, to diarrhea and vomiting. *Immunity* is the ability to resist infections; it involves several layers of protection to keep pathogens out.

The first layer of protection is broad antimicrobial strategies that fend off intruding pathogens. If the pathogen is able to bypass these barriers, the pathogen next encounters the branch of the immune response called *innate immunity.* Innate immune cells are the first responders to an infection and work to kill off the pathogen. These cells also send out signals to recruit the second branch of the immune response, called *adaptive immunity.* Adaptive immune cells are specialized for clearing certain pathogens; they learn from experience so that they're better prepared the next time an infection occurs.

The immune system involves many cell types, tissues, and organs, all of which work together to deal with the diversity of pathogens the immune system encounters, including bacteria, viruses, fungi, and parasites.

In this section, we cover all these elements of the immune response in greater detail.

Putting up barriers to infection

Regions of the body that come in direct contact with the environment are the ones most frequently exposed to pathogens. Here are the various parts of the body that are on constant guard to prevent pathogens from getting in:

- **Skin:** The skin is a thick layer of cells that routinely sheds to eliminate microorganisms on the skin. Oil glands secrete oily fats that are antimicrobial and lower the pH of skin to discourage microbial growth.
- **Secretions:** Earwax, saliva, tears, and perspiration contain antimicrobial factors including a low pH and enzymes such as lysozyme that damage microorganisms.
- **Stomach:** The stomach produces a highly acidic gastric juice that is strong enough to kill microorganisms and their toxins.
- **Respiratory, gastrointestinal, and urogenital tracts:** The respiratory tract, the gastrointestinal tract, and the urogenital tract are lined with mucous membranes that secrete a thick dense layer of mucous to trap microorganisms.

» **Microbes on and in your body:** The normal *microbiota* is the collection of microorganisms that live on your body and inside of places like your gut. They have an important role in competing with pathogens and preventing them from causing an infection.

Raising a red flag with inflammation

If a pathogen gets past the initial barriers and gains entry into the body, it starts to infect cells and injure tissue. The damaged tissue sends out chemical alarms that cause an immediate, local inflammatory response. The inflammatory response is what causes a cut to turn red and swell up. Inflammation helps to move immune cells from the blood to the injured tissue to prevent an infection.

The inflammatory response is important during the early stages of an infection to warn the immune system that something bad is happening. However, when immune cells are recruited, it's important that the inflammatory response is dampened to prevent an overreaction to injury. Inflammation that remains ongoing can cause further tissue damage and prevent repair. *Chronic* (continuous) inflammation contributes to autoimmune diseases (such as rheumatoid arthritis) and infectious diseases (such as tuberculosis).

REMEMBER

In the case of inflammation, too much of a good thing can be bad for the body. The inflammatory response must die down in order to allow repair of tissue following an infection.

Holding down the fort with innate immunity

Innate immune cells are the first cells of the immune system to contact pathogens and rapidly respond to initiate the clearance of infection. They're able to respond to a broad number of pathogens because they recognize features common to many microorganisms. These features are called *pathogen-associated molecular patterns* (PAMPs), and they include carbohydrates, proteins, DNA, or lipids only found on microorganisms.

Pattern recognition receptors (PRRs) are surface-bound proteins on innate immune cells that detect PAMPs and indicate the presence of an infection. An important class of PRRs includes the *toll-like receptors* (TLRs), each of which detects a certain class of PAMPs. The binding of a PAMP to a TLR calls the innate immune cell to action and initiates a response that aims to eliminate the pathogen (see Figure 15-1).

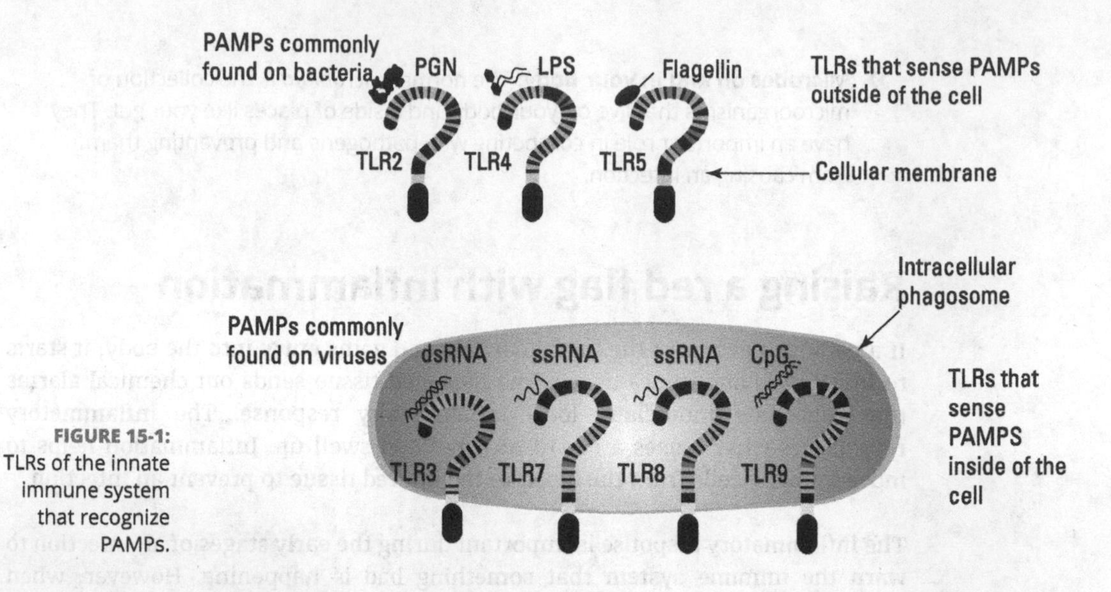

FIGURE 15-1: TLRs of the innate immune system that recognize PAMPs.

Most innate immune cells share a common mechanism of pathogen killing called *phagocytosis*. During phagocytosis, the cell takes up the pathogen and traps it within the *phagosome*, a specialized organelle that releases highly toxic substances that destroy the pathogen. Here are some phagocytic immune cells:

- Macrophages are "big eaters" and scavenge the environment for unwanted material. They reside is mostly every tissue in the body and ingest and kill pathogens.
- Dendritic cells are also found in many tissues and help remove pathogens. They play an important role in starting an adaptive immune response to an infection.
- Neutrophils circulate in the blood and are called to tissues when there's an infection.

Nonphagocytic innate immune cells are important for other branches of the immune response. Mast cells, basophils, and eosinophils release substances out of the cell to promote immune activation during allergic responses or parasitic infections. Natural killer (NK) cells destroy infected cells, as well as cancerous cells in tumors.

Sending out the troops for adaptive immunity

Several key features differentiate the adaptive immune system from the innate immune system:

- Ability to respond to specific microbial components called *antigens*
- *Tolerance,* or the inability to respond to molecules normally found in the body
- Memory of infection so that the response to reinfection is faster and more effective

Adaptive immune cells respond to *antigens* (*anti*body *gen*erators), which are molecules such as proteins, sugars, or lipids found in the pathogen. Each cell of the adaptive immune response expresses a unique surface receptor that is able to recognize a specific antigen. The immune system carefully chooses between the T cell or B cell receptor it expresses so that it responds only to microbial antigens and not to proteins, sugars, or lipids commonly found in the body. When a receptor binds to an antigen, it triggers the adaptive immune cells (like T cells, see "The tenacious T cell" section) to divide into a larger population and join the fight against the intruding pathogen. The population of adaptive immune cells (B cells, see "The beneficial B cell" section) that is activated during infection remains in the body as *memory cells* so that next time the same pathogen is encountered, the immune system responds more rapidly see Figure 15-2.

Different adaptive immune cells use different mechanisms to combat infection.

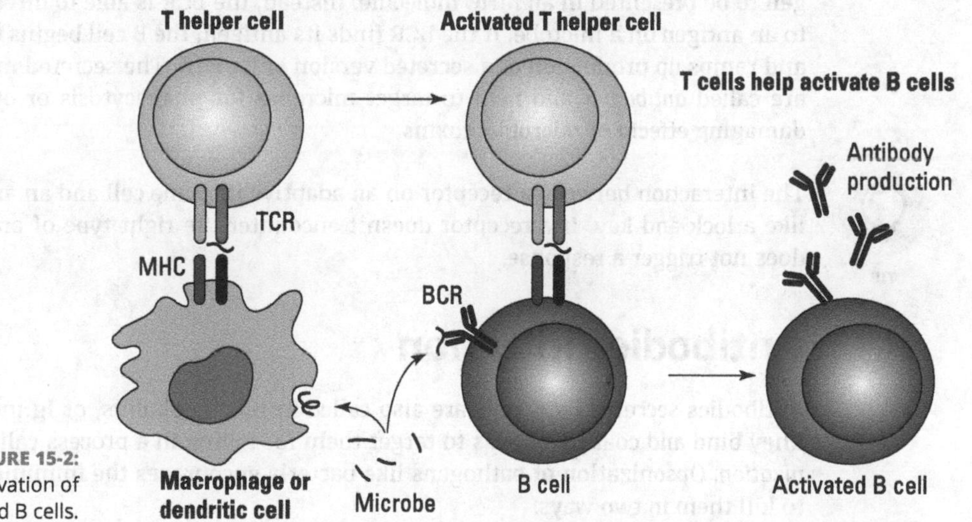

FIGURE 15-2: Activation of T cells and B cells.

The tenacious T cell

The receptor present on T cells is called the *T cell receptor* (TCR). The TCR is able to recognize antigens only when they're nested within a molecule called the *major histocompatibility complex* (MHC). When macrophages or dendritic cells phagocytose and destroy a pathogen, they load its chewed-up bits into the MHC molecule and display it on its surface to show to cells of the adaptive immune system. If a TCR is able to recognize the antigen shown, the T cell is recruited to fight. Different types of T cells help clear infections in different ways.

Cytotoxic T cells (T_C) are designed to hunt and kill infected cells. When they find their target cell, they release damaging molecules that trigger cell death, preventing the spread of an infection.

T helper cells (T_H) encourage recruitment and activation of other immune cells. During an infection, T_H cells help macrophages kill microbes they have taken up, activate B cells to produce antibodies, and recruit immune cells, like neutrophils to do damage control.

The beneficial B cell

The *B cell receptor* (BCR) is different from a TCR in that it doesn't require an antigen to be presented in an MHC molecule. Instead, the BCR is able to directly bind to an antigen on a microbe. If the BCR finds its antigen, the B cell begins to divide and ramps up production of a secreted version of its BCR. The secreted molecules are called *antibodies* and help to target microbes for phagocytosis or offset the damaging effects of microbial toxins.

TIP

The interaction between a receptor on an adaptive immune cell and an antigen is like a lock and key. If a receptor doesn't encounter the right type of antigen, it does not trigger a response.

Antibodies in action

Antibodies secreted by B cells are also called immunoglobulins, or Ig molecules. They bind and coat pathogens to target them for killing in a process called *opsonization*. Opsonization of pathogens like bacteria encourages the immune system to kill them in two ways:

- Macrophages and neutrophils have special receptors on their surface that recognize antibodies. If they come across a microbe coated in antibodies, they phagocytose it.
- The *complement cascade* is a series of proteins that assemble on the surface of a microbe when antibodies are present. Altogether, they form a hole on the microbe's surface and cause cell death.

Ig molecules are typically Y-shaped, with two identical arms that can bind to antigens (see Figure 15-3). Some antibodies have two Y-shaped Ig molecules but most have only one.

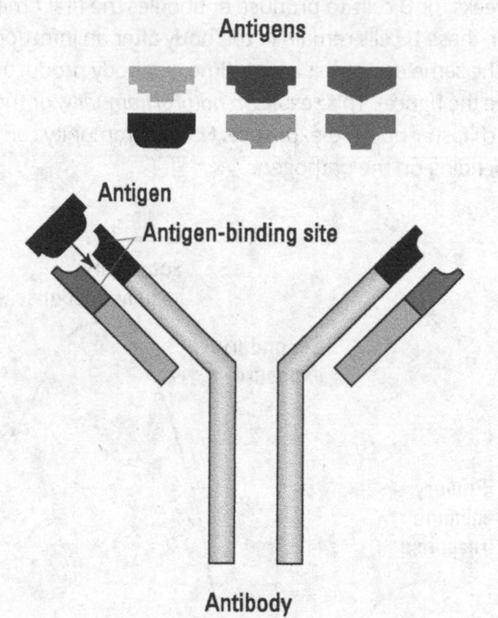

FIGURE 15-3: The structure of antibodies.

B cells are able to produce five different classes of antibodies, each of which serves a different function in the immune response (see Table 15-1).

TABLE 15-1 Functions of Antibody Classes

Antibody Class	Number of Ig Molecules	Main Location	Main Function
IgA	2	Tears, saliva, breast milk	Present at mucosal membranes
IgD	1	Blood	Activate B cells
IgE	1	Blood	Are part of the allergic response
IgG	1	Blood and tissue fluid	Coat the surface of a microbe, making it easier for the microbe to be phagocytosed (called opsonization)
IgM	5	Blood	Activate complement

INTERPRETING NATURAL IMMUNITY AND IMMUNIZATIONS

It can take up to two weeks for B cells to produce antibodies the first time they respond to an antigen. However, these B cells remain in the body after an infection is cleared, and if they encounter the same antigen at a later time, antibody production occurs much more quickly (see the figure). This results in *natural immunity*, or the ability to eliminate infection much faster upon re-exposure. Natural immunity can last for several years to a lifetime, depending on the pathogen.

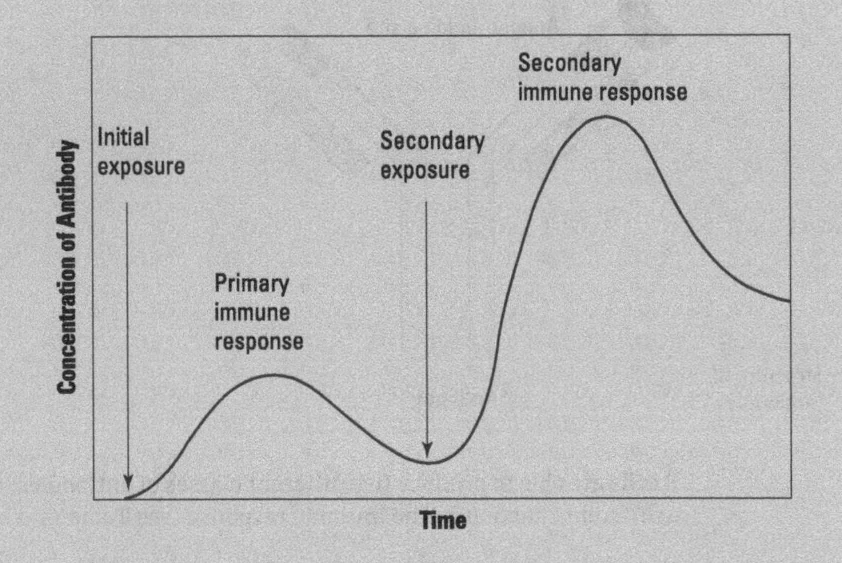

Newborn babies are highly susceptible to infections because their immune systems are not fully developed. Mothers transfer antibodies to babies in two ways to provide *passive* natural immunity.

- Antibodies of the IgG class cross the placental barrier and provide protection during and after pregnancy.
- Antibodies of the IgA class are present in breast milk, which newborns consume shortly after birth.

Passively transferred antibodies provide protection for only several weeks to months.

The specificity of antibodies for a pathogen and the speedy response the second time around is the basis of immunization. Vaccines deliver harmless antigens to the body in a controlled setting and prime B cells to produce antibodies. If the individual becomes infected with the pathogen they're vaccinated against, the immune system is prepared to stop infection early in its tracks.

Currently, close to 30 vaccines are available that protect against various infectious diseases, such as influenza, rabies, and meningococcal disease. Vaccines have eliminated many viral and bacterial infections (reduced the number of cases to near zero), and in the case of smallpox, have completely eradicated infections (no existing cases), saving many lives.

Relying on Antimicrobials for Treating Disease

Getting a bacterial infection is likely not a major worry for you. At the first signs of an infection, you head to the doctor to get a prescription of antibiotics and take them until the infection goes away.

It was a very different picture, however, just 80 years ago. People of every age died from untreatable bacterial infections. Antibiotics are a relatively modern advancement, one that has increased life expectancy and saved many lives. The use of one of the first antibiotics, penicillin, had such a significant impact on society that the scientists who discovered it were awarded the Nobel Prize.

Although we've come a long way in developing antibiotics to treat all kinds of bacterial infections, it's unfortunately not a battle that medicine has yet won. Drugs that once worked for treating infections are no longer effective because pathogens have developed *antibiotic resistance* (which involves changes in the genes of a pathogen such that the pathogen is no longer susceptible to the drug's action).

Pathogens are resistant to nearly every antibiotic developed to date. The current picture of antibiotic resistance worldwide is so grim that the World Health Organization (WHO) recently declared it a global health threat. There is an urgent need for more effective antibiotics in order to combat bacterial infections to prevent entry into what we fear will be a post-antibiotic era.

In this section, we fill you in on how antibiotics work and what they're used to treat.

AT THE FOREFRONT OF THE FIELD

Several key scientists made important contributions to the field of microbiology leading up to the development of antibiotics.

- In the early 1900s, Paul Ehrlich put forward the idea that in order to stop bacterial infections, a "magic bullet" was needed that would target and destroy a pathogen without harming the host. Following this theory, Ehrlich discovered the first antimicrobial agents, including Salvarsan, a drug used to treat syphilis infections in 1910. Ehrlich's "magic bullet" concept is now known as *selective toxicity* and is still important for drug discovery.
- In 1928, Sir Alexander Fleming accidentally discovered that a fungus, *Penicillium notatum,* inhibited the growth of the human pathogen, *S. aureus.* Scientists Howard Florey and Ernst Chain isolated the compound produced by the fungus, now known as penicillin, and found that it saved mice from infections with staphylococci and streptococci. Penicillin was first used as an antibiotic in 1940 and saved the lives of many soldiers during World War II.
- Gerhard Domagk researched the antimicrobial properties of a red dye and found that, although the red dye did not inhibit bacterial growth on a Petri dish, it was very effective at treating infected animals. His work uncovered that the body turns certain compounds into highly potent antimicrobials, and forged the way for the use of live animals for drug discovery. Domagk's work also led to the development of an entire class of antimicrobials called sulfa drugs.
- Selman Waksman recognized the idea that microorganisms themselves, such as the mold on Fleming's Petri dish, produce compounds with antibacterial properties. Waksman screened thousands of soil bacteria and fungi for their ability to inhibit the growth of pathogens. In 1944, his efforts led to the isolation of streptomycin from the *Streptomyces* bacteria, which resulted in a drug used to treat tuberculosis. Even today, environmental microorganisms are screened in hope of finding new antibiotics.

These discoveries had such an impact on society that they were recognized with the Nobel Prize. Fleming, Florey, and Chain received the Nobel Prize for their contributions to the discovery of penicillin in 1945. Waksman received the Nobel Prize in 1952.

Fundamental features of antibiotics

Antibiotics are a broad class of chemical compounds that are categorized based on some general properties. Some of these properties help doctors decide which antibiotic to use to treat an infection:

- **Spectrum:** The *spectrum of activity* describes, essentially, how many different types of bacteria a drug can kill. *Narrow-spectrum antibiotics* are able to kill only a select type of bacteria. For example, penicillin is effective at killing Gram-positive bacteria, but not Gram-negative bacteria. In contrast, *broad-spectrum antibiotics* are active against many types of bacteria, such as chloramphenicol, which is active against both Gram-positive and Gram-negative bacteria.
- **How they work:** Not all antibiotics kill bacteria on the spot. *Bacteriostatic* antibiotics inhibit the growth of bacteria and make it easier for the immune system to control infection. *Bactericidal* antibiotics are more potent and directly kill the targeted pathogen. Sometimes two bacteriastatic drugs can be combined to have a bacteriacidal effect, this is called *combination therapy.*
- **Where they're made:** Some antibiotics are isolated from environmental microbes, while others are made in a lab. Antibiotics produced by microorganisms are referred to as *natural* antibiotics, whereas antibiotics made in a lab are referred to as *artificial* or *synthetic* antibiotics.

REMEMBER

Whether antibiotics come from a natural source or are artificial has no impact on their effectiveness at killing microbes.

Antibiotics that target pathways unique to bacteria are great because they're unlikely to produce side effects in humans. Some antibiotics, however, affect human cells and can be harmful, especially if used in high doses. Individuals can develop a sensitivity and may have an allergic reaction to an antibiotic cannot be treated with it. This is the case for individuals with penicillin sensitivity.

WARNING

Don't confuse *sensitivity* to an antibiotic with *susceptibility* to an antibiotic. A person may be sensitive to an antibiotic (have an allergic reaction to it), whereas a microbe may be susceptible to the action of an antibiotic.

Targets of destruction

Antibiotics target several types of structures or pathways in bacterial cells that are absolutely necessary for survival (see Figure 15-4). Some of these targets are structures that are unique to bacterial cells, such as the peptidoglycan cell wall. Other antibiotics target pathways important in both bacterial and human cells but take advantage of slight differences in the structure of enzymes, making them more selective for bacteria. This section describes the cellular targets of antibiotics and the groups of antibiotics that target them.

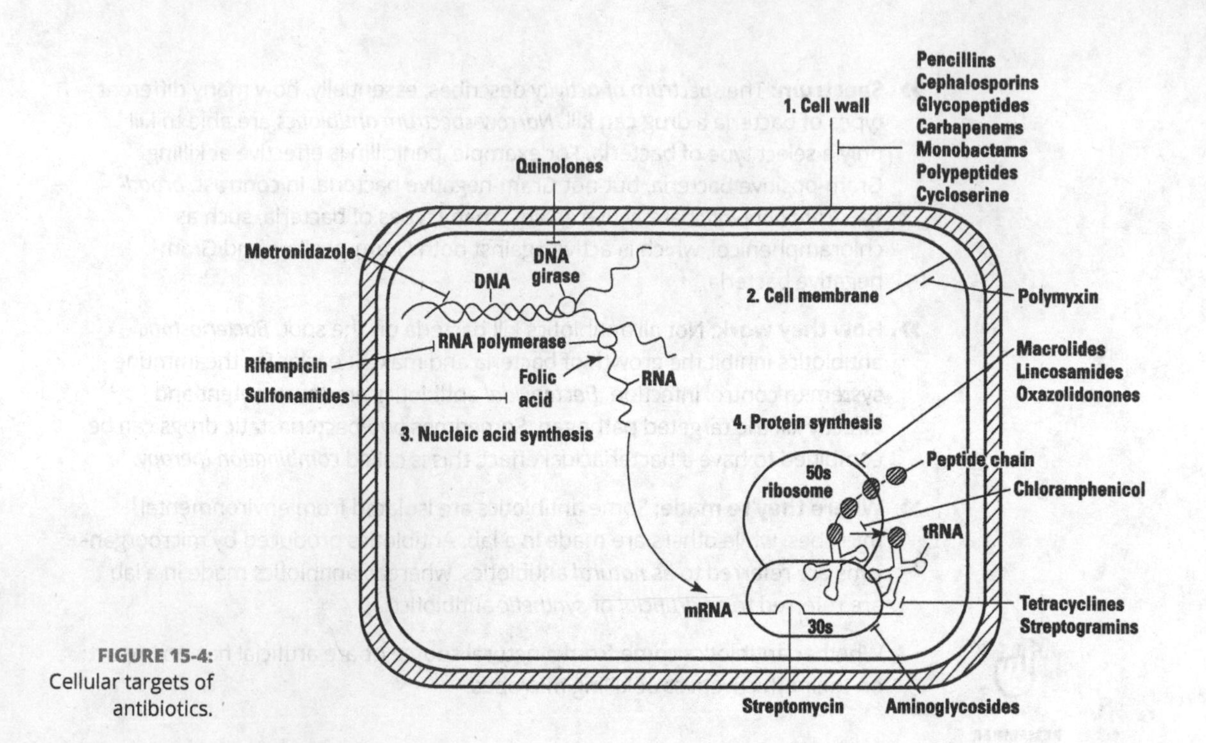

FIGURE 15-4: Cellular targets of antibiotics.

The cell wall is the target of several antibiotics. Antibiotics such as penicillins, cephalosporins, bacitracin, cycloserine, and vancomycin inhibit bacterial cell wall synthesis by interfering either with the synthesis or with the crosslinking of the subunits of peptidoglycan. Antibiotics like gramicidin and polymixin poke holes in the cell membrane causing bacterial cells to become leaky and die. This second type of antibiotic is used infrequently, however, because they also act against animal cells because they also have a cell membrane.

Antibiotics that target nucleic acid synthesis include

- Quinolones block bacterial DNA synthesis.
- Metronidazole is taken up by bacterial cells and randomly nicks the DNA, causing destruction of the bacterial genome.
- Rifampicin blocks the RNA polymerase enzyme needed for RNA synthesis.
- Sulfa drugs block the synthesis of folate, an essential component in the synthesis of RNA. Because mammals get all their folate from their food, these drugs are specific against bacteria.

Bacterial protein synthesis is a major target of antibiotics, all of which bind to various parts of the bacterial ribosome and interfere with translation of mRNA. Examples of drugs that inhibit bacterial protein synthesis include aminoglycosides (neomycin, gentamycin and kanamycin), tetracyclines, macrolides, lincosamides, and chloramphenicol.

Unraveling microbial drug resistance

Antibiotics are one of the major scientific advances in the past century that have undoubtedly improved the quality of life, yet the battle against infectious diseases has yet to be won. Bacteria are remarkably capable of *evolving* by altering their genetic code to make them less susceptible to the antimicrobial effects of antibiotics.

What does resistance look like?

Some bacteria acquire mutations that alter the target of an antibiotic enough that the antibiotic can no longer bind to its bacterial target. For example, some strains of *Staphylococcus aureus* are resistant to methicillin, a β-lactam antibiotic, which interferes with enzymes involved in peptidoglycan synthesis. To be methicillin-resistant, *S. aureus* must carry the *mec* gene, which allows the cell to continue making peptidoglycan as usual and not be affected by the antibiotic.

Bacteria use efflux pumps (membrane-bound proteins that work to pump things out of the cell) to get rid of unwanted waste and toxins. Pathogens make use of efflux pumps to pump out antibiotics (see Figure 15-5). Efflux pumps are not very selective for the substrates they expel — they're able to pump out many different drugs. *Multi-drug* efflux pumps are present in *Escherichia coli*, *Pseudomonas aeruginosa*, and *S. aureus*. These strains often have mutations that lead to increased expression of efflux pumps to pump antibiotics out of the cell right away before they exert their toxic effects.

The structure of an antibiotic is very important for its target specificity and antimicrobial activity. Some bacteria carry enzymes that chemically modify antibiotics to render them inactive. Another mechanism of drug inactivation includes degradation. The most well-known example is the β-lactamase enzyme. This enzyme cleaves the β-lactam ring in penicillin to prevent it from interfering with peptidoglycan synthesis.

Where does it come from?

Emergence of antibiotic-resistant bacteria is inevitable once an antibiotic is used to treat infection because bacteria that can resist the antibiotic have a greater chance of survival. Many bacteria are actually the producers of antibiotics and, therefore, must themselves be able to resist the toxic molecule they produce. These protective genes for some bacteria then move to pathogenic bacteria and make them resistant to the antibiotics. Initially, the rate of resistance is low, but with increased use of antibiotics, resistance becomes widespread.

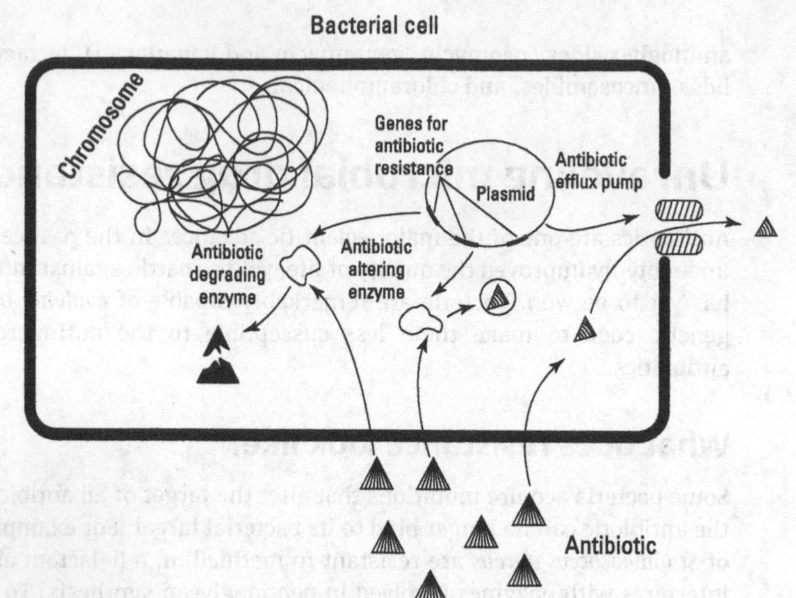

FIGURE 15-5: Mechanisms of antibiotic resistance.

Some bacteria are *naturally* resistant to antibiotics, meaning that the cells are innately unaffected by the drug. For example, some antibiotics like vancomycin are ineffective at treating Gram-negative bacterial infections because the drug isn't able to penetrate the outer membrane.

Other bacteria *acquire* antibiotic resistance either through mutation of existing genetic material or acquisition of new genetic material. Example of this include

- Spontaneous mutations can confer resistance to a drug, usually by changing the target to prevent binding of the drug. Such mutations are passed on directly to bacterial cell progeny during division in a process referred to as *vertical gene transfer*. Spontaneous mutations are not a major source of resistance because they're infrequent events and could have negative consequences on target gene function.
- Horizontal gene transfer (HGT) is a more prominent mechanism of acquiring antibiotic resistance. HGT involves the exchange of genetic information between bacteria. Mechanisms of HGT include conjugation, transformation, or transduction (see Chapter 8 for more detail). Resistance genes can be found on plasmids, referred to as *R (resistance) plasmids;* R plasmids exist separately from the chromosome or on mobile genetic elements, such as transposons.

Antibiotics can come from natural sources, such as soil bacteria or fungus, so some pathogens are exposed to antibiotics long before we use them to treat infections. Scientists suspect that nowadays bacteria experience increased exposure due to the inappropriate use of antibiotic treatment and overuse of antibiotics in the farming industry.

Discovering new antibiotics

Most antibiotics used today were discovered in the "golden age" of antibiotic discovery, from the 1940s through the 1960s. In the decades since, the "antibiotic pipeline" has run relatively dry and no new classes of broad-spectrum antibiotics have been developed despite an urgent need.

Drug discovery faces a variety of challenges, including the following:

- **Lack of new target discovery:** Few new bacterial targets, to which antibiotics act, have been recently discovered.
- **Discovery of drugs with only a narrow spectrum:** Discovering drugs to target Gram-negative bacteria is particularly difficult because of their outer membrane.
- **Rapid development of resistance:** Bacteria rapidly become resistant to antibiotics as these drugs are used more frequently. Because antibiotic therapy is short-term treatment, it doesn't bring in as much money for pharmaceutical companies as drugs that people have to be on long term (or even for life). For this reason, pharmaceutical companies aren't as eager to invest a lot of money in discovering new antibiotics.
- **Risk of toxicity:** As the safe drugs have been used up (meaning that more people have resistant infections and the antibiotics are useless against them), antibiotics that are more toxic to people are being used. Newer drugs are often more toxic to people because the antibiotics designed against the known bacterial targets have already been made.

With these challenges in mind, there are several approaches to antibiotic discovery today:

- **Generation of synthetic antibiotics:** This approach is attractive because the structure of the drug can be created to optimize crossing of the membrane and binding to the target protein. This approach involves *rational drug design,* in which a target gene (ideally one that is essential for bacterial growth) is selected to target and inhibit. Structural models are used to predict drug-target binding to optimize drug design.
- **High-throughput screening (HTS) of chemical compound libraries containing thousands of synthetic and natural compounds:** The compounds are screened in automated processes for their ability to inhibit bacterial growth. In contrast to rational drug design, HTS can result in identification of novel targets to inhibit.

THE DOWNSIDE OF ANTIBIOTICS

In recent years, scientists have learned a lot more about the *human microbiome*, the collection of bacteria that normally colonizes our bodies. Antibiotics can't distinguish these "good" bacteria from *pathogens* (disease-causing bacteria), so antibiotics can end up altering the microbiome, which may have implications for a person's health.

Children are the most common age group to be prescribed antibiotics. Exposure to antibiotics in early life may affect immune system development and increase susceptibility to allergic and metabolic diseases later in life. In order to minimize the effects on the microbiome, development of antibiotics that are specific to pathogens is an active area of research.

Alternatives to traditional antibiotics to treat bacterial infections include the following:

- **Targeting genes that are not necessarily required for growth, but contribute to the ability of a pathogen to cause infection:** Targeting such a factor may enable the host immune response to control infection more efficiently.
- **Antimicrobial peptides (AMPs) are produced by the innate immune system and are broadly active against many bacteria:** A boost of AMPs during infection could promote clearance of a pathogen.
- **Phage therapy:** This involves the use of *phage* (viruses that infect and destroy specific populations of bacteria) to target infections. Progress has been made in this field, but challenges in logistics still require attention.

Searching Out Superbugs

Antibiotics are of bacterial origin. They're part of the arsenal of bacterial chemical warfare against one another. So, it's no surprise that antibiotic resistance genes are common among bacteria, too. When antibiotics are used, they affect all the bacteria that are sensitive to the antibiotic. When a member of the bacterial community is resistant to the antibiotic used, it's given an opportunity to flourish. Superbugs are the result of antibiotic-resistant bacteria that been allowed to flourish and have become a problem for human health.

Different superbugs are a problem for one of these reasons:

- They cause disease and are resistant to the drugs we usually use to treat them.

» They're usually harmless microbes that are now impossible to treat if they do cause an infection (for instance, in someone with a compromised or weak immune system).

» They're resistant to several antibiotics.

Staying ahead of vancomycin-resistant enterococci

Enterococci are Gram-positive bacteria that normally colonize the gut and the female genital tract of healthy people and are generally not harmful, but hospital patients undergoing surgery or with open wounds are highly susceptible to enterococcal infections. The most common type of infection is the urinary tract infection; others include bacteremia, endocarditis, and intra-abdominal infections.

Infections spread directly from patient to patient, indirectly on the hands of healthcare workers, or on contaminated surfaces or equipment. The many sources of transmission and the many people at risk in a hospital make the enterococci a major cause of hospital-associated infections.

Enterococcal infections can be difficult to treat because strains are often antibiotic resistant. Enterococci are naturally resistant to β-lactams, aminoglycosides (low levels), macrolides, and sulfa drugs. The emergence of vancomycin-resistance enterococci (VRE) has had a major impact on the treatment of infections, leaving doctors with limited options.

An added threat of VRE is the possibility that the resistant genes could be passed to other pathogens, including *Staphylococcus aureus.* In 2002, a doctor reported the first case of a methicillin-resistant *Staphylococcus aureus* (MRSA) strain that was also vancomycin resistant and carried the VRE resistance genes. Fortunately, only 13 additional cases have since been reported to date. (See the next section for more on MRSA.)

Battling methicillin-resistant Staphylococcus aureus

Staphylococcus aureus is a prominent human pathogen capable of infecting nearly every site in the body and producing infections that range from mild to life threatening. Stopping *S. aureus* infections is difficult because it has a history of rapidly developing resistance to antibiotics. Penicillin-resistant strains emerged in 1944, just four years after the drug was introduced to market, and in the 1960s methicillin-resistant *S. aureus* (MRSA) appeared. Over the last 60 years, MRSA has spread to cause epidemics globally.

Initially, MRSA infections were primarily a problem in hospital patients undergoing surgery or with open wounds. Outbreaks of MRSA in hospitals were reported throughout the 1970s and 1980s (see Figure 15-6).

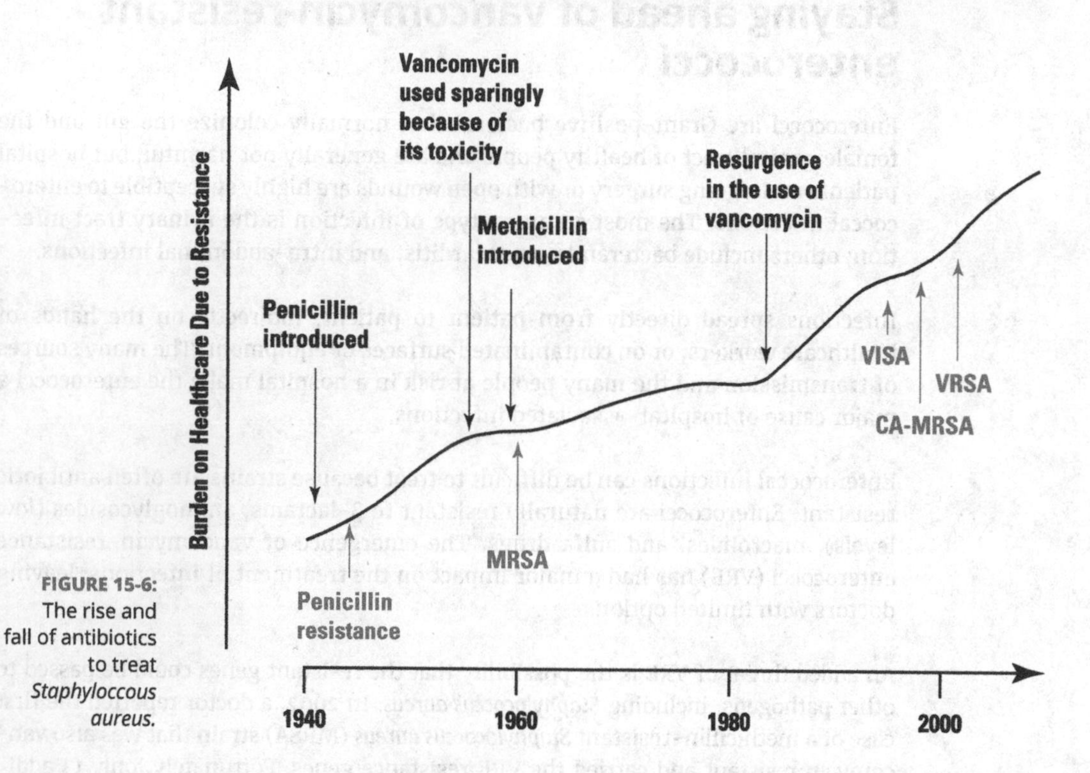

FIGURE 15-6: The rise and fall of antibiotics to treat *Staphyloccous aureus.*

In the late 1990s, MRSA spread out of hospitals and into communities. Community-associated (CA) MRSA affects healthy individuals with no known risk factors of infection, and commonly causes outbreaks in athletes, recreational facilities, prisons, and daycare centers. CA-MRSA primarily causes skin infections that are typically red, swollen, painful, and pus filled. Treatment of infections involves drainage of pus or, if more serious, use of antibiotics. If left untreated, the infection can spread to other body parts and cause a more severe infection.

CA-MRSA is spread by direct skin-to-skin contact or contact with contaminated objects. If you come into contact with CA-MRSA, you can spread it to other individuals if you don't wash your hands properly, or can infect yourself through an open wound.

The antibiotic vancomycin is used as the alternative to methicillin to treat MRSA. Although it was discovered before methicillin, vancomycin was not widely used because of its toxicity to human cells. With the rise of MRSA and other antibiotic-resistant bacteria, the use of vancomycin has increased by 100 times in the last 30 years and, not surprisingly, has led to the rise of vancomycin-tolerant (VISA) and vancomycin-resistant (VRSA) *Staphylococcus aureus*. With the loss of vancomycin to treat *Staph aureus* infections, treatment options for this nasty bacterium are slim indeed.

Outcompeting Clostridium difficile

Clostridium difficile is the most common cause of diarrheal infections in hospitals and long-term care facilities. Infections most often occur in the elderly or in a hospital setting when a patient is on continuous antibiotic treatment. Antibiotics can disturb the normal gut bacteria. Then *C. difficile* endospores can germinate and take advantage of the absence of other bacteria to grows rapidly.

The symptoms of a *C. difficile* infection include diarrhea, nausea, fever, and abdominal pain and can range from a mild form of infection to serious and possibly fatal cases. Mild cases don't require treatment, whereas more serious infections require medication or surgery. Recently, more harmful strains have emerged and have caused serious outbreaks in hospitals resulting in death of up to 20 percent of infected patients.

Because they are so hardy, *C. difficile* endospores can survive on surfaces and equipment in hospitals and can resist common disinfectants. This makes it difficult for hospitals to get rid of *C. difficile*. The spores spread from the feces of individuals who carry *C. difficile* in their gut. Hospitals enforce hand washing to prevent transfer of spores; in some cases, they use specialized machines that sterilize a room after a patient has had a *C. difficile* infection.

Often after a patient gets a *C. difficile* infection, that person is more likely to get another one in the future. The cause of recurrent infections is not clear, and without knowing the cause, it's difficult to find a cure. One treatment gaining appreciation is *fecal transplants* (literally, the transfer of stool from one individual to another). Although the idea isn't the most appealing, the results are too good to ignore — it boasts a 95 percent success rate over the approximately 50 years of its use.

Pressure from extended-spectrum beta-lactamases

New Delhi metallo-β-Lactamase (NDM-1) is a recently described broad-spectrum β-lactamase enzyme that inactivates all β-lactam antibiotics except aztreonam. It belongs to a group of acquired β-lactamases, called carbapenemases.

Gram-negative bacteria carrying carbapenemases have gradually increased over the years yet remain a relatively rare cause of human infections. The rapid emergence of infections caused by bacteria carrying NDM-1 in several countries has raised global alarm.

NDM-1 is encoded on a conjugative plasmid in isolates of primarily *Escherichia coli*, *Klebsiella pneumoniae*, as well as other *Enterobacteriaceae* species. The spread of NDM-1 across many unrelated bacteria is not common to resistance mechanisms and is another reason for raising concern. NDM-1 isolates cause a range of infections, including urinary tract infections, diarrhea, septicemia, pulmonary infections, and soft tissue infections. The infecting organisms are usually highly multidrug resistant, leaving few if any treatment options.

Most people acquire an infection with NDM-1 producing bacteria in a hospital setting. Currently, it appears that individuals who received medical care in India or Pakistan are most at risk of infection. With the amount of global travel in today's society, NDM-1 producing bacteria are a threat worldwide. Some hospitals have begun to screen patients carrying bacteria with the NDM-1 gene to detect sources of infection early and prevent spread. New antibiotics to target multidrug resistant Gram-negative bacteria are urgently needed to treat infections.

Knowing the Benefits of Prebiotics and Probiotics

Probiotics are live microorganisms with claimed health benefits when consumed in adequate amounts. They're components of fermentation products (such as yogurt) or packaged in a supplement form. Once consumed, probiotics colonize the gut for a brief period of time and exert their beneficial effect.

Prebiotics are nondigestible nutrients that are used as an energy source by beneficial bacteria normally found in the body. They stimulate the growth of these bacteria to promote a healthy gut microbial population.

Probiotic bacteria are thought to provide benefits to human health in three ways:

- Protecting against intrusion of pathogens
- Enhancing gut function to promote antimicrobial defense
- Altering immune cell populations and their function to strengthen nonspecific immunity against pathogens

Probiotic bacteria can prevent entry of pathogenic bacteria in the gut by competing for nutrients and surface colonization. They also secrete metabolites that inactivate toxins or are antimicrobial against other bacteria, called *bacteriocins.*

TECHNICAL STUFF

Cell surface components of probiotics act similarly to the PAMPs discussed earlier in this chapter — they bind to surface receptors on immune cells to induce an immune response. Yet in contrast to PAMPs, which induce an inflammatory response, probiotic microbial associated molecular patterns (MAMPs) can induce an anti-inflammatory response, as well as other beneficial changes in intestinal cells such as an increase in the mucous production. This anti-inflammatory response is thought to limit the amount of inflammation in the gut and is referred to as *immunomodulation.* It may help restore a healthy gut in people with weak immune systems, but whether this provides beneficial effects in healthy people isn't clear from the studies on probiotics.

With the increasing negativity surrounding antibiotic use, including their association with the development of antibiotic resistance, and possible long-term health effects due to disturbance of normal gut bacteria, probiotics are becoming appealing therapeutics. They're currently used to prevent antibiotic-associated diarrhea and acute infectious diarrhea in adults. They're also used to successfully treat necrotizing enterocolitis (NEC), a devastating infection of preterm or sick infants where tissue damage kills parts of the bowel.

Attacking Viruses with Antiviral Drugs

The infectious cycle of a virus involves the hijacking of a host cell and use of its machinery to replicate and produce progeny virus. Development of antivirals was initially a daunting task since the stages of viral replication depend on host cell machinery, making it difficult to develop selective drugs. Luckily, virologists discovered virus-specific pathways leading to the development of successful antiviral therapeutics.

In general, antivirals can target several steps of infection:

- **Viral entry:** The viral particle enters the host cell.
- **Virus particle uncoating:** An enveloped virus fuses with the host cell membrane, allowing entry of the viral capsid (containing the viral genes) into the host cell.
- **Nucleic acid synthesis:** After viral infection, the host cell machinery is hijacked to make copies of the viral genome by synthesizing viral nucleic acids.

- **Viral particle production:** After the host cell machinery is finished making viral genomes, it makes viral proteins and then assembles viral particles from the proteins and the genomes.

Most antivirals are specific to the type of virus they inhibit because not all pathways targeted are common to all types of viruses. Many antivirals target influenza and HIV. The development of anti-HIV drugs has been critical to the management of HIV infections and consequently AIDS. Currently no vaccine exists for HIV; anti-retroviral drugs, like AZT, have significantly improved the life expectancy of HIV-infected individuals.

Retroviruses use a special viral protein called reverse transcriptase to copy their RNA genome into DNA after they've entered the host cell. Drugs like azidothymidine (AZT) are nucleoside analogues. Nucleosides are the building blocks of nucleic acids and because AZT looks like a nucleoside (that's the analogue part), AZT works by taking the place of nucleotides as they are incorporated into the growing DNA molecule during reverse transcription. But because they're shaped slightly differently than the cellular nucleosides, they act to terminate the reaction (see Figure 15-7).

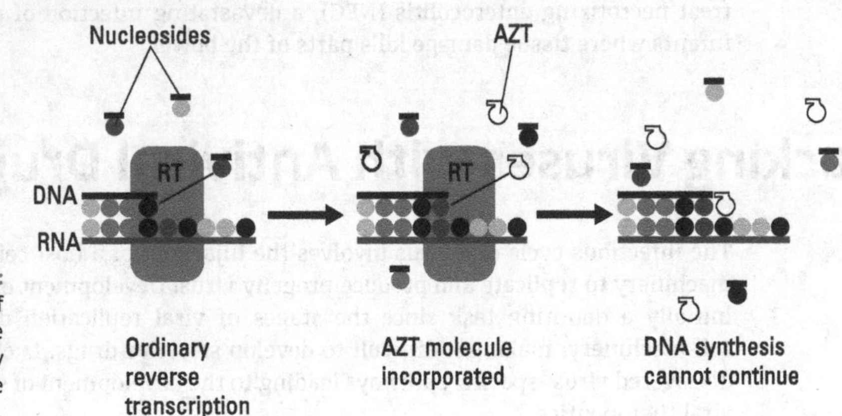

FIGURE 15-7: Inhibition of reverse transcription by nucleoside analogs.

IN THIS CHAPTER

» Finding out about recombinant DNA technology

» Giving traits to microorganisms

» Seeing how biotechnology works for you

Chapter 16
Putting Microbes to Work: Biotechnology

Microbial biotechnology is the manipulation of a microorganism so it can be used for a commercial purpose. Altering the genetics of a microorganism by inserting or deleting genes is called *genetic manipulation*, and the result is called a *genetically engineered* or *genetically modified* microorganism. Many decades of research have gone into developing the tools needed to be able to manipulate DNA, and the results of this hard work are a set of tools collectively referred to as *recombinant DNA technology*. The reason this is important to a book about microbiology is that many of the tools used for recombinant DNA technology were discovered in, come directly from, and are applied to microorganisms.

The expression of foreign DNA in microbial cells poses many challenges, so it may seem like each section of this chapter is filled with obstacles. The field of biotechnology has for years been addressing these challenges, and it has come up with some clever solutions to many of them, building tools that are used to either improve or solve a problem in the areas of research, medicine, food and alcohol, energy, mining, and environmental contamination.

Using Recombinant DNA Technology

Despite its fancy name, the concepts in recombinant DNA technology are quite simple:

1. Get a sequence of DNA that you're interested in.
2. Modify it (optional).
3. Cut it to fit and then join it to a cloning vector.
4. Insert it into a host cell.
5. Choose, or *select for,* the host cells that now carry your DNA of interest.
6. If you want the protein product, extract it from the host cell.

This, in a nutshell, is the process that has been used over the last several decades on countless genes of interest. In this section, we go into a bit of detail about how these steps are accomplished and talk about some of the molecular tools used.

Making the insert

The first step to any recombinant DNA strategy is choosing a gene or pathway of interest. This depends on the biotechnology application, but two broad categories may include

- Making a protein of interest to be purified and used without the organism (that is, an antibody, an antibiotic, or a therapeutic agent)
- Engineering an organism to do something, which usually involves an entire pathway (that is, multiple genes)

After you've chosen the pathway or protein of interest, you need to get the DNA sequence that corresponds to it. Until recently, going from function to DNA involved complicated screens to identify and then capture the genes of interest. Now, with recent advances in DNA sequencing, complete genomes are available for a large number of organisms. These genomes can be used to either predict the gene function computationally or compare the genes to genes from other organisms for which the function is known.

Two other relatively modern breakthroughs that make the process of generating a DNA insert easier are chemical DNA synthesis and the polymerase chain reaction (PCR). With these breakthroughs, it's possible to simply amplify the gene(s) and then use them in a cloning strategy.

If the genome of your organism of interest hasn't been sequenced or if the gene you're interested in hasn't been studied before, it's a bit more complicated to get the DNA you need to work with, but thanks to decades of work done prior to the advent of PCR and gene databases, this too is relatively straightforward. But that's the subject for a whole other book. . . .

In any case, in order to amplify your gene(s) of interest, you must first extract total DNA (or RNA) from cells of that organism. DNA extraction is essentially the same for both prokaryotic and eukaryotic cells and involves breaking open the cells, neutralizing the enzymes (that may degrade nucleic acids), getting rid of the cellular debris, and then cleaning away the impurities. Some types of cells — notably Gram-positive bacterial cells and tough plant tissues like seeds — require more work to break open, but other than that, the rest of the steps are pretty much the same.

THE POLYMERASE CHAIN REACTION

PCR, shown in the figure, is the process of creating a large amount of a particular DNA sequence from:

- A small amount of that sequence, called the *template*
- Some small pieces of DNA that match the ends of the sequence of interest, called *primers*
- *Nucleotides* (the building blocks of DNA)
- An enzyme that knits nucleotides into a sequence (called a *polymerase*)

This method, developed in the 1990s, changed everything for researchers working with DNA. In a few short years, they went from having to painstakingly extract large amounts of DNA to work with or find creative ways to work with tiny amounts of DNA, to being able to generate all the necessary DNA in a couple of hours.

A surfer/chemist named Dr. Kary Mullis developed the technique, for which he got a Nobel Prize, while struggling to get enough DNA to study. He reasoned that if he could just cycle through the steps of DNA synthesis enough times, it should be possible to keep copying a stretch of DNA, the template, over and over again. The copies end up becoming the templates and the number of DNA molecules increase at an exponential rate. The beauty of this method is that every copy is identical, except for the odd error made by the polymerase enzyme. From a handful of DNA molecules, you can make millions of copies of your DNA sequence of interest.

(continued)

(continued)

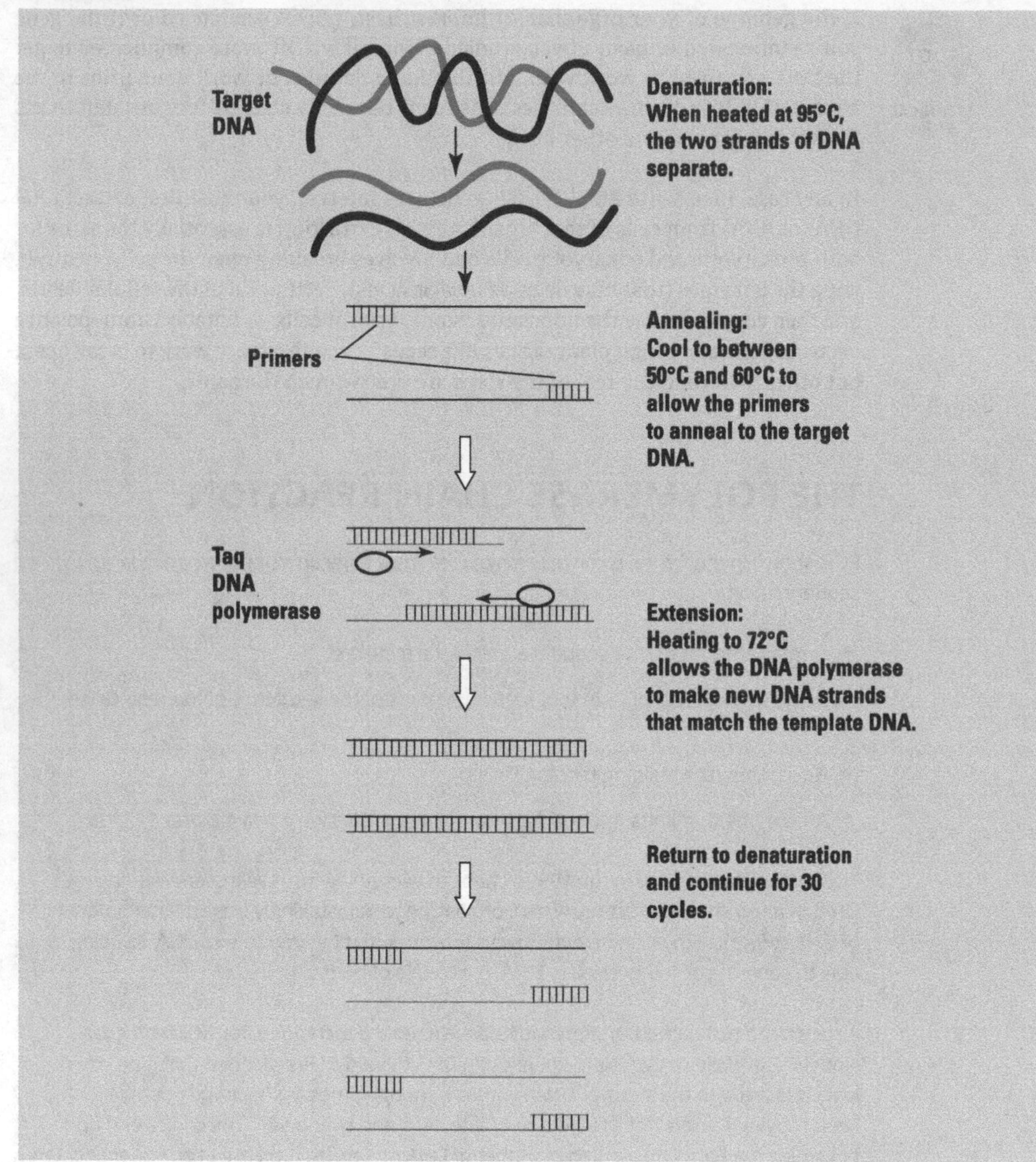

Target DNA
Denaturation: When heated at 95°C, the two strands of DNA separate.
Primers
Annealing: Cool to between 50°C and 60°C to allow the primers to anneal to the target DNA.
Taq DNA polymerase
Extension: Heating to 72°C allows the DNA polymerase to make new DNA strands that match the template DNA.
Return to denaturation and continue for 30 cycles.

After you have DNA, or complementary DNA (cDNA) made from RNA, from the organism whose genes you're interested in, you can either use them the way they are or modify them to improve or change the function or structure of the proteins they encode. Modifications are done either by *directed mutagenesis*, where a specific change is made to the DNA sequence, or by *random mutagenesis*, where errors are intentionally made during amplification of the genes and the resulting proteins are tested for the desired changes.

Now the sequence of interest, modified or otherwise, is ready to be used to genetically engineer a microorganism to do your bidding.

Employing plasmids

One of the most valuable assets to recombinant DNA technology has been the *plasmid*. These small, circular pieces of DNA that occur naturally in bacteria are not part of the main bacterial chromosome (thus, they're said to be extra-chromosomal). They replicate when the bacterial cell divides and can exist as many copies inside a cell.

Plasmids are very handy because you can manipulate them by adding DNA of interest and then put them back into a host cell, such as bacteria or yeast. Plasmids that have been modified to have certain features added to make them suitable for the transfer of genetic material to a host are called *cloning vectors*. Several features of cloning vectors make them handy, all of which exist within a plasmid backbone:

- **Origin of replication (ori):** This is the area where the enzyme complex of DNA replication binds, allowing the plasmid to be copied before cell division. If a plasmid has an ori that is not recognized by the DNA polymerase of the organism it's in, then the plasmid won't be replicated and won't be transferred to the new cell when cell division happens. The ori can be specific to the type of bacteria the plasmid originally came from, giving the plasmid a *narrow host range*, or be more general, giving it a *broad host range*.
- **Selectable marker:** Selectable markers are usually antibiotic-resistant genes that allow bacterial cells to survive in the presence of an antibiotic. Selection is important — without it, bacteria would lose the plasmid during successive rounds of cell division. Making the proteins encoded on the plasmid, as well as the plasmid itself, is metabolically expensive for a bacterial cell, so it'll try to do without it if possible.
- **Site of insertion:** The plasmid must have a site of insertion for the target gene. It's important that this be a restriction enzyme recognition site that doesn't occur anywhere else in the plasmid; otherwise, when digested, the plasmid will be cut into more than one piece, making recircularization inconvenient. For convenience, many cloning vectors have been engineered with a multiple cloning site (MCS), where sites for multiple different restriction enzymes have been put next to one another to make cloning easier.

» **Promoter:** If the gene of interest is to be expressed as a protein, this stretch of DNA contains all the binding sites for the enzyme RNA polymerase and all the accessory proteins needed to begin transcription. The promoter can either be put ahead of the site of insertion or be inserted as part of the target gene so that it can drive expression of the gene of interest, once inserted.

One great thing about cloning vectors is that the promoters, selectable markers, and plasmid backbones can be mixed and matched to create the right vector for the gene(s) of interest.

Cutting with restriction enzymes

Bacteria have a natural defense system involving enzymes that cut up foreign DNA. This is needed because viruses that infect bacteria, called *bacteriophage*, inject their DNA (or RNA) into the bacterial cell and then take over the cellular machinery, eventually killing the cells. Restriction enzymes work by cutting up any foreign DNA that is found in the cytoplasm. These enzymes are either *exonucleases*, which remove bases from the ends or *endonucealses*, which cut at sites within the DNA.

Restriction enzymes are one of the most useful tools in recombinant DNA technology for the following reasons:

» **They cut DNA in a specific and reproducible way, called a *digestion*.** This is important because it allows researchers to know the exact size of the fragments that they'll get when they put in a known sequence and a particular restriction enzyme. The way this works is that restriction enzymes recognize a particular sequence in the DNA and always cut in the same way when they've bound there.

» **The sequences can be four, six, or eight nucleotides long** (for many options and versatility) and always have the same sequence in either the 5′ or 3′ direction. An example of the six base sequence for EcoRI, a common restriction enzyme, is shown in Figure 16-1 with cut sites always occurring between the G and the A.

The naming convention for restriction enzymes is also shown in Figure 16-1, where the genus, species, and strain of the organism from which the enzyme was originally found make up the first few letters in its name and then the order of the discovery of each restriction endonuclease for that organism is shown by the roman numeral at the end of the name.

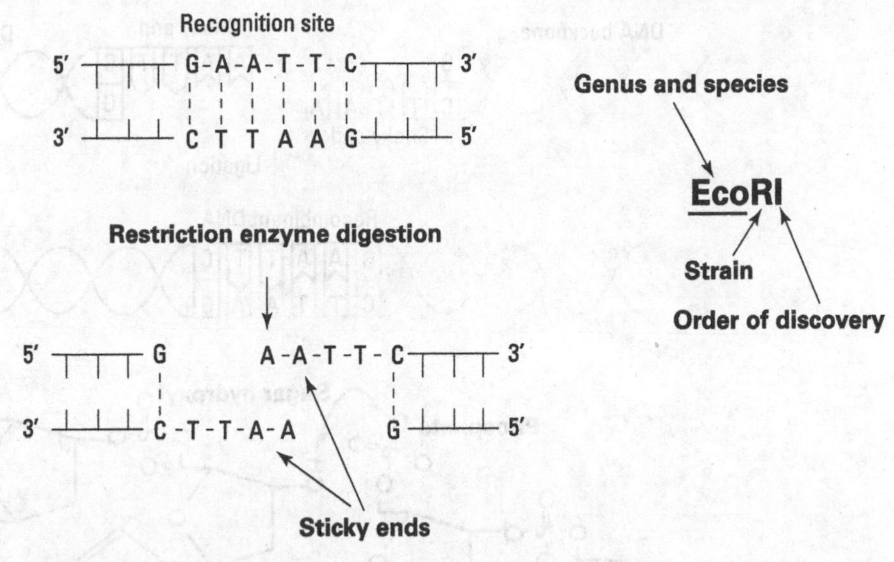

FIGURE 16-1: Recognition site and naming convention for the restriction enzyme EcoRI.

Today more than 3,500 restriction enzymes have been isolated that recognize around 250 different target sites. Shorter recognition sites are present more frequently in genetic sequences than are longer ones. Restriction enzymes that recognize four and six base sites are used most frequently, whereas those that recognize eight base sites are less commonly used.

After a restriction endonuclease is used to digest a piece of DNA, the two ends of the DNA made by the cut are staggered (see Figure 16-2) and said to be "sticky." Any other piece of DNA cut with the same enzyme will have the same sticky ends, and the two can be joined together through *ligation* (refer to Figure 16-2).

Ligation requires another enzyme called a *ligase*, which seals the nicks in the backbone by bonding the hydroxyl residue (–OH) on one end to the phosphate (–PO_3) on the other. What you're left with is a seamless sequence of DNA called a *construct* that can be cut and ligated again to add more DNA or put into a cloning vector. (A cloning vector containing an insert is sometimes called the *cloning vector–insert DNA construct*, but let's just say the *cloning vector–insert.*) For multi-gene constructs, a different restriction enzyme site is required for each addition, making it difficult to construct very large pieces of DNA or entire genomes. Recently this challenge has been overcome by the use of recombination (see the "Making long, multi-gene constructs" section, later in this chapter).

TIP

Not all restriction enzymes produce sticky ends. Some cut sequences leaving the two strands flush with one another, called *blunt ends.* Ligating pieces together that have blunt ends is not difficult, but the ligase has no way of knowing which way the ends should be glued together, so extra care has to be taken afterward to verify that the sequences are in the right order.

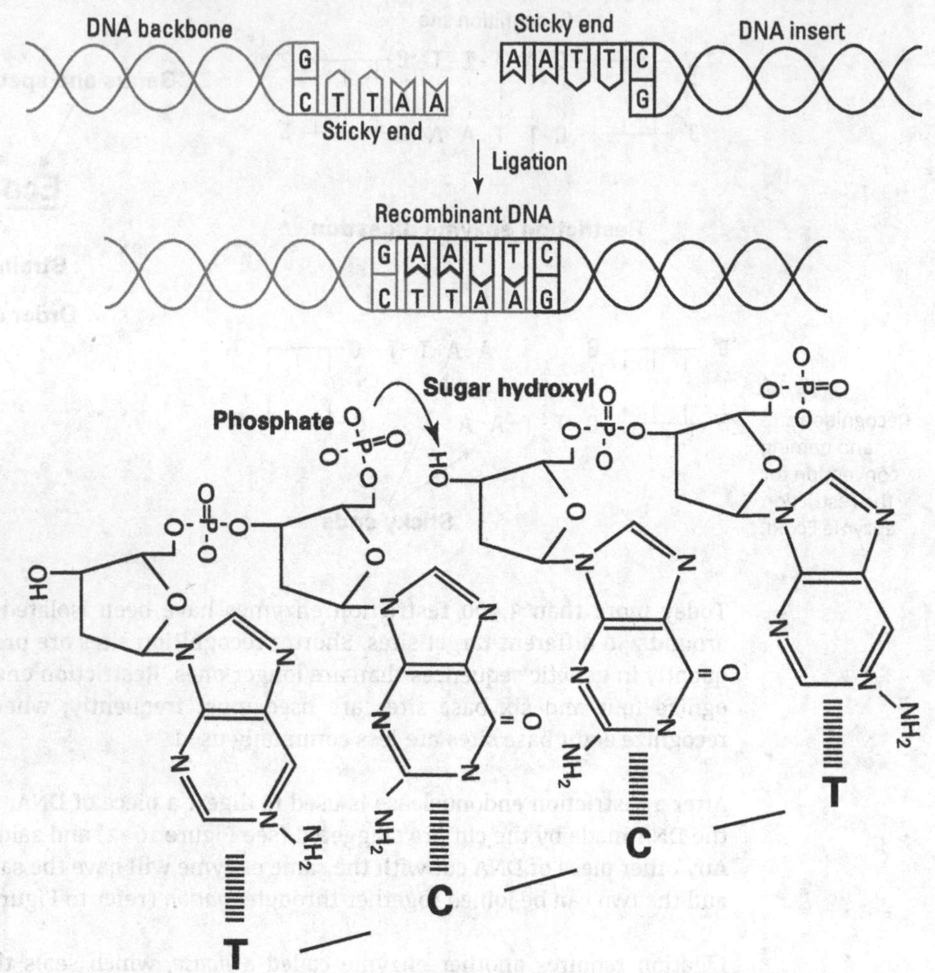

FIGURE 16-2: Ligation of two pieces of DNA cut with the same restriction enzyme.

Getting microbes to take up DNA

If a bacterial species is not naturally able to take up DNA, or not *competent*, it must be induced to take up the cloning vector-insert or DNA of interest. This can happen in a few different ways:

- **Heat shock:** Chemical competence can be induced by exposing a mid-log phase culture of the bacteria to cold calcium chloride and then exposing it to a high temperature (42°C) for two minutes. Plasmid DNA in solution along with the bacterium is taken up through transient holes in the cell wall.

» **Electroporation:** Electroporation is used to make cells take up DNA by exposing them to a short electrical pulse. Again, DNA mixed in with the culture just before the pulse is taken up through the cell wall via small openings that close shortly afterward. Although we know that it works, not much is really known about the mechanism involved in DNA uptake through electroporation.

» **Conjugation:** The genes necessary to form these cellular connections are not generally present on cloning vectors but can be supplied by a second plasmid called a mobilizable plasmid, which is supplied by a third bacterial strain called a helper strain. Figure 16-3 shows how the donor strain passes genetic information to the recipient strain, the helper strain is shown.

For conjugation the donor contains the cloning vector with the insert that you eventually want to be transferred into the recipient. The helper has the mobilizable plasmid with all the instructions for how to move plasmids between cells. After conjugation has happened, the trick is to select only for the cells you want, namely recipients with the plasmid of interest, and screen out all the others.

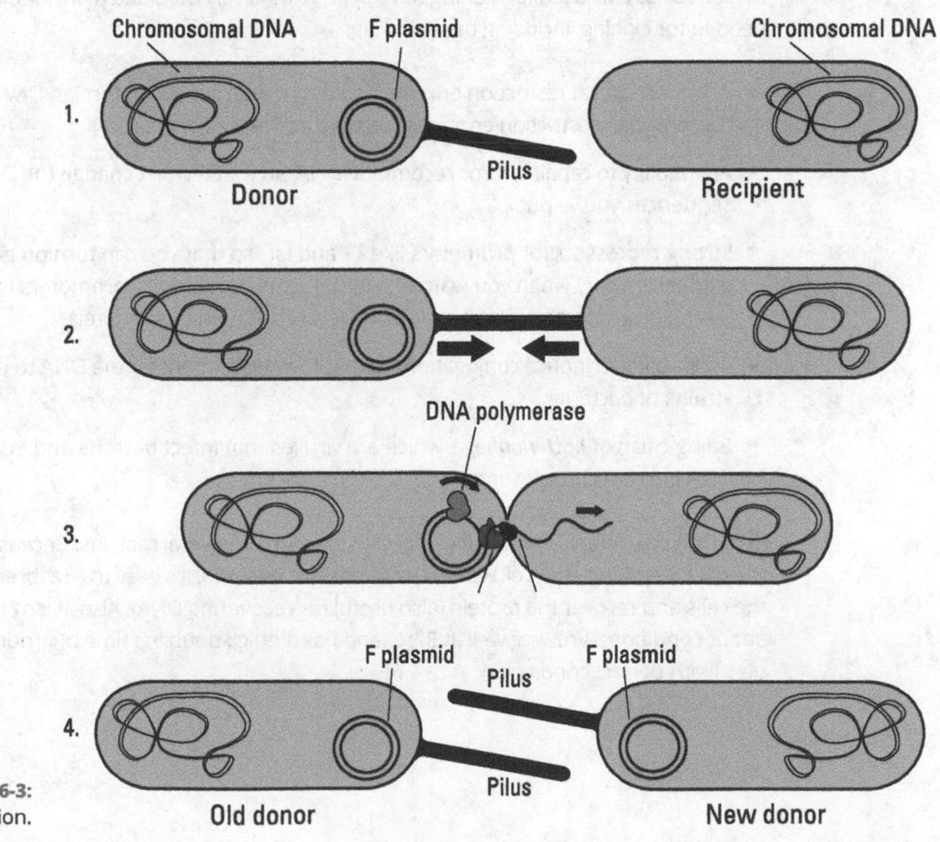

FIGURE 16-3: Conjugation.

Yeast cells can be transformed with DNA in ways that are similar to bacteria, such as using electroporation or methods involving calcium phosphate. When DNA is introduced into mammalian cells, the process is called *transfection* instead of transformation. For these eukaryotic cells that aren't so easily transfected, creative methods have been used to get the foreign DNA into the cells, like using viruses, shooting gold particles at them, or using magnets.

E. COLI: THE WORKHORSE OF MOLECULAR BIOTECHNOLOGY

Escherichia coli, otherwise known as *E. coli,* is a Gram-negative bacterium found in the gastrointestinal tract of animals and humans. Because it's easy to grow, it became the model organism for bacterial research on cell structure and function, as well as genetics. Because *E. coli* is Gram-negative, it's easily transformed to take up foreign DNA such as plasmids, allowing it to be used for recombinant DNA technology and protein work.

Most *E coli* strains used for cloning have been extensively modified to include traits needed for cloning, including the following:

- A lack of natural restriction enzymes, so that it won't chew up foreign DNA (see the "Cutting with restriction enzymes" section, earlier in this chapter)
- An inability to repair and/or recombine DNA, so that it won't change the DNA sequences you've put in
- Strong repressors for promoters like T7 and lac, so that you can turn on gene expression only when you want to, which means fewer leaky promoters (see the "Using promoters to drive expression" section, later in this chapter)
- An inability to induce conjugation, so that they won't transfer the DNA to other strains of bacteria
- Being clean of *bacteriophage,* which are viruses that infect bacteria and insert their DNA into the chromosome

E. coli is also useful for producing a foreign protein for several reasons. Because it lacks the thick peptidoglycan cell wall of Gram-positive bacteria, it's easy to *lyse* (break open) the cells and recover the protein (also useful for recovering DNA). Also, it isn't too picky about conditions, growing well at 37°C, and has a quick doubling time of around 20 minutes (with perfect conditions).

Using promoters to drive expression

A *promoter* is a region of DNA that contains sequences where the enzyme RNA polymerase and transcription factors bind to initiate transcription (see Chapter 6).

REMEMBER

Promoters are used to drive protein expression.

Two aspects of promoters are important to their function in the cell:

- **They can be strong and induce the production of a lot of protein, or they can be weak, producing less protein.** You want strong promotion to get the most protein possible, but sometimes when the protein product is toxic to the host cell or is very metabolically expensive to make, a weak promoter has to be used.
- **They can be either *inducible* (meaning that they're turned on only when needed, so as not to create a heavy metabolic burden on the cell) or *constitutive* (meaning that the cells continuously express the protein).**

Promoters also contain regions where repressors or inducers can bind. *Repressors* are molecules that interact with the DNA and/or the transcriptional machinery (RNA polymerase and transcription factors) and repress (turn off) transcription. When promoters are hard to repress, often because there are fewer repressor molecules than copies of the promoter, they're said to be leaky because gene expression happens despite the repressor being present.

Making use of expression vectors

Expression vectors are plasmids that have been modified specifically to help the genetically engineered cell make functional protein, which is usually harvested from the cells. Ahead of the actual protein coding sequence is the ribosome binding site (RBS) where the machinery that will make a growing peptide chain binds to the mRNA after it's made. Strong binding of the host cell's ribosomes to the transcript is important — it ensures that the protein will be made efficiently. Alterations in this sequence are especially important when using prokaryotic cells (like *E. coli*) to express eukaryotic proteins (like those from humans), because the RBS differs between the two organisms.

Another way to ensure that the protein of interest is efficiently made is to make sure that there isn't a codon bias in your DNA construct. *Codons* are the three base codes used to signal each amino acid during protein translation. They're the sites at which a transfer RNA, or *tRNA*, binds to add a specific amino acid to the growing peptide chain during protein synthesis. More than one codon — and hence, more than one specific tRNA — can code for the same amino acid, but most organisms preferentially use some over others.

When a protein with one codon preference is expressed in a cell with a different codon preference, it can create a shortage of tRNAs and slow down protein synthesis. Likewise, some codons are rare in many bacteria, so expression of proteins with these codons in *E. coli* creates bottlenecks and is too slow.

There are a couple of solutions to these problems:

- You can use strains of *E. coli* containing tRNAs for these rare codons.
- You can alter the codon usage of the DNA sequence so that it matches the host more closely.

Making sure that the final protein will be stable is important. Some amino acids, or combination of amino acids, create sites that are vulnerable to degradation by *proteases* (enzymes that degrade proteins). This is how proteins are naturally degraded in the cell under normal conditions. When a heterologous protein is being expressed in bacteria, however, it's good to have the least amount of degradation possible to maximize yields. A way of improving this is to change the amino acids to others that are unlikely to cause the same problem. This is easier said than done — the amino acid changes have to be chosen so as not to alter the protein's function.

Properly folding proteins

You can imagine that you've carefully chosen the protein, cleverly tweaked it to be more stable, and put it into an expression vector with the right promoter. Now *E. coli* is conveniently making a bunch of the protein for you in a test tube. But when you try to purify the protein, you find that most of it isn't functional and it's stuck in an insoluble fraction containing *inclusion bodies*. These are deposits of misfolded protein either in the cytoplasm or the periplasmic space. Foreign proteins don't fold properly for a variety of reasons but mainly because their disulfide bonds haven't been formed.

Disulfide bonds are bonds between the sulfur-containing thiol groups of cysteine residues that give protein their structure.

Gram-negative bacteria are not great at making disulfide bonds, so proteins that need a lot of them won't fold right and will end up in inclusion bodies. A few strategies have been tried to improve protein folding, but most have had limited success. The best ones include

- Overexpressing disulfide bond forming proteins along with the protein of interest.
- Lowering the temperature at which protein expression is turned on.

Being mindful of metabolic load

Sometimes a genetically engineered bacterial strain will not grow as quickly as you would expect and can even lose the gene(s) of interest despite selective pressure. This happens because the foreign DNA creates a metabolic change in the host cell, which can be detrimental to the host cell.

The amount of detrimental impact of the foreign gene(s) on the host is called *metabolic load.* When the metabolic load gets to be too high, the cells tend to not grow very well. Metabolic load can arise for several reasons:

- A high plasmid copy number or large plasmid size can be expensive for the host cell to maintain.
- An overproduction of a foreign protein can use up important or rare tRNAs and/or amino acids.
- Sometimes foreign products like enzymes have activities that interfere with host cell pathways by using up needed intermediates or by creating compounds that are toxic to the host.
- If the increased rate of metabolism of regular cellular processes plus foreign protein can't be sustained by the level of oxygen in the media, the cells won't grow well.

Along with decreased growth rates, other effects of metabolic load include the slowing down of other processes in the cell that need a lot of energy (for instance, nitrogen fixation), an increase in the number of errors made during translation, and the increase in secretion of polysaccharide substances making the cells much stickier than usual.

Making long, multi-gene constructs

The traditional ligation-based method for cloning in *E. coli* is useful for assembling relatively short sequences, but when the construct is very large with a size exceeding 20 kilobases (kb), difficulty amplifying and a lack of unique restriction sites creates insurmountable challenges. To overcome the 20 kb barrier, a number of strategies have been developed recently, all of which rely on the ability of a DNA molecule to pair with and combine with another piece of DNA with a similar sequence in a process called *homologous recombination* (see Figure 16-4).

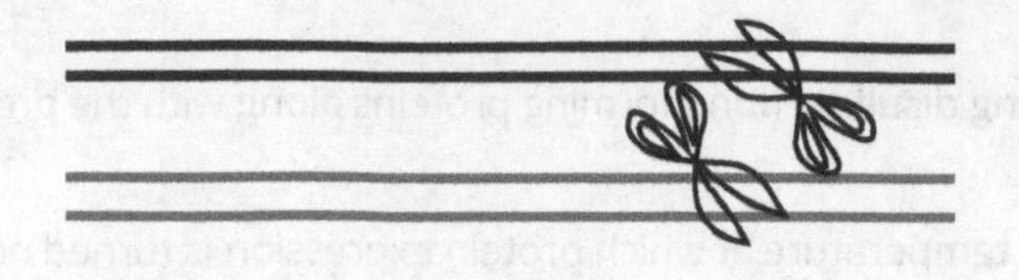

Two similar DNA strands are nicked.

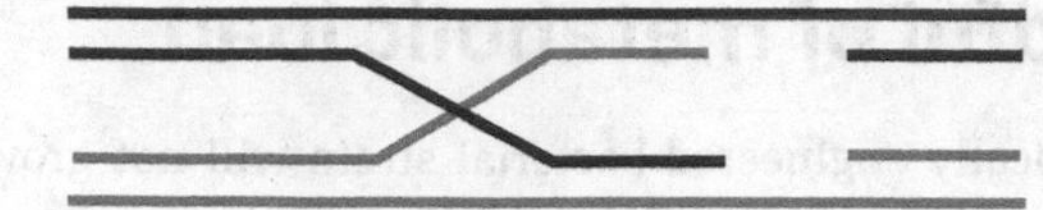

The similar strands base pair with each other in a process called crossing over.

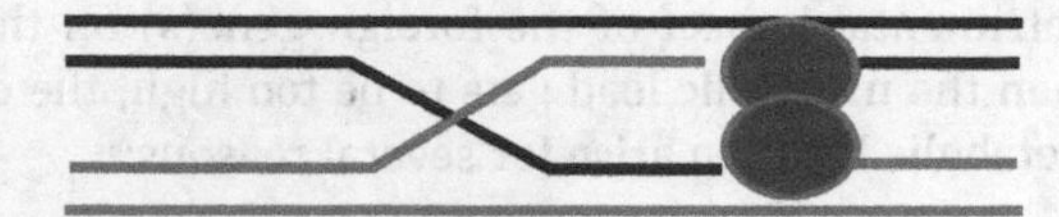

The nicks are sealed by DNA ligase.

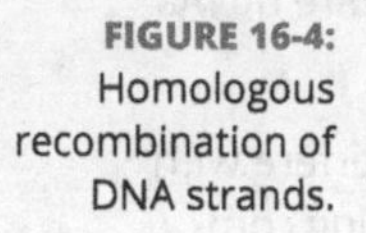

Recombinant DNA molecules are the final result.

FIGURE 16-4: Homologous recombination of DNA strands.

Recombination-based assembly (the construction of long pieces of DNA with homologous recombination) was used in 2007 to reconstruct both mitochondria and chloroplast genomes in the bacterium *Bacillus subtilis* and in 2010 to create a completely synthetic genome based on the bacterium *Mycoplasma.* This method involved using the yeast *Saccharomyces cerevisiae* to perform DNA recombination of over a thousand fragments of synthetic DNA in a process called *transformation-associated recombination* (TAR), shown in Figure 16-5.

Each 1 kb fragment was first chemically synthesized and designed to contain regions at each end that overlapped sequences to either side of it in the genome. When two or more such pieces of DNA are taken up by the yeast cell, they're naturally paired up, using these overlapping regions, and then combined using homologous recombination. Pools of ten sequences each were put into yeast and combined to make longer fragments (10 × 1 kb = 10 kb), which were then put into yeast again to make longer fragments. In three rounds, the complete 1 mega base (Mb) genome of the *Mycobacterium* was constructed.

FIGURE 16-5: Assembly of long DNA constructs by transformation-associated recombination (TAR).

REMEMBER

The advantages of this method can't be overstated. Making a DNA construct of this size with only restriction enzyme digestion and ligation would never have been possible — but thanks to yeast cells and homologous recombination, complex gene constructs can be assembled easily.

YEAST: THE NEW WORKHORSE OF MOLECULAR BIOTECHNOLOGY

Saccharomyces cerevisiae is not a newcomer to genetic engineering or research labs, but it is starting to gain fame from its unsurpassed ability to assemble recombinant sequences *in vivo* (in the cell). As a eukaryotic cell, yeast has for years been used:

- To study aspects of eukaryotic cellular biology and physiology
- In the food and beer industries and in biotechnology to produce proteins of interest
- As a model to study cancer

Like all eukaryotic cells, yeast carries its genetic information in chromosomes, but it can also use DNA in circular, plasmidlike molecules called yeast artificial chromosomes (YACs). Problems with instability of YACs have recently been overcome, so many of the cloning steps that used to be done exclusively in *E coli* are now being accomplished more easily in yeast.

Providing Therapies

Pharmaceuticals (such as insulin, interferon, and anticoagulants), hormones (like human growth hormone, epidermal growth factor, and fibroblast growth factor), and anti-cancer drugs are all made through microbial biotechnology. In many cases, only a tiny amount of these drugs could be extracted from animal tissues. Some of these therapeutic agents used to be made from biological systems, usually protein based, but they were expensive and hard to make. This meant that patients often didn't have access to them. Now many of these products are commonly used to treat people with diabetes, people with infectious diseases or cancer, and stroke patients, among others. A few examples are given in the this section.

Improving antibiotics

Many of the antibiotics that are used every day have been developed and produced through biotechnology. Although the discovery of naturally occurring antibiotics has slowed, attempts to improve current antibiotics have continued. One example of this is the attempt to design new modes of action for current antibiotics by genetically engineering changes in known steps of antibiotic synthesis genes.

Manipulation of the expression of these genes in organisms such as *Streptomyces* has been used to try to improve antibiotic yield. *Streptomyces* are a genus of bacteria that is commonly found in soil. They're an invaluable group of bacteria because they produce a lot of *bioactive compounds*, which are those that have effects on bacteria (as well as other organisms). More than 50 different antibiotics have been isolated from this group of bacteria.

One obstacle to high yield of antibiotics from strains of bacteria such as *Streptomyces* has been a limit to the amount of dissolved oxygen in the culture media. Antibiotics are complex to make, and producing large quantities of them requires more oxygen than can easily be dissolved in liquid media. One solution to this problem has been to engineer strains of *Streptomyces* with genes for a bacterial heme that can bind oxygen and deliver it to the growing cells.

Another problem is that organisms that produce antibiotics are often slow growing. A solution to this problem has been to identify all the genes involved in the biosynthesis of a particular antibiotic and engineer *E. coli* to produce the antibiotic of interest, since *E. coli* grows quickly in culture.

Developing vaccines

Current vaccines are produced in two ways: by culturing the pathogenic agent in a lab and then either killing it (killed vaccines) or inactivating its ability to cause disease (attenuated vaccines). These methods of vaccine production have some drawbacks:

- It's expensive.
- It produces low yields.
- It's dangerous to the people making the vaccine.
- It requires that the vaccine be tested rigorously to ensure that no live agent has contaminated the final vaccine.
- The vaccine has to be refrigerated once made.

Researchers have come up with several ways to improve vaccines:

- Deleting the virulence genes, ensuring that the strain won't revert and become virulent again.
- Using a benign microbe to express the bits of the pathogen that can illicit an immune response, called the *antigenic determinants.* This way the body makes a robust immune response and there is no risk of spreading the disease.
- Expressing the antigenic determinants in bulk in a fast-growing lab strain of bacteria like *E. coli* and then recovering them and using them in a vaccine.

The last two methods are especially useful for pathogens that can't be grown in culture and for which there are no current vaccines. A number of vaccines have been developed based on these ideas, but for the most part they're used in veterinary medicine. There is a real need to produce cost-effective and easily distributed vaccines for people in the developing world. Recently, there has been a resurgence in research in this area thanks in part to grants by the Bill & Melinda Gates Foundation.

Using Microbes Industrially

Many industries used to pump out waste products with a high carbohydrate load. Now that we know that the environment can't totally absorb all our waste, industries have had to find ways of recycling or disposing of high-carbohydrate waste, called *biomass.* Biomass has the potential to be used in many industrial processes as *feedstock* (the nutrient input) to fuel the microorganisms that are producing a product of interest. An example of this is the use of *whey* (the watery waste from the cheese industry) as input into the production of xanthan gum, which is used as a stabilizer in cosmetic products.

Commercial microbial production is called *industrial fermentation*, even when fermentation in the biochemical sense is not part of the process. These types of reactions happen in large vessels called *fermenters*, which can often hold more than 3,500 liters. The following things are important to successful industrial fermentation:

- Inside a fermenter the conditions are rigidly controlled to be optimal for the microbes so that they produce the highest yield of products.
- The input media has to be cheap, so it's often the waste from something else, like whey from cheese production.
- The microbial strains used have to be unstressed, healthy, free of pathogens (like viruses), and good at producing the product of interest. A lot of research goes into developing better microbial strains.
- The fermenter has to be protected from contamination with pathogens or competing microorganisms.
- Products should be easy to extract.
- The product has to be safe and free of bacterial substances that could be dangerous for the consumer.

Examples of food and alcohol fermentation include bread, cheese, yogurt, sauerkraut, pickles, beer, wine, and spirits. Food additives such as amino acids and thickening agents are produced from microorganisms, as well as organic acids and industrial alcohol.

Many of the additional steps involved in industrial fermentation are performed by enzymes derived from microbial sources. Typically, these enzymes are from genetically engineered strains. In the brewing industry, for instance, the yeast used to make the beer is rarely genetically engineered, but recombinant enzymes are used for other steps such as malting and post-fermentation steps like clarifying.

Protecting plants with microbial insecticides

During its spore-forming phase, the bacterium *Bacillus thuringiensis* (Bt) makes a group of proteins that are toxic to insects called the Cry proteins. *Cry proteins* are formed as crystals that, when ingested by certain insect larvae, are converted into the active toxin by the stomach pH and enzymes. The mature Cry toxin inserts itself into the cellular membrane of epithelial cells lining the intestinal wall of the insect, causing cell death and eventually dehydration as the intestinal cells are destroyed.

The bacterial Cry proteins have been engineered into crop plants to give them protection against insect pests. These transgenic plants are widely used in agriculture. Depending on the type of promoter used, the Cry proteins can be expressed either in all plant parts or only in the parts that are most affected by the insect. Target insects feeding on any part of the plant expressing the Cry genes will be exposed to the toxin. Other commercial preparations of the Cry proteins, used by organic farmers, include purified protein or Bt spores that can be applied to crops in the place of chemical pesticides.

WARNING

As with all other pesticides, insects have managed to form resistance to the Cry toxins, with at least five species of insect known to be insensitive to them. This serves as a warning about overuse of an inhibiting compound, which if put into the environment in large amounts will inevitably cause the development of resistance to it in the target organism.

Making biofuels

Biofuels are currently produced as alternatives to fossil fuels. To make biofuel, carbohydrates are fermented anaerobically by yeast to form ethanol. Currently, the yeast *Saccharomyces cerevisiae* is used to make biofuel from the carbohydrates derived from corn wheat and sugarcane. For the most part, the carbohydrates used come from the sugary and starchy parts of the plant, leaving the bulk of the plant biomass unused. This is because the rest of the plant (for instance, stalks and leaves) contain compounds called lignin and cellulose, which require many different enzymes (from many different microbes) to break down. It makes sense that these compounds would be resistant to microbial digestion — they provide the rigid support to plant structures in nature.

WARNING

As an alternative to fossil fuels, current methods for producing biofuels are not perfect. The energy input for crop production, the impact on food prices and greenhouse gas emission from farming all decrease the payoffs (renewable resource, less greenhouse gas emissions, and less energy put in).

In an effort to improve the energy yield of the process, an alternative microbe has been proposed for producing ethanol. One of the problems is that yeast uses some of the energy it gets from fermentation to grow, creating more microbial biomass. Ethanol is, after all, a waste product, so in nature microbes would want to limit the amount of energy lost as waste. Some microbes, however, like the bacterium *Zymomonas*, ferment sugars through a different pathway that produces less microbial biomass from sugars, releasing more ethanol in the process.

Another improvement is the use of more complex carbohydrate sources as input to the system because breaking down complex carbohydrates either chemically or enzymatically before feeding them into the fermentation process only makes the process less efficient. For this, genetic engineering of additional enzymatic pathways into both *Zymomonas* and *E. coli* has been used. Improvements to this process have been developed in the lab but aren't currently used in industry.

Bioleaching metals

Acidithiobacillus and *Leptospirillum* bacterium can clean up abandoned mine sites. The traditional way of extracting metals from rocks has been through smelting processes. These processed have huge environmental impacts in terms of the release of carbon dioxide, sulfur dioxide, arsenic, and mercury. Smelting is also pretty inefficient. It releases large quantities of waste called *slag* that still contain metals but at a concentration too low to be extracted economically. *Bioleaching* (the extraction of metals from rocks by microorganisms) is currently used as an alternative in about 20 percent of copper mines. Microbes can extract metals that are at lower concentrations and from types of rock that are difficult to smelt. They can also remove contaminating arsenic and convert it to a form that won't leach into waste water or contaminate the final product.

Bioleaching is overall less expensive and requires less infrastructure than smelting, making it more economical and environmentally friendly. The way it works is that a solution of bacteria is percolated through a slag pile, removing metals as it goes. When the solution drains out, the copper is extracted from the liquid through electrolysis. All the microbes known to do this are acid tolerant and some are *thermophilic* (they like to grow at temperatures above 45°C) members of the *Acidithiobacillus* group. Some are iron-oxidizing members of the *Leptospirillum* group of bacteria.

Microbes used in bioleaching to mine metals can also be used to clean up contamination in mine sites — specifically, uranium-contaminated sites, where the radioactive uranium has to be stabilized.

Cleaning up with microbes

Bioremediation is the degradation of organic contaminants by microorganisms. Contaminants from chemical spills, industrial waste, or other waste include

- Hydrocarbons
- Polycyclic aromatic hydrocarbons (PAHs)
- Polychlorinated biphenyls (PCBs)
- Hazardous material such as explosives

- Xenobiotics such as pesticides, herbicides, refrigerants, and solvents
- Inorganic compounds such as cyanide, ammonia, nitrates, and sulfates

There are two ways of inducing bioremediation to occur:

- **Bioaugmentation:** The addition of bacteria to a site where they don't naturally occur in order to induce breakdown of a contaminant
- **Biostimulation:** The addition of nutrients or growth stimulants to a site thought to already contain the bacteria needed

Stimulants can include adding or removing oxygen, water, nutrients, or electron donors/acceptors. Bioremediation can be done at the site, or the contaminated material can be removed and treated elsewhere and then returned to the site afterward.

Many soil bacteria from the genus *Pseudomonas* are able to degrade contaminants. The genes are often located on plasmids. One organism can sometimes use a wide range of compounds as its sole carbon source. Organic contaminants fall into a couple of groups based on whether they have aromatic structures and/or halogen atoms (see Figure 16-6). These features make them more difficult to degrade and more persistent in the environment.

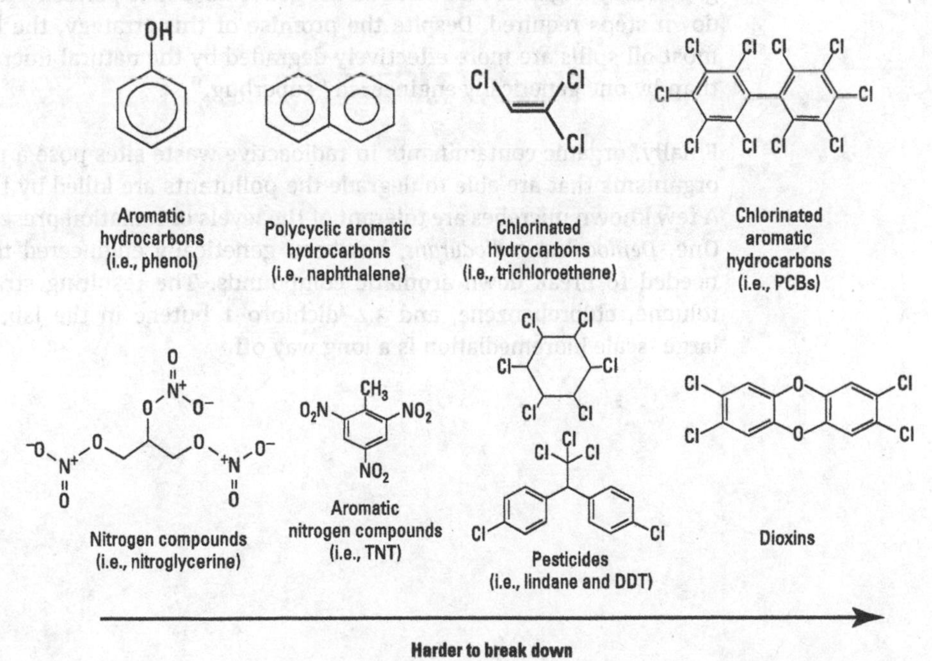

FIGURE 16-6: Classes of environmental contaminants in order of ease of breakdown.

Nonhalogenated compounds are for the most part all converted to catechol, which is cleaved to common compounds that all microbes can use.

Halogenated compounds are more difficult to break down. For each halogen atom, the rate of degradation is cut in half. In order to be degraded fully, the compound has to have the halogen group(s) removed in a process called reductive dehalogenation that has only been found in some anaerobic syntrophic communities that have yet to be completely defined. The microbes in these communities are *syntrophs* — that they depend on one other to complete redox reactions. For this reason, halogenated pesticides like dithiothreitol (DTT) that were widely used in the 1940s persist to this day in the environment, four decades after they were banned in North America.

Strategies to allow bioremediation of contaminated sites sometimes involve using a mix of microbes that can work together to perform all steps of the degradative process. Drawbacks of this approach are that contaminated sites often contain a mix of pollutants, some of which can inhibit the growth of some of the microbes in the mix. Another problem is that although a microorganism may have the pathways needed to break down a compound, it may not perform well under the environmental conditions at the contaminated site.

To overcome these problems, a strategy has been developed to choose one microbe, or a small set of microbes, known to survive well at a site of contamination and genetically engineer into them all the genes needed to perform the various breakdown steps required. Despite the promise of this strategy, the hydrocarbons in most oil spills are more effectively degraded by the natural microflora in the sea than by one genetically engineered "superbug."

Finally, organic contaminants in radioactive waste sites pose a problem because organisms that are able to degrade the pollutants are killed by the radioactivity. A few known microbes are tolerant of the levels of radiation present at these sites. One, *Deinococcus radiodurans,* has been genetically engineered to express genes needed to break down aromatic compounds. The resulting strain can degrade toluene, chlorobenzene, and 3,4-dichloro-1-butene in the lab, but its use for large-scale bioremediation is a long way off.

IN THIS CHAPTER

» Keeping an eye on public health

» Identifying a microbial pathogen

» Protecting ourselves with vaccines

Chapter 17

Fighting Microbial Diseases

In previous chapters, we discuss the variety of microbial life on earth, how we study microbes, and the role of microbes in the environment. The truth is, microbes would probably get much less attention if some of them didn't cause serious human and animal diseases. The proportion of all microorganisms on earth that make us sick is minuscule, but their impact has shaped human history and driven scientific discovery like almost nothing else.

Protecting Public Health: Epidemiology

Over the past century, there has been a drastic reduction in the burden of infectious disease. This is due to a combination of factors, including the use of vaccinations to prevent infections, the use of antibiotics to treat infections, overall improved nutrition and living standards, and water and food safety regulations.

The scientists who monitor the trends of infectious disease and the interventions that protect against them fall in the field of public health. Public health incorporates information from a range of fields — including science, medicine, sociology, statistics, politics, and education — to improve the health of an entire population.

Epidemiologists are central to the field of public health and are experts in identifying, tracking, and stopping outbreaks.

Tracking diseases

Epidemiologists collect information called *surveillance data* on the health of a population on a regular basis. Outbreaks of infectious disease are sporadic, usually unpredictable, and can sometimes spread rapidly through a population, so it's important that epidemiologists ask questions about the general trends of a particular infectious disease so that they can identify when something goes awry.

Several pieces of information help epidemiologists describe the pattern of an infectious disease:

- **Incidence of a disease:** The number of new cases in a population in a given time period
- **Prevalence of a disease:** The total number of new and existing cases in a population in a given time period
- **Reservoirs:** Sources of infection, such as animals, soil, or food
- **Carriers:** Individuals who carry a pathogen without showing signs of an infection
- **Morbidity:** The incidence of a disease in population that includes both fatal and nonfatal cases
- **Mortality:** The incidence of death in a population

Collecting surveillance data for each type of disease is important, because their occurrence, distribution, and impact on the population can vary quite a bit. For example, one case of the flu is not sufficient to sound the alarm, but one case of Ebola is an immediate threat.

Investigating outbreaks

Infectious disease epidemiologists are like detectives. When their data indicate that an outbreak could be occurring, they begin to investigate the cases of people infected. They look for commonalities in people's food consumption and social interactions to help identify the source of the infection and how it's transmitted. With these clues, they decide on the best way to intervene and reduce the spread of the infection.

HOSPITAL-ASSOCIATED INFECTIONS

Ironically, many outbreaks originate from hospitals. This is partly because hospital patients often have compromised immune systems and are at higher risk of getting an infection, and partly because healthcare personnel are in frequent contact with many patients and unintentionally transmit infections. Around 5 percent of patients admitted to a hospital will get an infection while there. Hospital-associated infections are called *nosocomial infections.* Common pathogens include *E. coli,* which causes urinary tract infections, *S. aureus,* which causes bloodstream infections, and *C. difficile,* which causes diarrhea. Hospitals take preventive measures to reduce transmission of infections by encouraging hand washing, isolating patients who show symptoms of an infectious illness, and thoroughly cleaning surfaces and surgical equipment.

Outbreaks are classified into three main categories:

- **Epidemic:** An outbreak is considered epidemic when an unusually high number of people are infected in a population.
- **Pandemic:** A pandemic is similar to an epidemic, but it indicates that cases are widespread around the world.
- **Endemic:** An endemic disease is always present in the population in low levels.

Catching the source

When pathogens aren't causing an infection, they reside in *reservoirs* (animals or inanimate objects that allow them to stay viable until they contact a susceptible host). Animals are common reservoirs of infections that can be transmitted to humans. Infections transmitted from animals to humans are considered *zoonotic.* For example, cows are a reservoir of a dangerous form of *E. coli* known as O157:H7. If cow manure containing O157:H7 contaminates soil or produce on a farm, the humans who consume that food are at risk of infection.

People who carry an infection without showing any symptoms, called *carriers,* are another major source of infections. Because they don't necessarily know they carry a pathogen, they don't take precaution against spreading it to susceptible individuals. Carriage can be short (such as unknowingly contracting a mild respiratory virus) or continuous (such as colonization for an extended period of time). Some examples of diseases where carriers spread infection include *Staphylococcus aureus* infections, hepatitis, tuberculosis, and typhoid fever.

Staying on top of transmission

Pathogen survival in a population depends on its transmission. To be successful at transmission, a pathogen must escape one host, travel to find another host, and then enter that host. Although it doesn't sound like a difficult task, many pathogens are used to the warm and moist environment of the body, and the outside environment can be a harsh place to survive.

Direct host-to-host transmission occurs when a pathogen does not pass through an intermediate. Some infections, like colds or the flu, are transmitted in air droplets after infected individuals sneeze or cough. Other infections, such as syphilis or HIV, are transmitted through exchange of bodily fluid or sexual intercourse.

Infections can also be passed indirectly through an intermediate. Intermediates can be inanimate objects (referred to as a *fomites*), including bedding, toys, or surfaces, or living things (referred to as *vectors*), such as insects or animals. Vectors of disease are organisms that pick up a pathogen without becoming infected and transfer it from host to host. Mosquitos are common culprits of indirect transmission of diseases, such as malaria.

Gaining control

Epidemiologists can implement several strategies to help stop an outbreak:

- Control of vehicles of infection such as water or food (for example, through water purification or milk pasteurization)
- Reduction of interaction with reservoirs of disease (for example, elimination of infected animals/insects)
- Quarantine and isolation of infected individuals
- Vaccination

Vaccination is the ideal method to reduce the spread of disease. The *coverage rate* (the number of people vaccinated in a population) does not necessarily have to reach 100 percent for a vaccine to be effective. A successful coverage rate is one that provides *herd immunity* in a population. Herd immunity occurs when a sufficient amount of the population is immune to a disease such that the whole population is protected. For example, the success of the measles vaccines is attributable to herd immunity.

In some cases, it's beneficial to reach 100 percent coverage, because this can lead to pathogen eradication. It's possible to eradicate an infectious disease if the only reservoirs of infections are humans. Such was the case for polio, a virus that infects children and causes paralysis. Jonas Salk invented the polio vaccine in the 1950s and, since then, global vaccination efforts have led to 99 percent eradication of polio, but efforts are still under way to eradicate polio completely.

USING DDT TO ELIMINATE INSECT-BORNE DISEASE

Dichloro-diphenyl-trichloroethane, commonly known as DDT, is an insecticide developed in the 1940s to eliminate insects that carry human diseases, such as malaria, typhus, and West Nile virus. In a global concerted effort in 1955, it was used in an attempt to eradicate malaria. Although it was a successful means of reducing mosquito populations and controlling malaria, its overuse led to the emergence of insecticide-resistant species. Scientists also found that DDT had a negative consequence on animal ecology and human health. It was banned in the United States in 1972, although it's still used today in developing countries where malaria is endemic, although its use is regulated. The benefits of protecting individuals against a deadly disease are thought to outweigh the health risks associated with DDT use.

Identifying a Microbial Pathogen

Many pathogens cause infections that share similar symptoms, such as a fever, pneumonia, or diarrhea. This makes it difficult to diagnose an infection based on a patient's symptoms. Clinical microbiologists have the task of identifying the causative microorganism in order to inform the doctor of how to treat the infection. The field of diagnostic microbiology uses genetic, biochemical, and immunological tests to narrow down the list of possible causative pathogens.

Traditional diagnostic tests first require culturing of the microorganism from infected tissue. This can take a few days — since some pathogens are slow growing or require very specific growth conditions. Although the traditional methods are effective, rapid identification is important for severe life-threatening infections. For example, the risk of death from septic shock increases 7 percent for every hour that passes. New technologies that allow rapid pathogen identification are making their way into diagnostic labs, making clinical microbiology a mixing pot of new and old tricks.

Characterizing morphology

A staple technique in every microbiology lab is the Gram stain (see Chapter 7). In the Gram stain, bacteria are categorized into two groups based on the results of the stain: Gram-negative or Gram-positive. The Gram stain also reveals the cell shape — for example, whether the cells are rod shaped or coccoid. From here, a clinical microbiologist has enough information to decide on the appropriate biochemical tests to use.

Mycobacterium, a group of important human pathogens, won't take up the crystal violet dye used in Gram staining because the cell wall contains high amounts of lipids. Instead, they're stained with a carbol-fuchsin dye that is red; they don't let go of the color when washed with acid, making them acid-fast positive.

Figure 17-1 gives an example of how to narrow down the identity of a bacterium based on staining and morphology.

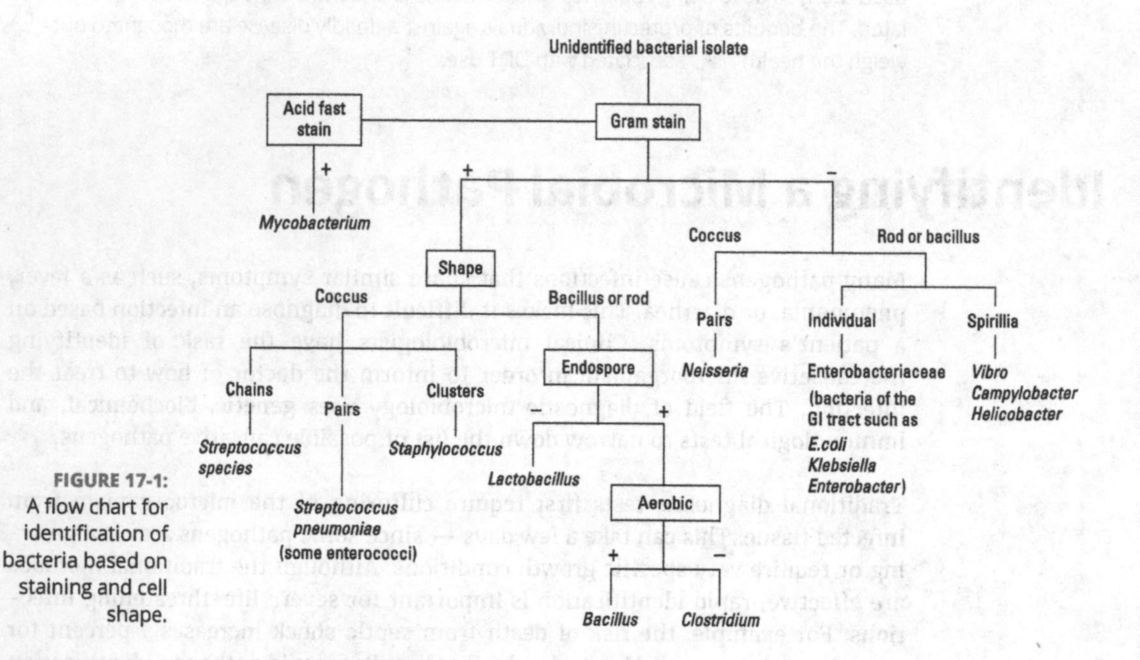

FIGURE 17-1: A flow chart for identification of bacteria based on staining and cell shape.

Using biochemical tests

The principle behind biochemical tests it that different species differ in metabolic capacities or enzymatic activities that allow them to be differentiated. These tests (outlined in Table 17-1 and Table 17-2) act as a process of elimination to reach a final conclusion of the pathogen's identity. There are many other tests but here are a few that illustrate the principle. Figure 17-2 is a flow chart for how these tests are run.

TABLE 17-1 Tests Used to Differentiate Gram-Positive Bacteria

Test	How It Works
Catalase test	Detects catalase enzymes, which break down hydrogen peroxide into water and oxygen gas.
Coagulase test	Detects coagulase, a virulence factor that clots blood plasma.
Mannitol salt agar	Selects for microbes that can grow at high concentrations of salt. pH indicator differentiates bacteria that are able to ferment mannitol (+ = yellow, – = pink).
Blood agar	Detects hemolysins, virulence factors that lyse red blood cells. Three hemolysis patterns can occur: complete hemolysis (beta), partial hemolysis (alpha), or no hemolysis (gamma).
Optochin sensitivity	Differential test that distinguishes microbes susceptible to optochin.
Bacitracin sensitivity	Differential test that distinguishes microbes susceptible to bacitracin.
CAMP test	CAMP factor is a protein that enhances the hemolysis of *S. aureus* on blood agar. It is also indicative of *S. pyogenes*.
Bile esculin agar	Selects for microbes that can grow in the presence of bile. Differentiates between microbes that can hydrolyze esculin.
Starch hydrolysis test	Differentiates microbes that can hydrolyze starch.
Motility agar	An agar tube is stabbed with the microbes. Diffusion away from the stab indicates the microbe is motile.

TABLE 17-2 Tests Used to Differentiate Gram-Negative Bacteria

Test	How It Works
Oxidase test	Identifies microbes that produce the cytochrome oxidase needed for respiration. If the enzyme is present, there is a change to a purple color.
Glucose broth with Durham tubes	Tests the ability of a microbe to ferment glucose and to convert pyruvic acid into gaseous byproducts.
MR/VP	Methyl red (MR) is a pH indicator that turns red if glucose is fermented into mixed acid products. If not, it turns yellow. The Voges-Proskauer test tests for production of acetoin, an intermediate in an alternative glucose fermentation pathway.
MacConkey agar	Selects for microbes that can grow in the presence of bile salts and crystal violet (inhibitory to Gram-positives). pH indicator differentiates microbes that can ferment lactose.
Urease test	pH indicator differentiates microbes that are able to hydrolyze urea.
Sulfur indole motility medium	Differentiates microbes that produce H_2S (black precipitate), convert tryptophan to indole (red indicator), and are motile.
Citrate agar	Differentiates microbes capable of using citrate as a sole carbon source. If it can, it produces a basic compound that turns the media from green to blue.

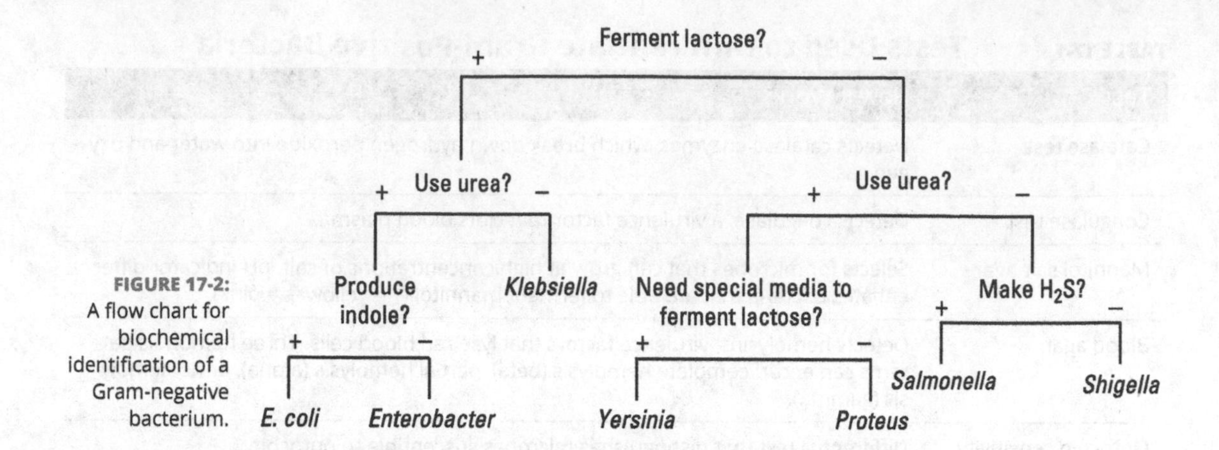

FIGURE 17-2: A flow chart for biochemical identification of a Gram-negative bacterium.

Typing strains with phage

Bacteriophage, often just called *phage*, are viruses that infect bacteria (see Chapter 14 for more information about these viruses). The fact that phage are very picky about their host has been used in diagnostic labs to identify which strain of a particular bacteria is present, a process called *phage typing*. In practice, enough of the unidentified bacterium in question is plated on solid media so that a thin layer of cells, called a *lawn of bacteria*, covers the entire plate. Onto this lawn are placed drops of bacteriophage that will infect and lyse their target bacteria (see Figure 17-3), creating clear circles where the bacteria have died.

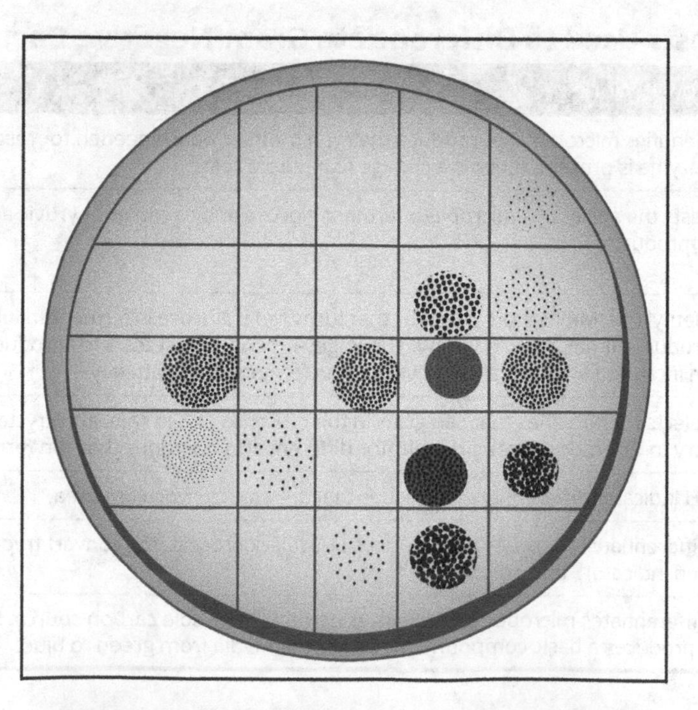

FIGURE 17-3: Phage typing of a bacterial strain. Clear areas are shown in black here.

SORTING THE STREPTOCOCCI BY LANCEFIELD GROUPS

The genus *Streptococcus* is made up of many different, closely related bacteria that have a range of impacts on human health, from innocuous colonization to nasty diseases like flesh-eating disease. The mystery behind this group of bacteria was that an astonishing array of diseases seemed to be caused by a group of bacteria that doctors couldn't tell apart.

Dr. Rebecca Lancefield's great knowledge about these bacteria, and about serology, enabled her to come up with a typing system for them. Using antisera against different *Strep* pathogens, Dr Lancefield divided them into groups and then worked on identifying the antigenic bacterial components — capsule polysaccharide and M protein — in order to subdivide them further.

Today the Lancefield system is essential to our understanding of the diseases caused by the streptococci. This also explained why Koch's postulates, stating that a single organism causes a single disease, didn't work to describe this group of bacteria — many streptococci cause several different diseases.

One bacterial strain can be vulnerable to more than one phage, but most strains can be distinguished from one another based on the combination of which phage kill it. Phage typing is useful for finding the source of a bacterial infection because samples from the site and a few other sources can be tested and then compared.

Using serology

Microorganisms are antigenic, meaning that they induce an immune response in a host that involves producing antibodies against the invader. Serology is the study of the serum of animals, and the antibodies present there and is a valuable tool for pathogen identification. The antibodies produced against an organism are very specific and can discern between closely related organisms, which are then called *serotypes* or *serovars*.

Serological testing involves the use of serum-containing antibodies produced in animals against known pathogens, called *antiserum*. Serological testing can be done in several ways:

- **Agglutination test:** Uses antisera, added to a sample of unknown bacteria, to test for the presence of a known pathogen. When the antisera contains the right antibodies, clumping of the bacteria, or agglutination, is visible.

- **Enzyme-linked immunosorbent assay (ELISA):** Uses antibodies to bind and then visualize the presence of an antigen in a sample. The antigen can be either bacterial or viral, so these methods have been used in the quick detection of known pathogens including viruses. The antibodies are attached to a solid surface, the sample containing the suspected pathogen is added, if present the pathogen binds to the antibodies, and the rest is washed away. Another antibody to the pathogen is tagged so that it either emits light or changes color, indicating that the pathogen was, in fact, present in the sample.
- **Western blotting:** Takes advantage of the same principals as ELISA, but the antigenic compound is bound to a matrix and the antibodies are added to it.

Testing antibiotic susceptibility

Knowing what organism is causing a disease, as well as the strain of that organism, is useful information. Most treatments are based solely on that information. However, it's often important to know the antibiotic sensitivity profile of an organism in order to treat the patient effectively. Also, a pathogen can develop antibiotic resistance over the course of treatment, which makes it necessary to test which other drugs can be used against it.

To do this, several different methods are used, although only the broth dilution test is truly an accurate measure of how much antibiotic is needed to kill a bacterial strain.

- **Disk diffusion assay:** Indicates the inhibitory effect of different concentrations of antibiotic. It works by first spreading a specific concentration of bacterial cells on solid media and then adding disks of antibiotics then giving the bacterial cells time to grow and form a lawn. The zone of clearing around each disk gives an indication of the effect of the antibiotic. This test is less precise than other tests because it relies on the ability of the antibiotic to diffuse through the media.
- **Minimum inhibitory concentration (MIC):** Can be calculated for each antibiotic in a similar way to the disk diffusion assay, except that a commercially produced strip is added to the plate; the strip has precise markings on it that indicate the concentration of the drug being delivered to the media.
- **Broth dilution test:** The preceding two tests only give an indication of the inhibitory effect of an antibiotic. To test the concentration of the drug needed to kill the bacterial strain, the minimum bactericidal concentration is calculated. This is done with the broth dilution test, where concentrations of antibiotic are added to tubes of culture media, which are then inoculated with the bacterial strain. The amount of growth of bacterial cells from each tube is monitored on culture media without any antibiotic. The broth dilution test is commercially available with wells of a microtiter plate loaded with the precise concentrations of antibiotics necessary.

MICROBIAL DISEASES: THE LAUNDRY LIST

An assortment of microbial diseases have afflicted humans throughout history and to this day. They range in severity from the minor inconvenience of a cold to the rapid deterioration and death that happens during sepsis. Individual species or strains of bacteria, viruses, fungi, and protozoans can be the causes of disease, each with their own method of transmission, target tissue in the body and disease symptoms.

We don't have space to catalog all known microbial diseases, but you can find a complete discussion of all microbial diseases at the following sites:

- **National Institute of Allergy and Infectious Diseases:** `www.niaid.nih.gov/topics/microbes/pages/diseases.aspx`
- **The Centers for Disease Control and Prevention:** `www.cdc.gov`

Understanding Vaccines

In the fight against infectious diseases, we have three main weapons: sanitation, antibiotics, and vaccines.

Although proper sanitation has gone a long way toward stemming the spread of some diseases, in the developing world this is still a struggle. Education about preventive practices has reduced the spread of communicable diseases, but prevention isn't always possible.

Antibiotics are very effective at killing or stopping the growth of bacterial infections, but they have several serious flaws. (See the article at `www.dummies.com/extras/microbiology` for a complete discussion of the problems with the overuse of antibiotics, as well as the sidebar in Chapter 20 and the section on superbugs in Chapter 15.) Bacteria can and eventually will develop resistance to most antibiotics, leading to the development of many antibiotic-resistant strains that are impossible to treat. Multi-drug resistant bacteria (superbugs) are a major concern even today, because some are no longer sensitive to any of the available antibiotics.

In the case of viral infections, preventive behaviors and vaccination are our only defense, because antibiotics are useless to treat viruses and some can't be treated at all once contracted. It's especially important for young children, the elderly, and people with a compromised immune system, in which vaccination is not effective, to not be exposed to infectious diseases because they're at the highest risk of both contracting a pathogen and having the most severe outcomes from an infection. For these reasons, it's important for everyone to be vaccinated, a process that limits the number of circulating pathogenic organisms in society, thereby protecting the vulnerable from exposure.

TIP

For a list of recommended vaccination schedules see www.cdc.gov/vaccines/schedules. For a list of vaccines needed if traveling outside North America, see www.nc.cdc.gov/travel/destinations/list.

Understanding how vaccines work

Pathogenic microorganisms are seen by the body because they're made of nucleic acids, proteins, and sugars, called *antigens*, which are foreign and provoke an immune response. The first time a pathogen is encountered, a primary immune response occurs that causes antibodies to be made against the antigens and memory cells to be formed. If the same pathogen is seen a second time, it provokes a quicker and much stronger immune response thanks to these memory cells.

REMEMBER

Epitopes are the specific structures on the antigen that are bound by antibodies. Antigens are made of protein or sugars and the parts of the molecules that are on the surface act as epitopes.

Vaccination takes advantage of this response to antigens by exposing a person to a harmless, or at least less harmful, version of a pathogen in order to induce protection from the real thing. Vaccination is extremely effective at reducing the number of deaths from a disease-causing microbe with the bonus that if enough people are protected, the amount of the virus or bacteria can no longer circulate in the population, eventually making vaccination against it unnecessary.

Two discoveries helped improve the safety and effectiveness of vaccines:

- **Adjuvants:** Adjuvants can boost the immune response, allowing both the use of smaller doses and the vaccination of people who normally only mount a weak response. Adjuvants were discovered by accident, but research into them found several compounds — namely, aluminum salts — that act to increase the primary immune response in ways that are not fully understood.
- **Similar but harmless relatives:** A harmless relative of a nasty pathogen can sometimes be similar enough that an immune response to it can provide protection against the pathogen itself. This was the case with one of the first-ever vaccines for smallpox based first on a similar virus that caused cowpox and later on the Vaccinia virus. Smallpox has been eradicated worldwide in large part due to the success of Vaccinia at infecting animal cells and causing a robust immune response that will protect against many of the pox viruses.

Ranking the types of vaccines

Although not the safest way, the most effective way to gain lifelong protection from a pathogen is to recover from an infection by it. For many pathogens, the mortality rate is very high, complications are unpleasant and sometimes permanent, and individuals at risk, like small children and the elderly, are disproportionately affected.

A safer way of protecting people from an infectious microbe is to give them a vaccine for it. There are several types of vaccines based on what parts of the pathogenic organism are used to stimulate an immune response. They vary in their effectiveness in providing long-term immunity.

- **Live attenuated vaccines:** The most effective vaccines are those that contain live attenuated virus or bacteria. They're called *attenuated* because their ability to cause disease has been disabled by either removing or disrupting one of their virulence genes. These vaccines produce immunity that lasts a lifetime. The risk that an attenuated pathogen will revert back to its virulent state is always there, making these types of vaccines riskier than the others.
- **Killed bacteria or viruses:** The immune system reacts to the pathogen but in a weaker way than it does to a live vaccine. Future exposures to the killed microbe in the form of boosters are required to keep immunity high.
- **Subunit vaccines:** Made by genetically engineering another, nonpathogenic microorganism, to trigger an immune response similar to what would happen with the real pathogen. Because the vaccine contains only one part of the original pathogen, a subunit vaccine gives a shorter period of protection than does the live vaccine.
- **Toxoids:** When our immune system encounters a bacterial toxin it produces antibodies that bind to it and inactivate it so that it can't bind to its target. Toxoids are inactive bacterial toxins that, when used as a vaccine, can provide protection again the bacteria that make them. Toxoids are not as effective as live vaccines.
- **Conjugated vaccines:** The immature immune system of young children can't respond strongly enough to some vaccines to give them protection from the pathogen later. For this reason, conjugated vaccines were developed that combine parts of the pathogen organism with an inactivated toxin protein. Because the immune system reacts more strongly to the toxin, it also remembers the pathogen by association.
- **DNA vaccines:** Work by letting muscle cells make pathogen proteins that the immune system then reacts to. Because there is no pathogen present to set up an infection, there is no risk of real illness. DNA is injected into muscle tissue where it's taken up by the nuclei of muscle cells that then transcribe and translate it.

WEIGHING THE RISK OF A VACCINE

The rôle of a vaccine is to protect the population from the disease in order to make the disease caused by the pathogen so rare that it's almost never seen. Because of their success, some vaccines now represent a higher risk to people than the diseases they were designed against. Examples of this include the polio vaccine and the rabies vaccine. In the case of rabies, only high-risk individuals and those who suspect that they have been exposed to the virus are vaccinated. The polio virus is transmitted through fecal contamination of drinking water and infects the throat and intestines moving to the central nervous system. There it attaches to peripheral neurons, moving up to attack and kill the motor neurons in the upper spinal cord. The oral vaccine, a live attenuated virus, was so effective that it eliminated polio and drove one of the serotypes to extinction. However, because it is live, it can revert to a virulent state in 1 in 750 000 doses, causing the disease again. Because the threat of the wild virus is now gone, this is an unacceptable threat to public health. As an alternative, a less-effective but completely safe killed polio vaccine is used in areas where it has been eradicated. Managing these risks is an important part of public health policy.

6 New Frontiers in Microbiology

IN THIS PART . . .

Identify the methods used to survey microbial communities and figure out what the members of each community are doing.

Learn about cutting-edge methods used to enhance classic microbiology, like DNA and RNA sequencing.

Get an introduction to synthetic biology, where the processes in bacteria and yeast are used like the parts of simple machines.

See how bacterial gene regulation signals can be used to drive the simple synthetic machines.

IN THIS CHAPTER

» **Becoming familiar with the study of microbes in their natural environments**

» **Getting creative with ways to see microbial communities**

» **Taking the mystery out of words ending in *-omics***

» **Counting the uncultivable**

Chapter 18 Teasing Apart Communities

Microbial ecology is the study of the interactions of microorganisms with one another, with other organisms, and with their environment. Because microbes are everywhere and are too small to see unaided, scientists have developed some clever methods for studying microbes in their natural environments, including soil, aquatic environments, the surfaces of plants, and even within humans.

Studying Microbial Communities

Microorganisms live in communities, in mixed populations with many different species coexisting together. How these communities interact affects the ecosystem and nutrient cycling on an even larger scale (see Chapter 11). To study these communities, scientists consider two things:

» **The biodiversity of microorganisms:** The number and kinds of species within the community

- **The activities of microorganisms:** Metabolic activities as well as activities involved in communication and competition

In this section, we explain how scientists study microbial communities.

Borrowing from ecology

The study of ecology focuses on all organisms and how they interact with their environments, not just microorganisms. The science of ecology began before microorganisms were really known to exist, but many of the main themes from ecology are applied to microbial communities as well. For instance:

- Both *abiotic* (nonliving) and biological processes have large impacts on nutrient cycles in nature.
- The evolution and genetics of the organisms in an ecosystem are important to the understanding of the ecology of that system.
- The pattern of how communities interact can't be known or predicted just by understanding each species alone.

Seeing what sets microbial communities apart from plants and animals

Communities of microbes are similar in many respects to communities of plants and animals, but they differ in two important ways:

- Microorganisms move around much more than plants and animals do, so they get to try out a lot of new niches.
- The rate of *speciation* (development of new species due to mutation) is much higher in microorganisms than it is in plants and animals, partly because of microorganisms' higher rates of mutation and because generation times can be faster than they are for larger organisms.

Both of these factors make the study of microbial ecology unique from macro ecology (the ecology of everything else).

Observing Communities: Microbial Ecology Methods

Microbial ecology is not new. Scientists have been studying bacteria in their natural environment since the 19th century. Some traditional techniques for studying microbes in the environment include enrichment culture and isolation of microbes. Advancements using new technologies, combined with important traditional methods, can be powerful ways of teasing apart microbial communities.

In this section, we cover some of the methods used to study communities. This isn't a complete list of every tool available — think of it more as a sampling of some fundamental tools.

Selecting something special with enrichment

Enrichment is when bacterial growth media contains specific supplements that are required for specific organisms to grow. These supplements vary from organism to organism to help the one of interest grow best. Scientists strive to re-create as best as they can the exact growth conditions of the bacteria as in nature such as temperature and mineral or nutrient supplements. Then the media is inoculated with a sample from the environment of interest. If the organism is there *and* the conditions are right, you may get growth of the organism of interest. Getting the right conditions is sometimes near impossible and is often very challenging. For this reason, if your organism of interest doesn't grow, you can't know for sure that it wasn't in the inoculum that you added — it may be that the conditions weren't right.

In order to get only the organisms that you want, you need to select against the organisms that you know are there but that you *don't* want. For example, if you're trying to grow bacteria A, but you know that bacteria B might be present, you can include an unfavorable supplement for bacteria B to inhibit its growth but still allow bacteria A to thrive. One problem is that some microbes can grow more quickly than others when given a chance. If another microbe in your sample can take advantage of the extra resources provided in your enrichment culture more efficiently than your microbe of interest, it may outcompete it and you'll see only the unwanted one. This is a problem with fungi, which are not easily selected against in some enrichment cultures and are experts at growing like crazy with favorable conditions.

After a successful enrichment culture has been achieved, the next step is to get a pure culture of the strain of interest using isolation techniques such as

- **Streak plate:** Using a sterile tool (loop, toothpick, stick, or swab), touch some bacteria. On a new agar plate, make a line to streak out the bacteria. Then, with a new sterile tool, streak again through the first streak to draw out fewer bacteria. Repeat again and allow the bacteria to grow. By the third streak, single colonies normally grow.
- **Dilution series:** From a liquid bacterial culture, dilute a small amount into clean liquid media. From this dilution, take a small amount and make a second dilution. As more dilutions are made into new tubes, the number of bacterial cells gets smaller and smaller. These dilutions can then be plated on agar and allowed to grow to look for isolated colonies.
- **Agar shake:** A dilution of bacteria is made as described above. Instead of growing on agar plates, a small amount of each dilution is added to molten semisolid agar media and allowed to grow in a tube while shaking. The semisolid nature of the media allows the bacteria to migrate through it, thereby making single colonies.

All these techniques rely on diluting microbial cells enough that single colonies can form and be chosen. To get a pure culture of the microbe of interest, these methods may need to be repeated several times on the same culture to make sure that it in fact only contains a single organism.

Seeing cells through lenses

Microscopy can be a powerful method of looking at and making observations about microbial cells. Regular light microscopy can be enhanced by the addition of fluorescent dyes that bind to DNA or to cell walls, giving contrast to samples, especially environmental ones where the background material can obscure the view. Another nonspecific dye can indicate which cells are alive and which are dead (see Table 18-1).

TABLE 18-1 Fluorescent Dyes Used to Label Microbial Cells for Microscopy

Dye	How It's Used	Excitation Light	Color
4',6-diamidino-2-phenylindole (DAPI)	Binds to DNA	UV	Blue
SYBR Green	Binds to DNA	Red light	Green
Propidium iodide	Binds to DNA, can't enter live cells	Green light	Red
Green fluorescent protein (GFP)	Probe label	Red light	Green
Rhodamine	Probe label	Green light	Red

TECHNICAL STUFF

Fluorescence is the light emitted from a protein after it has been excited by specific kinds of energy. In Chapter 9, we explain the different wavelengths of light and how each carries with it a certain amount of energy. When excited by light such as ultraviolet (UV) light, fluorescent proteins emit light of a different wavelength, giving them a different color.

Probe labels are specific proteins that will bind to a known target and then fluoresce, or light up, when exposed to the proper wavelength of light. The proteins that can be used as probe labels can also be used directly by genetically engineering microbial cells to express them (see Chapter 16 for a discussion of genetically engineering microbes). These proteins are handy for strains that are studied in the lab but are obviously not useful for environmental samples.

TIP

These dyes work because, when added to a sample, they only fluoresce when bound to DNA. Most can enter the cells easily, but others, like the vital stain propidium iodide used for live/dead staining, can't enter cells, so they'll only bind to DNA outside a cell or enter through holes in the membranes of dead cells.

Another useful technique involves labeling only the cells based on *phylogeny*, or their evolutionary relationship. In essence this method, called *fluorescent in situ hybridization* (FISH), will light up cells of a desired microbial group based on their genetic similarity.

TECHNICAL STUFF

Ribosomes are the essential machinery for making protein in the cell. They themselves contain proteins but also have a core of RNA called the ribosomal RNA (rRNA). Because all cells need their ribosomes to survive, they're conserved through evolution and don't change that quickly. Organisms that are closely related will have a similar rRNA gene sequence. The more distantly related two organisms are is related to how different their rRNA gene sequences are. There are exceptions to these simple rules, but by and large, they hold true, making rRNAs and their genes useful tools for the identification of microbial groups and the study of microbial evolution (see Chapter 8).

FISH probes work because of the tendency of complementary nucleic acid sequences to bind to each other. The target sequence is the one carried inside the microbial cell, the sequence that is added to bind to the target sequence is the probe. Probe sequences are chemically attached to a fluorescent protein, called a *tag*, making them visible microscopically.

Measuring microbial activity

When researching functioning of an ecosystem it's also important to know what the microorganisms are doing. There are several ways to measure microbial activity in a sample; some are direct measurements and others are indirect measurements.

Direct measurement of activity involves measuring the products of chemical reactions after adding the required substrates. When the products can't be measured directly or are made in very low amounts, then indirect methods are used. These indirect methods include the following:

- **Measuring gene expression:** Gene expression can be measured to see which genes are transcribed. This gives an indication of which proteins are likely being made. Methods for measuring gene expression are discussed in the next section.
- **Using radioisotopes:** Radioisotopes can be used to trace the fate of an element or the rate of conversion of an added radioactive compound. For instance, radioactive sulfur can be added in the form of sulfate to see how quickly sulfate is released as hydrogen sulfide.
- **Probing stable isotopes:** Stable isotope probing is similar to using radioisotopes, except that heavy elements are used instead of radioactive elements. Most elements in nature have heavy and light versions present in the environment. When glucose made from heavy carbon (C13) is fed to an environment expected to contain a microbe that will break down glucose, then the C13 eventually makes its way into the DNA of that organism. Because heavy DNA can be separated from light DNA and sequenced, the identity of the glucose consuming microorganism can be deduced.

Identifying species using marker genes

Sequencing can be used to identify microbial species within a sample. This technique is done most often with bacteria, but it can also be done with any other type of microorganism including viruses, as long as there is a database available against which to compare the sequencing results. As mentioned earlier in the "Seeing cells through lenses" section, the 16S rRNA gene is often used to tell species apart. Marker genes, including 16S rRNA genes, are handy because

- They're quick to amplify using the polymerase chain reaction (PCR).
- They're less expensive to sequence than the entire genome.
- They can give pretty good taxonomic resolution.

A *phylotype* is a sequence from a molecular profile that differs from all other sequences in that survey by greater than 3 percent. A phylogenetic tree can be made based on the differences found in the 16S rRNA gene in bacteria or archaea. Phylogenetic trees are useful not only for identifying microorganisms and thinking about their evolutionary history but also for calculating the amount of biodiversity in a community or ecosystem (see Chapter 8).

Getting the Hang of Microbial Genetics and Systematics

During the molecular biology revolution, techniques like sequencing and genetic manipulation were great tools to study microorganisms and their genes and processes. The recent generation of sequencing technologies have provided even more tools to complement those from microbial ecology. Next-generation sequencing has made sequencing millions of bases possible in a fraction of the time it used to take and at a fraction of the cost. Fast computers and more advanced computational algorithms have made the study of the DNA, RNA, and even protein of microorganisms from mixed communities much more feasible than ever before.

Sequencing whole genomes

Prokaryotic genomes, from bacteria and archaea, are simpler than those from eukaryotes, namely because they're shorter and they don't contain *intergenic regions* (noncoding regions of DNA that interrupt genes). For this reason, the sequencing and assembly of bacterial and archaeal genomes can reasonably be done in a short timeframe for a reasonable amount of money. Sequencing the genome of an organism can give you information about its core metabolic functions, its virulence and communication genes, and how closely related it is to other known genomes.

The steps involved in sequencing a bacterial genome, for instance, are as follows:

1. Isolating the genomic DNA from a culture of the organism of interest
2. Sequencing the isolated genomic DNA
3. Assembling the sequences
4. Annotating the genes

The effort required to accomplish this last step can sometimes be enormous. For this reason, many genomes are published as *draft genomes*, where a rough idea of how the genome looks is known but maybe not all the pieces are there and not all the genes are annotated. A complete genome is decidedly more work, but the end result is a reliable map of an organism's entire genome that is very helpful when studying genes.

If the organism you want to study isn't readily grown in culture, there are some creative ways of getting a genome sequence. From a mixed population of bacteria or archaea, single cells can be isolated that either resemble or are shown to be from the group of interest using FISH (see "Seeing cells through lenses," earlier in this chapter).

TIP

JUST SAY *-OME*

Don't be intimidated by all of the *-ome* words. These days, scientists will take any opportunity to make their research sound sexy and complicated by adding an *-omic* to the end of it. The suffix *-ome* means "all of," and the suffix *-omic* means "the study of all of." Some *-ome* words are entrenched in the English language and mean something specific (think *genome*); but other *-ome* words are fluid and can mean a couple different things. In this chapter, we walk you through some ways of studying microorganisms alone or in their natural communities using all the genes in one organism (genome), all the genes from all the microbes in a community (metagenome), all the RNA being expressed (transcriptome), all the protein in a cell (proteome), or all of the small molecules floating about a cell (metabolome).

Not everyone uses these words the same way. For example, to some people, the word *proteome* means "all the proteins that can be expressed by an organism." To others, *proteome* means "all the proteins measured during an experiment." Sometimes these *-ome* words are used to refer to something "at one time" and other times they're used to refer to "always." For instance, because RNA is sometimes expressed and other times is not expressed, if you measured a cell's transcriptome, you would be getting a picture of the RNA expressed under the conditions at the time you collected the RNA. As you read through the rest of this chapter, you become more familiar with and understand more clearly the different types of *-omics* studies used and their purposes.

Once they're isolated, single or small numbers of cells can be used for genome sequencing. The challenge here is that the small number of cells often don't provide enough DNA for sequencing. To overcome this challenge, a whole genome amplification method can be used to increase the copies of the genome present. Multiple displacement application (MDA) relies on a viral enzyme that starts copying DNA at random places and displaces the other strand as it goes (see Figure 18-1).

MDA is quicker than PCR (which has to be cycled through a high temperature to denature the two DNA strands) and generates hundreds of copies of the genome of interest within minutes. One area of concern when doing MDA is looking out for artifacts from non-specific base pairing.

Using metagenomics to study microbial communities

The word *metagenomics* literally means "a high-level view of genomics." In practice, it means the high throughput sequencing of all the microbial genes in a sample. The word *metagenomics* is also used to mean the sequencing of any microbial gene(s) from DNA extracted from a sample. Metagenomics as the sequencing of all microbial genes in a sample differs from the whole-genome sequencing of an organism in that it's applied to a community of microbes and not just a single microorganism, and there is no genome assembly step.

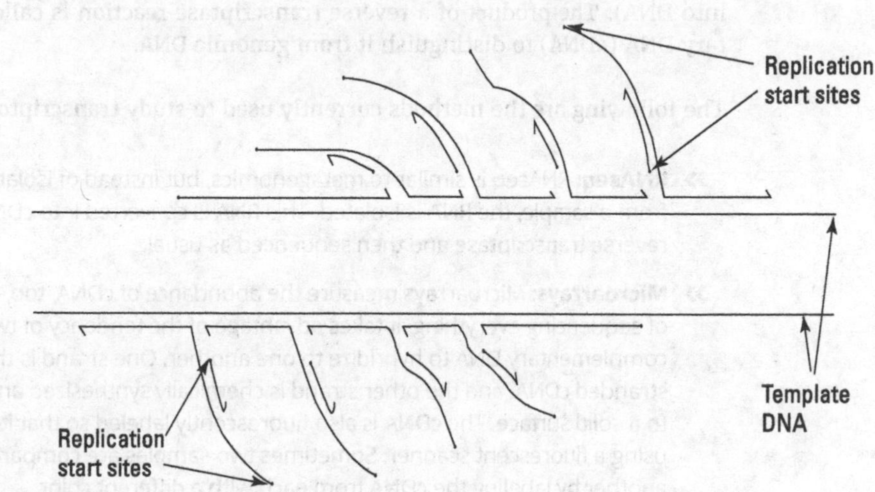

FIGURE 18-1: Multiple displacement amplification.

The list of genes present in a sample provides an estimate to how the community is functioning. It can indicate which enzymatic processes were likely taking place in that community and what kinds of competition and cooperation were taking place as well. One feature that microbiologists often look for when analyzing the metagenomic data for a bacterial community is the presence of genes that encode antibiotic-resistance. Recently metagenomic studies have found that antibiotic resistance is found nearly everywhere we look. It's enriched in bacterial communities that are exposed often to antibiotics like hospitals and farm animals.

Also, identifying organisms is a lot easier when you have more than just one marker gene. Although this is true for simple communities where the members are well known, with many environmental samples the microbes in them are often unknown so representatives of these communities won't be in a reference database. Metagenomic analysis is the only way of knowing about the presence of these organisms and how similar or different they are from known bacteria or archaea.

Reading microbial transcriptomics

Transcription is the process of making mRNA from DNA for the purpose of making proteins in the cell. *Transcriptomics,* then, is the sequencing of all the RNA in a sample for the purpose of measuring how much of each gene in a community is being expressed. Again, this is different from measuring the levels of RNA from one strain — instead, it's a measurement of all the RNA made by all the microbes in a sample.

Transcriptomics is different from metagenomics because it tells you which genes are being expressed and at what levels. There are a few different technologies that measure transcriptional differences between groups of microbes, each of which takes advantage of *reverse transcriptase* (the ability of a viral enzyme to copy RNA into DNA). The product of a reverse transcriptase reaction is called complementary DNA (cDNA) to distinguish it from genomic DNA.

The following are the methods currently used to study transcriptomics:

- **RNAseq:** RNAseq is similar to metagenomics, but instead of isolating the DNA from a sample, the RNA is isolated. This RNA is converted into cDNA using a reverse transcriptase and then sequenced as usual.
- **Microarrays:** Microarrays measure the abundance of cDNA, too, but instead of sequencing everything, it takes advantage of the tendency of two strands of complementary DNA to hybridize to one another. One strand is the single-stranded cDNA, and the other strand is chemically synthesized and attached to a solid surface. The cDNA is also fluorescently labeled so that it can be read using a fluorescent scanner. Sometimes two samples are compared to one another by labeling the cDNA from each with a different color.
- **Quantitative RT-PCR (qPCR):** qPCR uses PCR to amplify a signal from a small amount of template RNA. The *RT* part of the name stands for *reverse transcriptase*, and the *quantitative* part means that as the copies of DNA are made during each cycle of PCR, their numbers are measured. By calculating back from the number of copies of DNA made, you can get a fairly accurate estimate of how much RNA you started with. This method is sensitive and fairly quick, but it has to be done on the transcript of each gene individually.
- **Nanostring:** Nanostring is a brand-new technology that has the sensitivity of qPCR and the capability of doing many genes like microarrays or RNAseq. You have to know which genes you're interested in ahead of time and have probes designed for each one, but unlike in a microarray, the probes are not attached to a surface. Instead, the mRNA and the nanostring probe hybridize directly and only those that form the hybridized complex are kept and counted.

Figuring out proteomics and metabolomics

Microorganisms change the proteins they make depending on their environment and how they're interacting with it. In order to look at which proteins are being made under certain conditions, proteomic techniques are used.

The genome can tell you a lot about which proteins a microorganism has. Because each gene encodes a protein, you may assume that by listing all the genes you could get a sense of the processes going on in the cell, but it's a bit more complicated than that. Genes are transcribed into mRNA at different rates and only when needed. Then not all mRNA molecules are made into protein. Measuring all the protein present in a cell at one time gives you information about how the cell is interacting with its environment under those specific conditions.

Two main proteomic techniques work well when used hand in hand (see Figure 18-2):

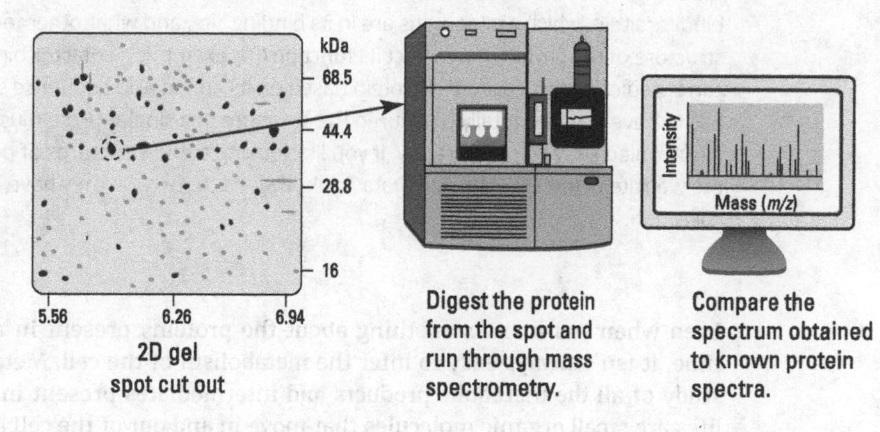

FIGURE 18-2: The use of 2D gels and mass spectrometry in proteomics.

- **2D gels:** The proteins are extracted from a pure culture of a microbial strain and separated by both their charge and their size. The different amino acids used to assemble a protein sequence have different charges that impart on the final protein an overall charge (see Chapter 6). Also, because proteins vary greatly in length, they have a range of sizes as well. When finished, a 2D gel looks like the side of a Dalmatian, and patterns between gels can be compared. Each spot can be extracted from the gel and used for mass spectrometry.
- **Mass spectrometry:** A protein sample is first digested into small peptides. Then the amino acid sequence of these peptides is calculated with mass spectrometry, which uses very sensitive mass calculations for each of the elements in each amino acid and gives a spectrum, or pattern, that is unique to each peptide. To identify the protein of interest, the spectra produced are compared to those for a standard set of known proteins.

MODELING PROTEINS

Based on the genetic sequence, assumptions can be made about the function of the proteins that the genes code for. These assumptions are made based on how similar a gene is to genes that are known from other organisms and where the activity has been tested. If the gene is very similar, like more than 70 percent identical, it's likely that they have the same function. If the gene is not that similar, its structure can still be modeled based on the structure of proteins that are known with much less than 70 percent identity.

Modeling a protein can give important information about a protein, such as where its binding site is, which amino acids are in its binding site, and what other features of the structure of the protein may affect its function (for example, a cofactor binding site). We can't predict the structure of a protein based on its amino acid sequence so someone has to have experimentally measured the structure of a similar protein using a technique called x-ray crystallography. If you like looking at the structures of proteins visit the website of the RCSB Protein Data Bank (`www.rcsb.org`) — they have an amazing collection.

Even when you know something about the proteins present in a cell at a given time, it isn't always easy to infer the metabolism of the cell. Metabolomics is the study of all the metabolic products and intermediates present in a cell. Metabolites are small organic molecules that move in and out of the cell and are the substrates, the products, or the intermediates of chemical reactions going on. There are many more metabolites than there are proteins, RNA, or DNA molecules, so metabolomics is the trickiest thing to study about microorganisms.

Looking for Microbial Dark Matter

With the many tools at our disposal to study microbial communities in nature, you'd think that we'd know everything there is to know about the bacteria, archaea, and eukaryotic microbes that we share the planet with. The truth is, there are still many species of bacteria and archaea that have yet to be discovered. Based on *culture-independent methods* (methods based on sequencing the DNA from microorganisms alone), scientists have found that there are far more species of prokaryotes left to be described than originally estimated. Shockingly, it seems that only the tip of the iceberg in terms of bacterial and archaeal species have been identified, and fewer still have been cultured.

IN THIS CHAPTER

» **Understanding the regulation of gene expression**

» **Seeing how bacteria systems can be used as switches**

» **Putting pieces together to make more complex systems**

Chapter 19
Synthesizing Life

The applied use of microorganisms goes back to the early days of agriculture and the use of fermentation for food production. Through a process of trial and error, humans have selected for microbes that are useful to us — for example, microbes that impart special flavors to our favorite cheeses, beers, and wines. At the same time, scientists have realized that microbes are perfect for studying the inner wiring of the cell. The facts that they're unicellular, they're easy to grow in a lab, and they have fewer genes than eukaryotic cells make them great for learning things like how and when DNA is turned into protein and how these proteins allow a cell to grow and divide. From painstaking genetic experiments that mapped how each gene in a cell functioned was born the molecular cloning era, when it became routine to isolate and make copies of DNA fragments in the lab, move the fragments from one cell to another, and even tweak the DNA sequences to see the consequence of mutations.

***Synthetic biology* is the merging of the area of applied microbiology and molecular cloning. Instead of looking for organisms that can do specific functions, why not just *make* an organism that does it? Synthetic biology began with the first experiments where we could move DNA between organisms and later evolved into isolating and cloning DNA. As tools for manipulating and synthesizing large fragments of DNA became available, synthetic biology really began to take off.**

In this chapter, we examine how simple systems to control gene expression in bacteria can be *rewired* to make the cells work for us by making specific proteins. Using parts from these simple systems, more elaborate systems are built and the field of synthetic biology has emerged.

Regulating Genes: The lac Operon

Because cellular functions are directly linked to the proteins made by a cell, control of gene expression gives you control over how a cell behaves. Another term for the control over when and how genes are expressed is the *regulation of gene expression* or simply *regulation*. Regulation of gene expression in bacteria was first discovered for the lactose degrading system, in a group of genes called the *lac* operon. The *lac* operon is the poster child for gene regulation systems because it has been extensively studied, is well understood, and has been intentionally altered to change how it behaves. The *lac* operon contains genes for the import and breakdown of the sugar lactose, and its expression is regulated by the *lac* promoter.

REMEMBER

All the genes needed to import and use lactose are grouped together into what is called an *operon*, where they're all transcribed together from a single promoter (P_{lac}).

Using a good natural system

Lactose is a sugar found in the environment. Bacteria take lactose into the cell to use for energy. It's a fairly large molecule, so bacteria must use a transport protein to get it across the membrane. The transport protein is called a *permease*. As lactose trickles into the cell through the permease, an enzyme called β-galactosidase converts it into two other sugars that are easier for the cell to digest: glucose and galactose.

At the same time, a small amount of the lactose is converted to allolactose. Allolactose is not used for energy; instead, it acts as a signal of how much lactose is in the cell. It does this by binding to a protein called the *lac* repressor (see Figure 19-1) and inactivating it. The *lac* repressor (LacI) is a protein that acts as an on/off switch for the expression of genes involved in the utilization of lactose.

The *lac* repressor binds to the *lac* promoter and blocks the RNA polymerase from transcribing the permease and β-galactosidase genes. When levels of lactose are high, allolactose is made and binds to the *lac* repressor, which stops it from repressing the *lac* operon. The system is then induced and the genes in the *lac* operon are transcribed. This system allows the cell to control the expression of the genes for lactose utilization.

TECHNICAL STUFF

François Jacob and Jacques Monod, French scientists, were pioneers in the study of the regulation of gene expression. To help with their experiments, they wanted to turn the system on without needing lactose or its permease. They screened through a collection of small molecule compounds and came across isopropyl β-D-1-thiogalactopyranoside (IPTG), which diffuses into *E. coli* without the need for the permease and binds to the *lac* repressor, turning on the *lac* operon.

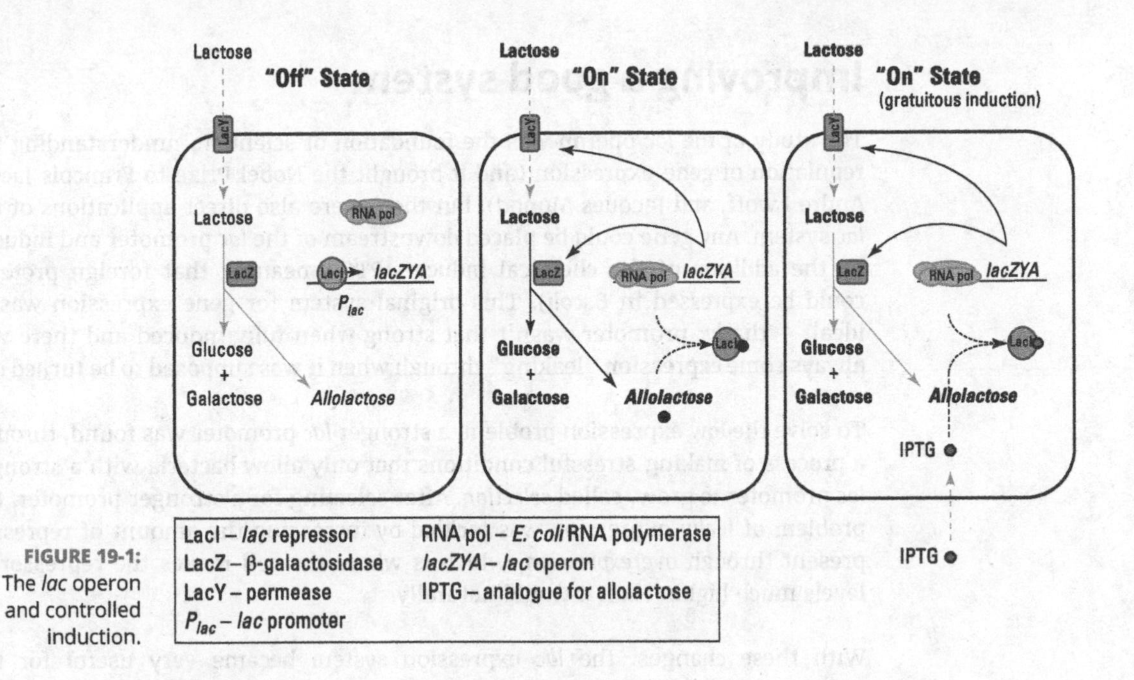

FIGURE 19-1: The *lac* operon and controlled induction.

There are a few features of the *lac* operon worth noting from a design perspective:

- **Leakiness of the *lac* operon is essential.** You need some low-level expression to have a little permease in the membrane to get some lactose into the cell and then some β-galactosidase to make some allolactose in order to get the system to turn on in the presence of lactose. If the system was really off, it wouldn't be able to be induced because lactose would never get into the cell. So, if you wanted to build a system with a true off state, the natural *lac* system would not be a good choice.
- **The system behaves in a positive feedback loop.** A little lactose gets in the cell initially through a few *lac* permeases, which then turn on the *lac* operon and make more *lac* permease and more allolactose and more *lac* permease and so on. This positive feedback loop means you can't get levels of expression of the *lac* operon that are directly proportional to the levels of lactose: It's an all-or-nothing system.
- **The *lac* promoter is not all that strong and is not good if you want to make lots of a product.**

REMEMBER

It's important to note that these properties are not "design flaws" in the *lac* operon but essential features that make the *lac* operon work well in *E. coli* in its natural environment.

Improving a good system

The study of the *lac* operon was the foundation of scientists' understanding the regulation of gene expression (and it brought the Nobel Prize to François Jacob, Andre Lwoff, and Jacques Monod), but there were also direct applications of the *lac* system. Any gene could be placed downstream of the *lac* promoter and induced by the addition of the chemical inducer IPTG (meaning that foreign proteins could be expressed in *E. coli*). This original system for gene expression wasn't ideal — the *lac* promoter wasn't that strong when fully induced and there was always some expression "leaking" through when it was supposed to be turned off.

To solve the low expression problem, a stronger *lac* promoter was found, through a process of making stressful conditions that only allow bacteria with a stronger *lac* promoter to grow, called *selection*. After selecting for a stronger promoter, the problem of leaky expression was tackled by increasing the amount of repressor present through *overexpression*, which is where the cell makes the repressor at levels much higher than it would naturally.

With these changes, the *lac* expression system became very useful for the overexpressing proteins in *E. coli* needed to purify and study them, as well as in biotechnology (see Chapter 16).

With the ideas from the *lac* system in mind, scientists designed an overexpression system in the 1980s that is still widely used today. The expression system was based not on bacterial genes but on the genes from the T7 bacteriophage. This bacteriophage encodes its own RNA polymerase, and when the phage infects a host bacterial cell, this polymerase is so active that soon almost all the mRNA in the cell is from the virus, ensuring that only T7 proteins are made.

Here are the features of T7 bacteriophage that make it so useful for protein expression:

- **T7 RNA polymerase is fast** — faster than the native *E. coli* RNA polymerase acting in the host cell — so it quickly outcompetes the *E. coli* RNA polymerase for resources and makes a lot of mRNA quickly.
- **T7 RNA polymerase only recognizes T7 promoters,** making it very specific for its target.
- **T7 promoters (P_{T7}) are very different from *E. coli* promoters,** so the T7 RNA polymerase doesn't ever transcribe from *E. coli* promoters by mistake.

WARNING

When the gene for a protein of interest is put behind a T7 promoter in a cell expressing the T7 RNA polymerase, the cell makes that protein almost exclusively. In the same way that expressing all phage proteins is bad for the host cell, over-expressing a protein of interest with the T7 RNA polymerase system is very toxic to the cell. To stop this from happening, the system has to be controlled so that expression of the target gene can be turned off until it's deliberately turned on.

To make the T7 system controllable, the expression of the T7 RNA polymerase was placed under control of the *lac* promoter and repressed by the *lac* repressor. In this way, target protein expression could be turned on only after cells had grown normally to a high cell density. The on switch was provided by IPTG added to the now large batch of cells, inducing the expression of T7 RNA polymerase that soon made lots of mRNA and subsequently large amounts of target protein.

The use of a combination of the *lac* operon and T7 genes, summarized in Figure 19-2, revolutionized protein expression and purification in bacteria.

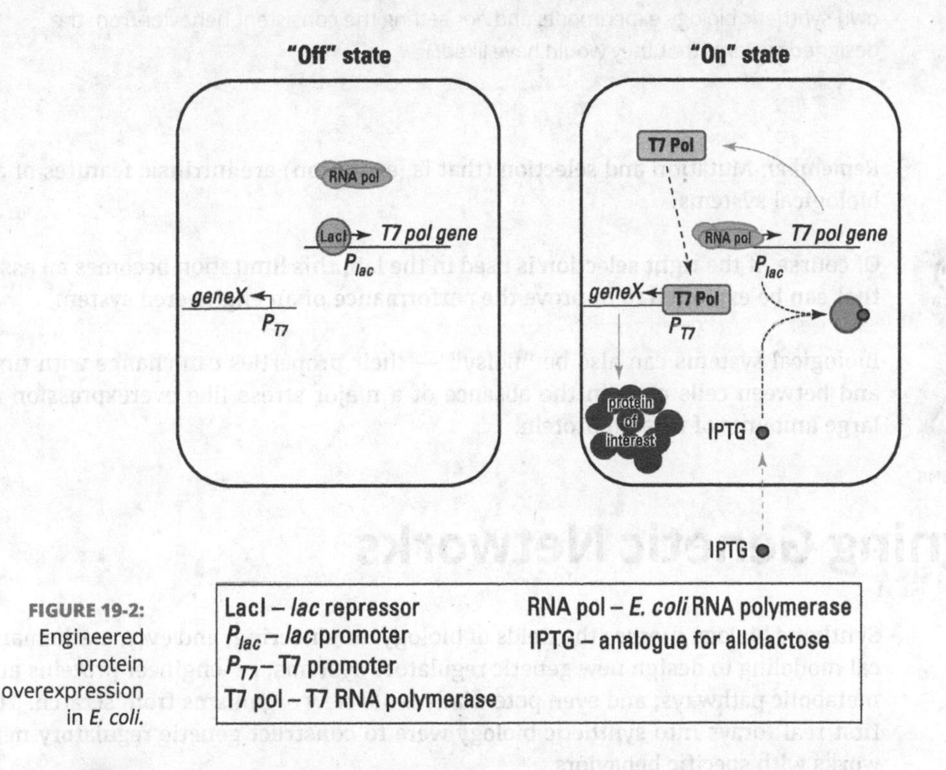

FIGURE 19-2: Engineered protein overexpression in *E. coli.*

WARNING

Once turned on, this system is very toxic, so *E. coli* cells will mutate to overcome the detrimental effects, effectively breaking the system. This is an important limitation to synthetic biology — biological systems are inherently unstable.

NOT SO ALIKE AFTER ALL

A tube of culture in the lab, containing billions of bacteria, is often thought of as completely homogeneous. That is, all the cells within it are thought of as identical little clones of each other, behaving alike. In theory, they *should* be, because they presumably all come from one cell. But as it turns out, this isn't entirely true. Cells in a seemingly homogeneous population will have different levels of gene expression — from a synthetic biology perspective this is an important point. Perhaps not surprisingly, the *lac* operon and synthetic biology are featured in the history of understanding cell-to-cell variability in populations.

In an elegant but often forgotten paper by Aaron Novick and Milton Weiner in 1958, they show heterogeneity in single cells within a population of *E. coli* with respect to the *lac* operon and all-or-none induction of the *lac* operon. And in more contemporary times, Michael Elowitz and colleagues have used green fluorescent protein (GFP)–based reporters to explore heterogeneity in *E. coli* cells, in part motivated by results of their own synthetic biology experiments and not getting the consistent behavior from the designed systems that they would have liked.

Remember: Mutation and selection (that is, evolution) are intrinsic features of all biological systems.

Of course, if the right selection is used in the lab, this limitation becomes an asset that can be exploited to improve the performance of an engineered system.

Biological systems can also be "noisy" — their properties can change with time and between cells even in the absence of a major stress like overexpression of large amounts of foreign protein.

Designing Genetic Networks

Synthetic biology merges the fields of biology, engineering, and even mathematical modeling to design new genetic regulatory systems, re-engineer proteins and metabolic pathways, and even potentially build new organisms from scratch. The first real forays into synthetic biology were to construct genetic regulatory networks with specific behaviors.

The examples in this section define simple systems or modules that can be used to design more elaborate systems. They serve to illustrate the basics and show how some simple designs are part of more complex circuits.

Switching from one state to another

One of the first synthetic genetic networks designed was inspired by bacteriophage lambda (λ). This phage can integrate into the genome of its host cell, called a *temperate phage*, and exist as a stable lysogen (see Chapter 14). A lysogen may be maintained for generations or induced into a lytic state, where the phage DNA is excised, replicated, and packaged into new phage particles. This system of lysogeny versus lytic states is not controlled by a simple on/off regulation but by a system of dual repressors (CRO and cI repressors) each inhibiting the expression of the other. The result is a bistable switch that can be stable in one state or another and switches between them.

A synthetic system built on this idea uses a temperature sensitive version of the λ cI repressor and the *lac* repressor (LacI). The cI repressor gene also has a GFP gene as part of the same operon, so when the system is on, the cells turn fluorescent green. This system uses two signals to switch states, shown in Figure 19-3:

- **Temperature:** The temperature-sensitive cI repressor is inactivated at 37°C.
- **IPTG:** When added, this compound inactivates the *lac* repressor.

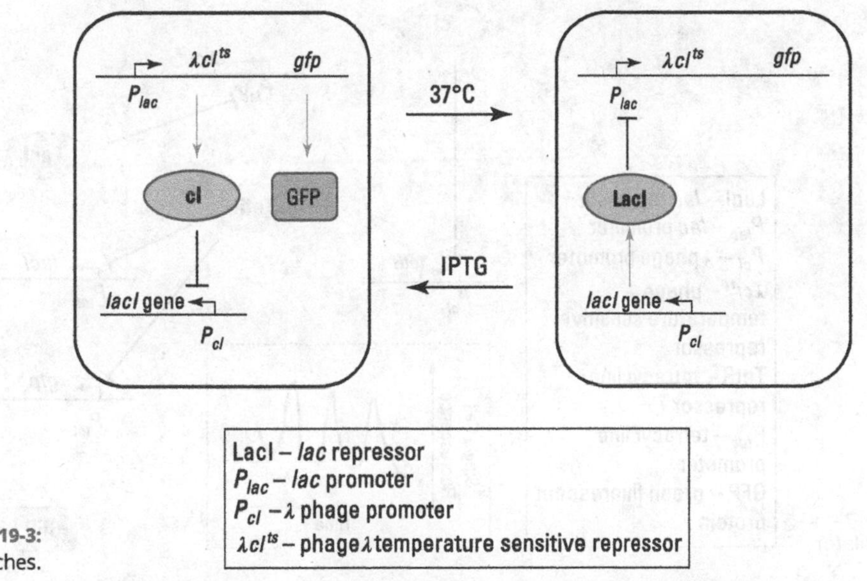

FIGURE 19-3: Bistable switches.

This simple system can be coupled to other genes to control cell functions, and scientists can switch the cells from one state to the other quickly by either changing the temperature or adding IPTG.

Oscillating between states

A more elaborate synthetic regulator network that was one of the early systems built was the *repressilator.* Many biological systems have oscillatory behavior — for example, circadian clocks that control gene expression on a 24-hour cycle.

TIP

Intriguingly, most circadian clocks — from simple cyanobacteria to our own clocks — are not regulated simply by light and darkness. That's why getting over jet lag is so difficult when we travel across many time zones — our internal clock is not easily reset by the new light/dark cycle.

Michael Elowitz and colleagues set out to design a simple oscillating genetic network. Whereas a two-repressor network can exist in bistable states, a three-repressor (or any odd number) system can oscillate if the parameters are just right. The repressilator (see Figure 19-4) was constructed with a gene for GFP under the control of the tetracycline repressor (TetR) that also repressed the synthesis of the *lac* repressor (LacI), which repressed the synthesis of the bacteriophage lambda repressor (λ cI), which in turn repressed the synthesis of TetR. The cells alternatively expressed and repressed the expression of GFP, creating an oscillation of fluorescence.

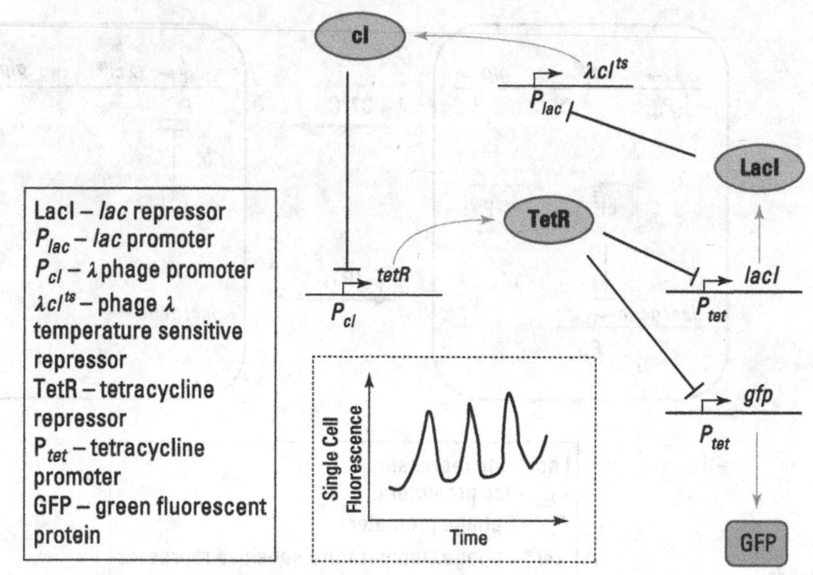

FIGURE 19-4: The repressilator.

One design feature that was critical for this system to work is that the repressors had to have short half-lives. Once they're made, repressor proteins usually hang around a while, causing long periods of repression of their target promoter. For the repressilator to work, the repressor proteins have to be broken down quickly so that the genes can be de-repressed. To make the repressors less stable, ssrA degradation tags can be added to them.

Keeping signals short

In bacteria, if there is a problem during the formation of a protein — for instance, if the ribosome stalls — there is a system in place to get rid of the unfinished protein. The peptide can't be released from the ribosome until a short amino acid sequence, called a *tag*, is added to the end of the peptide chain. Afterward, the tagged peptide is directed to the proteolytic machinery and quickly destroyed. Although the natural tag is 11 amino acids long, it turns out that just adding the last 3 amino acids (LAA) works just as well. This is called an ssrA tag (short for *small stable RNA A*) and is a handy tool in synthetic biology. By changing the tag sequence (to AAV or LVA), the amount of time it takes to degrade the protein can be tweaked, giving you control over how long a repressor will be around after it has been made, for instance.

The Synthetic Biologist's Toolbox

Engineers love to build things. Imagine that instead of a toolbox full of wires, resistors, switches, and other bins and bobs, they have one filled with transcriptional regulators, promoters, and engineered genetic networks, each with a function similar to a switch, a blinking light, or other functions not possible with inanimate objects made from metal and plastic. These biological parts can be used to *rewire* cells to carry out all sorts of novel functions. Synthetic biology has expanded into more complex eukaryotic systems such as yeast with a toolbox similar to that developed for bacteria.

Making it modular

Although on paper these systems seemed straightforward to build, it takes quite a bit of effort to build them from scratch, and early synthetic biology systems were done by trial and error, with lots of blood, sweat, and tears! For this reason, a collection of synthetic biology "parts" — also called BioBricks — and protocols is available and can be distributed to anyone looking to build things from living parts.

Several resources exist to help you get started:

» **The iGem Registry of Standard Biological Parts (`http://parts.igem.org/Frequently_Used_Parts`):** A synthetic biology toolkit of well-characterized parts, including operators, promoters, and regulators, that continues to grow.

» **The BioBricks Foundation (www.biobricks.org):** An organization dedicated to making synthetic biology open to everyone and to keeping it ethical and safe.

» **OpenWetWare (www.openwetware.org):** A place to share information, look for help, and get inspired to build things.

With a toolkit in hand, you can start to design all sorts of interesting networks. This inspired Drew Endy and colleagues to start an undergraduate competition to do just that, the iGEM competition.

Participating in the iGEM competition

The International Genetically Engineered Machine (iGEM) competition is an international synthetic biology competition that gives university students the opportunity to use the tools of synthetic biology to design and operate novel genetic systems in microbes. Teams from universities around the world are given BioBricks from the iGEM repository, or tasked with creating their own building blocks, and then they combine them in a way such that they perform a new function. Teams present their projects at the annual meeting and compete for recognition of executing the best project.

An iGEM project involves selecting a problem to tackle, brainstorming solutions using synthetic biology, and then executing the science. Along the way, teams are encouraged to participate in community outreach and consider the public's perception of their development. Teams must ask themselves, "Would the public consider their inventions safe?" and "Would they approve of its use?"

The iGEM competition is by no means your typical science fair. Students flex their scientific muscles and show up with high-quality creations. By tweaking genetic networks and moving genes around between microbes, students have used synthetic biology to provide solutions to a variety of fields. The following are examples of projects presented at iGEM competitions:

» The use of microbes to create building material, in a process called biocementation, in space so that we don't have to transport large and heavy bricks to habitable planets in the future (Brown and Stanford universities).

» Engineered worms that sense and migrate to environmental pollutants, like toxic material in oil sands, and then take up and digest the toxic material a way that safely removes it from the environment, a process called *bioremediation* (Queens University, Canada).

- A "smart" drug delivery system that links engineered *E. coli* with drug-containing nanoparticles, such that *E. coli* senses when it arrives in a certain environment and produces the enzyme needed to release the drug from the nanoparticle (École Polytechnique Fédérale de Lausanne, Switzerland).
- The creation of light-sensitive bacteria that act as photographic film. The team engineered typical *E. coli* to produce a black chemical unless it's exposed to light. Shining a patterned light onto a plate of bacteria created an image with a resolution of 100 megapixels (University of California, San Francisco, and the University of Texas).

7
The Part of Tens

IN THIS PART . . .

Get a view of some very unfriendly microbes — from nasty viruses like Ebola and influenza to the bacteria that cause things like cholera and tuberculosis.

See the many ways that microorganisms are used every day, from making wine to treating sewage.

Learn about the ways that microbiology approaches apply to your everyday life, including keeping you and your pets healthy and protecting the environment.

IN THIS CHAPTER

» Finding out about diseases with severe symptoms

» Uncovering new microbial threats to human health

» Seeing how antibiotic resistance is creating nasty pathogens

Chapter 20
Ten (or So) Diseases Caused by Microbes

Microbes play critical roles in human and animal health, maintaining ecosystems, and driving major geochemical processes. They inhabit every environment on earth, from the bland to the extreme. However, most people equate microbes or "germs" with disease. There is no shortage of microbes that cause disease; some are notable for the number of people they infect, and others for the nastiness of the infections they cause.

The interaction of people with pathogens has shaped the course of human history. Although it's tempting to think of them as the bad old days, many of the same things that plagued people hundreds of years ago remain potential threats today. Travel between continents was responsible for bringing devastating infectious diseases to the New World, and it's still a risk today as unprecedented numbers of people move around in our highly connected world. Air travel means we can reach any corner of the globe in 24 to 48 hours. It also means that microbes can reach us from anywhere on earth just as quickly. The threat of great epidemics is still here, despite great advances in treating and preventing infectious disease.

As we push deeper into remote areas of the world, the risk of new pathogens jumping from other species to humans, called *zoonotic infection*, also increases. Zoonotic infections can be nasty — the more balanced relationship that was there between the natural host and the pathogen is gone.

In this chapter, we list ten nasty microbes, including

» Pathogens that have been with mankind for thousands of years

» Pathogens tamed with vaccines and antimicrobials that are reemerging threats due to resistance

» Emerging pathogens that are new threats to man, often as zoonotic infections that have jumped from other host species

Ebola

Ebola is one of the most lethal viruses to infect humans, with a mortality rate reaching 90 percent in some outbreaks. Ebola is one of two known filoviruses that cause severe hemorrhagic fever in humans. First identified in 1976 in the Congo near the Ebola River, infections by this filovirus have been confined to small outbreaks in isolated regions of central west Africa. Fruit bats are considered the natural host, but contaminated wild meat may be the source of initial infection. The virus is spread through contact with bodily fluids of infected people or animals.

What makes Ebola so nasty is that the virus can be in a person unnoticed for around 21 days. But when symptoms begin to show, the disease progresses rapidly, starting with fever, fatigue, muscle pain, and headache, and ending with vomiting, diarrhea, internal bleeding, and often external bleeding. Because of the virus's long incubation period, containing an outbreak can be challenging. Apparently healthy people travel into new areas where they eventually get sick, and if they aren't isolated immediately, others can become infected. Cultural practices of gathering groups of people around the deceased at burial ceremonies haven't helped matters — infection is spread most quickly during the last stages of the disease.

Anthrax

Bacillus anthracis, which causes the disease known as anthrax, is a Gram-positive sporulating bacteria that gets its name from the distinct coal-like, black tissue that is shed from a skin infection (*anthrakis* is Greek for coal). It's primarily associated with infections of grazing animals — spores can persist in the soil for decades. In humans, three types of infections are observed:

- **Cutaneous anthrax** is easily diagnosed by its characteristic skin lesions and is easily treated.
- **Gastrointestinal anthrax** arises from eating infected undercooked meat. This form of anthrax is uncommon and can be effectively treated. However, because it presents initially like typical food poisoning, it's often not identified in time and consequently has a high mortality rate.
- **Inhalation anthrax**, from breathing in spores, was once confined to the tanning industry when spores would be released from the hides of infected animals. Patients initially present with flulike symptoms, but the disease rapidly progresses after about 48 hours to an acute illness followed by coma and death in up to 95 percent of infected individuals.

The potential to weaponize anthrax spores was realized early and is a major concern. The largest outbreak of inhalation anthrax occurred in 1979 in Sverdlovsk, in the former Soviet Union, downwind of a military complex; it resulted in up to 105 deaths. In late 2001, up to 6 letters containing anthrax spores were mailed to U.S. politicians, resulting in 11 cases of inhalation anthrax (5 deaths) and 11 cutaneous anthrax cases. The potential impact of aerosolized spores on a concentrated population could be devastating, and the threat of anthrax as a biological weapon is real.

Influenza

How often do you hear someone complain about being sick a couple days with the flu? More often, this is probably just the common cold. The real flu is a serious illness with fever, headaches, and fatigue lasting for many days. The flu is caused by the influenza virus and occurs seasonally throughout the world — it's always flu season somewhere! Each year, the seasonal flu is responsible for up 5 million severe cases and half a million deaths. It can be particularly hard on the young and the elderly, although a recent flu was more severe in young adults. New flu viruses arise each year, so one year's flu vaccine is usually not effective the following year.

About three times each century, a flu pandemic occurs, which spreads more effectively and is more virulent. The 1918–1920 flu pandemic, also called the Spanish flu, was thought to have killed 20 million to 100 million people.

TECHNICAL STUFF

The way that we keep track of influenza strains is by naming them for the different variants of two viral coat proteins — hemagglutinin (H) and neuraminidase (N) — to which numbers are assigned. The 1918 pandemic, as well as illness caused in 2009–2010, for example, were both caused by H1N1 viruses.

The flu virus usually spreads through contact or aerosols from coughing or sneezing patients, but some, like the avian flu (H5N1), can be carried by birds with the obvious potential to spread very rapidly around the world. Although it did not turn into the next big pandemic, H5N1 was highly pathogenic in humans and birds.

Tuberculosis

Tuberculosis (TB) is a global epidemic and a leading cause of morbidity and mortality throughout the world. Each year, it infects up to 9 million people and is responsible for over 1.5 million deaths. Once infected with *Mycobacterium tuberculosis,* about 5 percent of people will develop active disease within five years; the remaining 95 percent will have a latent infection that can persist for decades, later becoming activated in about 5 percent of these individuals. With treatment, the mortality rate is less than 2 percent, but without treatment mortality rates are much higher.

A major change that has increased the severity of TB is the emergence of multidrug-resistant strains. These strains of *M. tuberculosis* are resistant to the antibiotics that were previously most effective against it, rifampicin and isoniazid. Extensively drug-resistant TB (XDR-TB) comes from strains that are resistant to all the most potent antibiotics, making them the most difficult to treat. Almost half a million multidrug-resistant TB cases and an increasing number of XDR-TB strains are occurring each year, making an untreatable epidemic of TB a real possibility if new drugs are not discovered.

HIV

As its name suggests, the human immunodeficiency virus (HIV) infects immune cells. The immune suppression caused by HIV infection makes patients vulnerable to secondary infections from other bacteria, viruses, fungi, or protozoa. It's these secondary infections that cause problems for those infected with HIV — they become hard to treat as the patient's immune system deteriorates. Acquired immunodeficiency syndrome (AIDS) applies to the most advanced stages of HIV infection where secondary infections are often fatal.

In 2012, it was estimated that 35 million people were living with HIV, with more than 2 million new infections and 1.6 million deaths per year. Secondary TB infections are attributed to about 25 percent of all HIV-related deaths. The virus is spread through sexual contact, bodily fluids, and contaminated needles.

AIDS first emerged in the 1980s as a disease with an unknown cause, spreading through the homosexual male community before scientists identified the causative agent. Today, HIV infection is not restricted to gay men; in fact, women are the fastest-growing group of newly infected individuals.

Anti-retroviral therapy is effective in managing HIV infections, but it's costly and widely available only in the developed world. Although these drugs have made HIV a largely manageable disease in the developed world, AIDS remains a devastating epidemic in many parts of the world.

Cholera

Cholera is an acute diarrheal disease, caused by the waterborne bacterium *Vibrio cholerae*. Every year, it affects 3 million to 5 million people and leads to up to 200,000 deaths. The rapid dehydration can be lethal within hours, but it can be effectively treated with oral rehydration therapy. Of course, rehydration therapy can be challenging because the disease is associated with poor sanitation and poor water quality. There have been 7 major cholera pandemics around the world in the last 200 years that have resulted in tens of millions of deaths.

John Snow, a London physician, is credited with starting the science of epidemiology. He mapped cholera cases in London in the 1850s and observed that they all clustered around the Broad Street public water pump. The pump handle was removed, and this was attributed to the decline of cholera in this epidemic. In places where *V. cholerae* lives naturally, outbreaks can often follow natural disasters where damage to infrastructure can prevent access to clean water.

Smallpox

The smallpox virus, *Variola major*, has been a significant pathogen throughout human history and has been responsible for major epidemics in the past. It's particularly noteworthy, however, because it represents a remarkable success story in the eradication of infectious disease. A concerted international effort that started in the 1960s was declared successful in 1980. Smallpox was the first and only major infectious human disease to be eradicated from natural transmission.

This was the culmination of the pioneering work of Edward Jenner, who demonstrated the efficacy of a cowpox virus as an effective vaccine for smallpox virus at the end of the 18th century. Smallpox virus is now thought to exist only in laboratory collections. However, the threat of smallpox as a biological weapon has renewed interest in continuing to develop newer vaccines in the event of a man-made outbreak of small pox.

Primary Amoebic Menigoencephalitis

Bacteria and viruses are most of what comes to mind when people think of nasty microbes, but there are many examples of other types of organisms that can be particularly nasty. Of the small number of opportunistic amoeba pathogens, *Naegleria fowleri* deserves special attention because it's known as the "brain-eating amoeba." This free-living freshwater amoeba finds its way into the nasal passages of its host and then eats its way along the olfactory nerve and makes its way along the nerve fibers into the brain, where it feasts. It causes a condition called primary amoebic menigoencephalitis (PAM), a very rare disease but one that can affect any individual and is acquired from fresh water, including swimming pools.

Symptoms appear within a week of exposure and start out as the loss of the senses of smell and taste, followed by generic symptoms of headache, vomiting, and fever, and ending with confusion, hallucinations, and seizures. Infections with *N. fowleri* are almost always lethal; death occurs one to two weeks after the initial symptoms appear.

MULTIDRUG-RESISTANT SUPERBUGS

Although not a specific pathogen, an increasing concern is the emergence of multidrug-resistant bacteria, also called *superbugs*. The sharp decline of infectious disease in the last half of the 20th century can be attributed to improved hygiene and water quality, vaccines, and antimicrobial drugs. These drugs were so effective, in fact, that in the late 1960s it was widely thought that infectious diseases — and bacterial infections in particular — had been defeated. Yet the victory didn't last.

Most antimicrobial drugs are made by other microbes, or modifications of them. Microbes have been in chemical warfare with each other for hundreds of millions of years, and while we have exploited these microbial chemical weapons to target pathogens, the pathogens have gained resistance genes that make the drugs ineffective. Widespread use of antibiotics in agriculture and misuse in medicine have accelerated the spread of resistance. Early in 2014, the World Health Organization declared that antimicrobial resistance is a threat to global public health. Resistance may soon move us into a post-antibiotic era where common everyday infections and minor injuries will be untreatable and death rates not seen for five or six generations will return.

The Unknown

Perhaps one of the most frightening and nastiest of the microbes may be the ones we don't know about and can't prepare for. Organizations like the World Health Organization and the U.S. Centers for Disease Control are constantly on the alert for new emerging pathogens, yet these pathogens can still appear without warning. The lack of vaccines or therapies to deal with these new threats to human health makes them very dangerous. Two recent examples of zoonotic viral outbreaks are good examples of this.

In 2002, an outbreak of an unusual pneumonia, which became known as severe acute respiratory syndrome (SARS), infected more than 8,700 people and killed almost 800 in southern China. It spread within weeks to 37 countries. Many of the deaths were in medical staff treating patients before the nature of the epidemic had been established. Infections from the SARS-coronavirus had a mortality of almost 10 percent. Almost as quickly as it emerged, the epidemic was contained; it's now considered a rare disease. This zoonotic virus likely made the jump to humans associated with the wild animal trade and has been found in civets, raccoon dogs, and ferret badgers.

In 2012, the first case of Middle East respiratory syndrome (MERS) was described in Saudi Arabia, caused by the MERS-coronavirus. It is believed to have originated in camels. The numbers of MERS cases are slowly increasing (unlike the rapid emergence of SARS). Although MERS may not become as severe a threat as SARS was, another SARS or much worse is out there waiting for an opportunity to jump to a new host.

IN THIS CHAPTER

» Seeing microbial processes in your everyday life

» Benefiting from microbial fermentation

» Maintaining a balance in your garden and your belly

Chapter 21

Ten Great Uses for Microbes

Because we usually can't see the microorganisms all around us every day, it's easy to overlook them. But in this chapter, we fill you in on ten ways that microbes affect our lives in important ways.

Making Delicious Foods

The yeast *Saccharomyces cerevisiae* has been used for millennia because it ferments sugars and makes carbon dioxide, which causes bread to rise. Commercial yeast, which is sold at the grocery store, is convenient because it acts quickly and predictably every time. One tasty kind of bread called sourdough is made using natural or wild yeasts instead of commercial yeast. Wild yeasts are present everywhere — they need only a little encouragement to start the fermentation process. Wild yeasts prefer more acidic conditions and grow more slowly than commercial yeast, so the bread they produce has a more acidic, complex, and (some would say) delicious flavor than other breads.

The lactic acid bacteria, such as *Lactobacillus* and *Bifidobacteria*, are used in making fermented milk products. Buttermilk, sour cream, cheese, and yogurt are made by letting these bacteria ferment milk to different degrees. After the initial stages of milk fermentation, several other kinds of microbes can be added to give cheeses their distinct flavor. Blue cheeses are inoculated with species of fungi to give them their sharp flavor and colorful veining, but another example is the use of *Proprionibacteria* to make Swiss cheese.

Microorganisms are also responsible for the pickling process. Species of *Lactobacillus* are happy in very acidic conditions, so they're almost always there at the final stages of pickling.

Lactic acid bacteria are also used to cure meats such as salami, pepperoni, and summer sausage that are then dried and/or smoked after fermentation.

Soy sauce is a fermentation product of soybean and wheat paste. The lactic acid bacteria and fungi such as *Aspergillus* produce the distinctive taste of soy sauce.

Vinegar is made not by lactic acid but by the acetic acid bacteria from wine, apple cider, or dilute ethanol.

Growing Legumes

Leguminous plants, such as soybeans, peas, clover, alfalfa, and beans, form intimate associations with nitrogen-fixing bacteria like *Rhizobium.* The bacteria essentially move into the root cells and form little homes for themselves called *nodules.* Within these nodules, the bacteria change completely from their free-living form and produce nitrogen compounds that are used by the plant in exchange for shelter and organic compounds made by the plant. This setup is great for both the plant and bacteria in question, but it also serves to replenish nitrate levels in the soil, which is why farmers often alternate these crops with other nonleguminous crops from one planting season to the next to prolong soil fertility.

Brewing Beer, Liquor, and Wine

Along with bread-making, the yeast *Saccharomyces cerevisiae* is probably best known for its use in making alcoholic beverages. This process involves mixing a sugar source — barley for beer, fruit juice for wine, or rice or potato starch for spirits — with yeast and water. The yeast first consumes the sugars aerobically until the oxygen runs out; then it ferments the sugars anaerobically, producing ethanol and carbon dioxide as waste. When the concentration of ethanol gets to be around 20 percent, the yeast stops fermenting, so distilling is used to make liquor with a higher alcohol concentration.

Killing Insect Pests

The bacterium *Bacillus thuringiensis* (also known as Bt) produces a toxin that is specific to certain types of insects, many of which are pests encountered by gardeners and farmers. The toxin is so effective at killing the target insect that for years it has been used as an insecticide. A concentrated solution of Bt cells is sold as a spray that can be applied directly to plant tissues. Scientists have isolated the genes for the Bt toxin and have engineered plants to express it, producing transgenic Bt crops.

There are a couple of drawbacks to this method of pest control, however. As with any insecticide, Bt toxin can kill nontarget insects that are actually helpful in the garden. In addition, some pest insects have developed an immunity to it.

Another group of insect pathogens uses small worms called *nematodes* to get inside of their insect host. Members of the bacterial genera *Photorhabdus* and *Xenorhabdus* hitch a ride inside soil-dwelling nematodes without harming them. It's only when the nematode crawls inside of insect grubs that the bacteria emerge and produce toxins that kill the grub, giving the nematode a lot of nutrients to use for reproduction. Gardeners wanting to reduce the number of grubs in lawns and gardens have taken advantage of this natural lifecycle.

Treating Sewage

Microorganisms are an important part of wastewater treatment during a few different steps of the process. After the sewage has been physically filtered to remove solid waste, called *primary treatment*, microbes break down the insoluble organic matter during *secondary treatment*. This process is especially necessary for wastes that are high in plant matter, such as those from industrial food processing sites. Huge communities of microorganisms present in bioreactors (or sludge digesters) act together to anaerobically break down the organic matter in the water.

Another type of secondary treatment is aerobic and involves microbes that form clumps called *flocks* that the remaining organic matter and other microorganisms bind to. The flocks are then allowed to settle, and the remaining water is much clearer than it was in the beginning.

Microorganisms in sewage treatment plants also convert nitrate in the wastewater to nitrogen gas that is lost to the air. This process, called *denitrification*, is important because water with a high nitrate content can stimulate huge blooms of algae that foul lakes, rivers, and streams.

Contributing to Medicine

Microbes have evolved over the last 3.5 billion years to compete with one another mostly through chemical means. It's not surprising, then, that the vast majority of the thousands of antibiotics known to us are made by microorganisms. Species of the bacterial genus *Streptomyces* alone produce more than 500 different antibiotics, including cycloheximide, cycloserine, erythromycin, kanamycin, lincomycin, neomycin, nystatin, streptomycin, and tetracycline. Other antibiotics are made by *Bacillus* bacteria and fungi, such as *Penecillium*.

Because of this long history of competition and survival in the face of these chemicals, bacteria are naturally suited to develop antibiotic resistance. This means that despite having such an impressive arsenal of antibiotics, we're still struggling to control the growth of many strains of bacteria that have become resistant to them.

Botox, the toxin from the *Clostridium botulinum* bacteria, is best known for its cosmetic use to decrease wrinkles by paralyzing the muscles of the face. It works by blocking release of the neurotransmitter acetylcholine from neurons, effectively stopping them from signaling the muscles. Recently, Botox has been used in many other medical applications, such as treating overactive muscle and twitching disorders, as well as treating headaches and chronic pain in joints, muscles, and connective tissue.

Setting Up Your Aquarium

In nature, fish live in complex environments filled with plants, insects, and microorganisms. Because of their biological complexity, natural habitats are good at absorbing and recycling fish waste, such as ammonia, so that it doesn't build up. In a closed container, however, like an aquarium, ammonia released in fish waste can build up quickly, making the water inhospitable and toxic.

To stop this buildup of ammonia from happening, aquarium owners have to do two things:

- About once a week, change one-quarter of the water in the tank.
- Encourage the establishment in the tank of ammonia-oxidizing bacteria and archaea that live in the filter and the tank sediment and convert ammonia to less harmful products in a process called *nitrification*.

TIP

Pet stores sell an aquarium starter mix, but this usually doesn't contain all the necessary ammonia oxidizers. You're much better off seeding your aquarium from another well-established tank by taking a small amount of the filter. Once they're established, the microbes will lead to a well-balanced tank ecosystem.

Making and Breaking Down Biodegradable Plastics

Most of the plastic used today is synthetically made from petroleum and is extremely resistant to degradation. Some attempts have been made to make biodegradable plastics synthetically by adding starch or using different polymers, but most of these plastics still aren't completely broken down. The best kind of biodegradable plastics are the ones made by bacteria because they can also be broken down by bacteria. These contain either polylactide (PLA), made as a product of fermentation, or polyhydroxyalkanoates (PHAs), made by bacteria as storage compounds.

PLAs are slower to degrade and aren't as good as PHAs that have many of the same properties of synthetically made polymers used in plastics. PHA polymers such as poly-β-hydroxybutyrate (PHB) and poly-β-hydroxyvalerate (PHV) have been combined to make containers for shampoo and other personal care products in Europe.

Unfortunately, biodegradable bacterially produced plastics haven't yet been able to compete with the synthetically made ones because oil is still cheaper than the sugar needed to feed the bacteria.

Turning Over Compostable Waste

The ability of the compost pile in your backyard or in your municipal waste treatment center to turn kitchen and yard waste into what looks like dirt depends directly on microorganisms. The process of composting starts with the initial breakdown of simple organic matter by microorganisms that produce, among other things, carbon dioxide, heat, and *humus* (broken down organic matter).

Next, heat-tolerant bacteria called *thermophiles* continue breaking things down and increasing the temperature up to between 140°F and 158°F (60°C–70°C). This phase is important because the increased temperature helps break down complex organic matter and kills most animal and plant pathogens. It's also at this point that the pile is mixed to provide oxygen and to keep the temperature from going too high and killing all the composting microbes.

Finally, the compost pile cools and matures to include a complex community of microorganisms and insects that are beneficial to the environment.

Maintaining a Balance

The complex microbial communities on and in the human body can sometimes get out of balance. Probiotics have been designed to administer bacterial species that are beneficial to the human body. They're thought to help the body's microbial community return to a balanced state. Widely used probiotics mainly include the lactic acid bacteria. They can be taken as a pill or in a fermented milk product, such as yogurt. Probiotics are thought to help with everything from inflammatory bowel disease and resistance against pathogenic bacteria and viruses to cancer treatment. Drawbacks to probiotics are that they aren't always effective and they need to be administered daily to have an effect.

Fecal biotherapy (sometimes called *fecal transplant*) is the insertion of feces from a healthy person (the donor) into the colon of a person suffering from a gastrointestinal disorder or disease. The idea is that the patient's microbial community is unbalanced and the healthy community of microbes from the donor sample can help restore a healthy balanced microbial community. The best example of its use has been in treating people suffering from a *Clostridium difficile* (or *C. diff*) infection, which is a nasty infection of the gastrointestinal tract that is often contracted in the hospital after antibiotic treatment. *C. diff* is very difficult to get rid of because the bacterium produces antibiotic-resistant endospores; it's often fatal if unresolved. Fecal biotherapy has been used very successfully to treat *C. diff* infections because the bacterium doesn't compete well with other microbes. Although fecal biotherapy is still controversial, a few clinical trials aimed at treating complex gastrointestinal disorders with fecal biotherapy are underway.

IN THIS CHAPTER

» Treating and preventing infectious disease

» Protecting the environment

» Minimizing microbial contamination

» Manufacturing natural products

Chapter 22
Ten Great Uses for Microbiology

There are many professions where knowing about microbes, either pathogenic or benign, is handy. There are also plenty of situations where applying a microbiology approach to a problem will get you out of a jam. In this chapter, we provide ten examples. Some are industries that employ microbiologists, and others are situations where a knowledge of microbiology is essential to getting the job done.

Medical Care: Keeping People Healthy

One of the places where microbiology is essential is in the healthcare industry. Here are just some of the healthcare professionals who think critically about microbes every day:

» **Nurses and doctors:** As the front line of the hospital healthcare system, infection control nurses and doctors are constantly on guard against transmission of microbial pathogens. Protocols exist to protect both patients and staff from viruses, bacteria, and fungi. Simple hand washing can do wonders to protect people. Also instituted are routine screening for antibiotic-resistant bacteria and strict rules surrounding the intensive care and surgical areas of the hospital.

- **Pharmacists:** Pharmacists dispense a variety of medications, many of which have microbial origins. Even synthetic drugs are often derived from a modified natural product.
- **Clinical microbiologists:** Identifying pathogenic microorganisms in the many patient samples collected every day is the responsibility of clinical laboratories. Identification of microorganisms is performed with selective media, polymerase chain reaction (PCR), and biochemical and metabolic tests.
- **Obstetricians and midwives:** Newborn babies are completely vulnerable to infection by microbial pathogens, which is why obstetricians and midwives screen mothers for some of the pathogens that can be passed on to babies during delivery. Born relatively sterile, babies start getting their natural healthy bacteria almost immediately after birth, so parents are encouraged to expose babies to good microbes, from breast milk and skin contact with parents and siblings, and keep babies away from microbes that could make them sick, like from people who are contagious.
- **Public health officials:** These folks monitor the spread of infectious agents in the population and provide education about disease transmission and vaccination schedules. Public health officials are also responsible for monitoring rates of foodborne and waterborne illnesses to track sources of contamination.

Dental Care: Keeping Those Pearly Whites Shining Bright

The dentist's office is a place where smiles get polished and nagging toothaches get attended to. But a big part of dentistry is knowing about the microorganisms that live in the mouth. The oral cavity has the most microbial diversity of any site in the human body, with several hundred species of bacteria alone.

Plaque, which accumulates around the area where the gums meet the teeth, is a hard substance made up mainly of polysaccharide and other extracellular products of bacteria. After every dental cleaning, bacterial biofilms reform in the mouth through an organized process that involves different species of bacteria working together to form a matrix that they share as a community. This is part of the reason that it's difficult to get rid of the bacteria thought to be responsible for *halitosis* (bad breath). It's hard to identify which microbe is the culprit and it's hard to get rid of just the ones you want to, because they live in such a tightly knit community.

Periodontitis is the inflammation of the tissue surrounding teeth because of a heightened immune reaction against pathogenic microorganisms living below the gum surface. This inflamed state within the gum tissue is thought by some scientists to be linked with many other negative health outcomes like cardiovascular disease. Others disagree with this connection, pointing out that people who are not well often don't practice good oral care. The issue is still hotly debated, but dentists aim to keep gums happy and free of disease for the benefit of the teeth with the possible side effect of contributing to their patients' overall well-being.

Veterinary Care: Helping Fido and Fluffy to Feel Their Best

As doctors with special knowledge about many different types of animals, veterinarians keep pets, livestock, and even wild animals healthy. The types of microbiology used in veterinary medicine are much like those used in human medicine — the pathogens just have different names.

When pets get sick with mystery illnesses, vets use some fundamental microbiology to figure out if a microbe is the cause. This includes microscopy and Gram staining of body fluids and stool, as well as culturing swabs. They administer vaccines for viruses like rabies and feline leukemia and give out other preventive treatments for parasites like heartworm.

Vets on the farm take care of infectious livestock diseases like *mastitis* (inflammation of the mammary glands and udder tissue) in cows and viruses in pigs and poultry. Lots of research is done on preventing and dealing with livestock infections, mainly because of the tendency for infections to spread quickly through animal populations and because loss of animal life is costly to the agriculture industry.

Many animal pathogens have been successfully eliminated from livestock populations only to remain in the wild animal population, which is much more difficult to treat. An example of treating the wild population is in the case of rabies. An oral vaccine has been developed that is safe for the animals and works well if ingested. Wild populations of raccoon, coyote, and fox have been immunized by lacing bait with the vaccine, stemming the spread of this dangerous virus.

Monitoring the Environment

National environmental monitoring programs, like the U.S. Geological Survey, use microbiology, both to impact environmental conservation and for basic research reasons. Examples of such activities include the following:

- **Monitoring the impacts of climate change on microbial populations:** Rising temperatures can cause blooms of some microbes, like cyanobacteria, impacting animal and plant populations. Increasing temperatures can increase soil microbe metabolic rates leading to more carbon cycling along with the release of carbon dioxide, a greenhouse gas.
- **Conserving wildlife:** Scientists monitor aquatic and land animal diseases and research ways to treat or protect them from infectious microbes.
- **Studying the interaction of microorganisms and their environments:** How microorganisms interact with plants, corals, and in soils are important aspects of ecosystem functioning. This field is called *microbial ecology*.
- **Recording and publishing water quality data:** Scientists study water quality to keep the public informed of the closure of recreational beaches and warn them of poor water quality in lakes and streams.
- **Studying how microorganisms interact with nonliving parts of the environment:** These include sediments, minerals, rocks, and the atmosphere. This field is called *geomicrobiology*.
- **Combining all of the above activities to measure overall ecosystem functioning:** This includes research into the stability of different ecosystems in the face of stress, as well as how to restore natural habitats. Because microorganisms play a huge role in ecosystem functioning, scientists study their presence and how they change.

NOWHERE TO HIDE

Populations of birds found nowhere else on earth are in danger of extinction on the Hawaiian Islands because of avian malaria. This nasty bird pathogen is transmitted by mosquito and decimates bird populations every year. Mosquitos only arrived on the islands in 1827, so species of birds that evolved without avian malaria were completely unprotected. In the past, the cooler temperatures at higher altitudes have acted as a refuge for birds looking to escape from the biting mosquitos. However, recent increases in temperatures due to global warming have reduced the area of each island that gets cool enough to stop mosquito breeding. For this reason, avian malaria in Hawaii's birds can act as an indicator of global warming. Because of avian malaria, bird species like the Hawaiian honeycreeper are going extinct at an unprecedented rate.

Making Plants Happy

Farmers, horticulturalists, and gardeners use microbiology every day to keep plants protected from plant pathogens. A slew of microorganisms make a living from infecting plant tissues, and it's sometimes hard to stay ahead of them. Plants infected with microbial pathogens are less healthy, often have unsightly blemishes, and can succumb quickly despite treatment. Prevention is the best medicine, especially because many pathogens get inside of plant tissues, making it hard to get rid of them before losing the plants.

Microbiology approaches used to combat plant pathogens involve the following:

- **Know the enemy.** It's important to identify the culprit using microbiology techniques such as culture, fungal identification, and bacterial staining. Also, many nasty plant colonizers have a lifecycle that spans over the winter, so it's important to clean up all infected material left over from the previous season; otherwise, the spores will germinate in the spring. And don't forget that many have more than one host so although it's the apples you care about, for example, you also have to get rid of all infected blackberry bush leaves (because this is where the fungus hides).
- **The enemy of your enemy is your friend.** Several microorganisms compete with and actively kill other microorganisms, so it helps to know which these are and how to encourage them to grow near, on, or in your plants. Examples include free-living bacteria in the soil that are bioprotective.
- **The best protection is healthy plants.** Plant pathogens often prey on sick or stressed plants that otherwise would not be susceptible to infection. Maintaining good soil moisture and drainage, having the appropriate amount of light, not crowding plants too close together, and making sure that the soil is not being drained of essential nutrients can help to keep plants healthy.

Keeping Fish Swimming Strong

The fishing industry works together with wildlife protection agencies to make sure that captive and wild fish populations are healthy and free of disease. To do this, a number of different microbial pathogens and hazards have to be considered:

- A huge part of the fishing industry involves keeping **fish farms** free of microbial pathogens. Farmed fish are particularly vulnerable to disease, and countries that share an ocean notify one another when they have a nasty fish virus or bacterial pathogen within their fish farms. Fish vaccines have been developed that help in this regard as well.

- Bacterial, viral and fungal fish pathogens are also a major problem for **natural lake populations of fish.** It's important to collect data on fish pathogens in wild fish populations in national parks to make sure that fish stocking of streams upstream aren't impacting wild fish populations.
- Large-scale deaths in young fish can happen because of ingestion of **toxin-producing cyanobacteria.** Invertebrates consume the bacteria and are, in turn, food for fish. The more invertebrates a fish eats, the more toxin it gets as well.
- **Invasive fish species** can impact local fish populations in a negative way, by preying on them, as well as competing for food and habitats. Another negative impact of invasive species is that they bring with them pathogens that native fish have no immunity against.
- **Zoonotic diseases** that can pass from one species to another.
- Some bacteria can cause **vitamin B1 deficiency** in trout eggs leading to death in early stages of life.

Producing Food, Wine, and Beer

Although bacteria and yeast are used extensively in the food industry to ferment a number of food products, microbiology is also used to keep food manufacturing processes safe from microorganisms that will either foul the products or make people sick. Clever ways to keep microbes off of surfaces in the food processing industry have been developed that include inhibiting bacterial biofilms in order to stop bacteria like *Listeria* from growing on food-processing equipment.

An important question is how to determine that the equipment is free of contaminating microbes. To answer this question, environmental monitoring is used, including culture plates and swabs to assess cleanliness. Culture plates are inoculated by touching them to the surface in question or swabbing; these are then incubated to allow bacteria to grow, and then the organisms, if grown, are identified. Even very low numbers of bacteria can be problematic, but these small numbers of organisms can be hard to detect by plating. Taking advantage of the fact that adenosine triphosphate (ATP) doesn't hang around long after cells die, special swabs are designed to detect minute amounts of ATP, indicating that a living cell is present. The test uses firefly luciferase as an indicator; it lights up in the presence of ATP.

Canned foods are heated to kill pathogenic organisms, but this process doesn't make the food inside sterile — temperatures used to sterilize would destroy much of the nutritional value and alter the quality of the food. Many heat-tolerant organisms still remain, but they aren't likely to make people sick because they aren't adapted to the human body and don't make compounds that are toxic for people.

Uses of microbiology in the beer and wine industry include monitoring the fermentation process at different stages to look for lactic fermenters that will spoil the product. The wine industry also relies heavily on agricultural microbiologists who can keep the grape vines free of plant pathogens.

Science Hacking

Twitter is awash with the hashtag #ScienceHack (`www.twitter.com/hashtag/sciencehack`) calling all amateur scientists, or rather all science enthusiasts not affiliated with a university, to participate in events aimed at making an impact on research and medicine. One example involves genetically engineering *E. coli* to produce an antimicrobial and possibly anticancer compound called *violacein*. This bright purple, aromatic compound is produced naturally by a Gram-negative microbe, similar to *Neisseria*, called *Chromobacterium violaceum*. If *E. coli* were engineered to both produce and excrete violacein, the cost of this useful drug would be greatly reduced. If you're more of a lone-wolf scientist, you can order a kit from Synbiota (`http://blog.synbiota.com`) and get to work in your kitchen, using biotechnology to change the world in months rather than years.

Looking for Microbes in Clean Rooms

Clean rooms are used in manufacturing and special research centers where limiting all sources of contamination from dust and microorganisms is important. In these rooms, the incoming air is filtered, people are covered completely to avoid bringing in contaminants from outside, and all material going into the rooms is carefully cleaned and/or sterilized. Surfaces are cleaned with harsh chemicals that are thought to eliminate all living microorganisms, and people monitor the area frequently to look for live microbes that may have slipped through the cracks.

It was precisely these precautions that selected for and allowed one rare microorganism to grow on surfaces in a clean room at NASA used for preparing the Mars lander *Phoenix*. A newly discovered bacteria, now named *Tersicoccus phoenicis*, was found during a routine check of the facility. Because it can withstand the harsh chemicals and high temperatures used to disinfect the clean rooms, this bacterium remained after all other microorganisms were removed.

Researchers think that *T. phoenicis* hasn't been seen before because there are so many other organisms in nature that it's hard to isolate and identify each and every one. The same microbe could be hiding in common dirt outside the doorstep of the facility and they'd never see it — it's only because they work so hard at getting rid of bacteria that the ones able to withstand their sanitation measures are found on surfaces.

Producing Pharmaceuticals

Secondary metabolism is the term used to describe any nonessential products made by an organism. Many times these are waste products that are excreted from the cell, but other times these products have an important activity on live cells and so are called *bioactive* and can be excreted or stored within the cell. Bioactive properties of secondary metabolites include the following:

- Attracting or repelling other organisms.
- Killing or halting the growth of bacteria. Antibiotics are the most well-known group in this class.
- Toxicity to eukaryotic cells, including chemotherapeutic drugs used to kill cancerous cells.

Many other functions exist and the spectrum of activities is still largely unknown.

Many compounds have more than one function, which often only comes to light after extensive testing and research. Plants, animals, and microorganisms make bioactive secondary metabolites that they use to interact with their environment. When these bioactive compounds are purified for the pharmaceutical industry, they're called *natural products*. In fact, the natural products industry is responsible for a large proportion of all the pharmaceuticals used today. Almost all natural products used today are from microbial sources, two-thirds from bacteria (notably actinomycetes like *Streptomyces*) and one-third from fungi. For this reason, the pharmaceutical industry employs a large number of microbiologists who work to discover, isolate, and study natural products from microbes, essentially trying to harness the power of these compounds for human use.

Index

Numerics

A

B

C

D

E

F

G

H

I

J

K

N

O

P

Q

R

S

T

U

V

W

X

Y

Z

About the Authors

Jennifer C. Stearns, PhD: Jennifer is a postdoctoral fellow in the Department of Medicine at McMaster University, where, along with Dr. Michael Surette, she pushes back the boundaries of medicinal microbiology every day. She currently researches how the usually benign bacteria in the respiratory tract can sometimes make people sick. Jennifer was captivated by the images of microbes in her mother's nursing textbooks as a child and eagerly soaked up knowledge about microbiology everywhere she could. Seeing her interest in all things micro, her high school biology teacher, Mr. Tunnicliffe, lent her a copy of the novel *The Hot Zone*, which forever made her love deadly viruses. She has harnessed the potential of microbes to improve crop plant stress and applied the principles of microbial ecology to the bacteria living in the human GI tract. She is currently inspired by the awesome diversity of microorganisms in nature and in our everyday lives. You can follow her musings on the microbiology of the human body on the *Human Microbiome Journal Club* blog at `http://hmjournalclub.wordpress.com`.

Michael G. Surette, PhD: Michael is currently appointed to both the Department of Medicine and the Department of Biochemistry and Biomedical Sciences at McMaster University, where he's unraveling some of the dynamic bacterial interactions inside the complex microbial communities of the human airways and gastrointestinal tract. He earned his bachelor of science in biochemistry at Memorial University of Newfoundland and his PhD, also in biochemistry, at the University of Western Ontario. His post-graduate research at Princeton University was on bacterial chemotaxis; this was followed by a faculty position in the Department of Microbiology and Infectious Disease at the University of Calgary, where he is currently an adjunct professor. Michael holds the Canada Research Chair in Interdisciplinary Microbiome Research, has been on the editorial boards of several microbiology journals, and is a member of both the Canadian Society of Microbiologists and the American Society for Microbiology. Michael has published over 100 peer-reviewed publications on bacterial sensing and communication, antibiotic resistance, genetics, infectious disease, and microbiological methods. He has been invited to give countless seminars on bacterial genetics, behavior, biochemistry, and infectious disease.

Julienne C. Kaiser, MSc: Julie is currently a PhD student in the Department of Microbiology and Immunology at Western University in London, Ontario. She completed her bachelor of science at McGill University and her master of science at McMaster University and over the years has studied various human pathogens including *E. coli*, *Salmonella*, *Streptococci*, and deadly MRSA. She has earned numerous awards in communication for presenting her research at scientific meetings and currently outlets her thoughts on microbiology on *The Human Microbiome Journal Club* blog.

Dedication

For Ben and Lily, you turn all the lead sleeping in my head to gold.

— Jennifer Stearns

For Matt, Ben, and Carolyn, for patience, support, and wonderful questions!

— Michael Surette

To Steve, for his encouragement and patience, and to my parents, for buying me my first microscope.

— Julie Kaiser

Authors' Acknowledgments

Because this is our first-ever book, we have many people to thank for both technical and moral support. We're grateful to Matt Wagner at Fresh Books and Lindsay Lefevere at John Wiley & Sons for the opportunity to write this book and for their valuable encouragement throughout the process. Thanks to Elizabeth Kuball for keeping us on track.

Several people contributed material to this book and we're grateful for their contributions. Thank you to Kayla Cyr for contributing most of the online article "Ten Reasons You May Not Need Antibiotics" and to Josie Libertucci for contributing to the online article "Fecal Transplants: What They Are and What They're Doing."

A special thanks to the entire Surette lab at McMaster University who hunt elusive microbes on and in the human body every day and to the anonymous asthma patient who donated the sample that appears on the cover.

We'd especially like to thank our technical editor Laura Rossi, in whose talented hands so many challenges melt away. We'd also like to thank our families for their unwavering support and encouragement throughout the writing of this book.

Publisher's Acknowledgments

Executive Editor: Lindsay Sandman Lefevere
Project Editor: Elizabeth Kuball
Copy Editor: Elizabeth Kuball
Technical Editor: Laura Rossi

Production Editor: G. Vasanth Koilraj
Cover Image: © KATERYNA KON/SCIENCE PHOTO LIBRARY/Getty Images

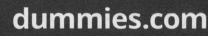

Leverage the power

Dummies is the global leader in the reference category and one of the most trusted and highly regarded brands in the world. No longer just focused on books, customers now have access to the dummies content they need in the format they want. Together we'll craft a solution that engages your customers, stands out from the competition, and helps you meet your goals.

Advertising & Sponsorships

Connect with an engaged audience on a powerful multimedia site, and position your message alongside expert how-to content. Dummies.com is a one-stop shop for free, online information and know-how curated by a team of experts.

- Targeted ads
- Video
- Email Marketing
- Microsites
- Sweepstakes sponsorship

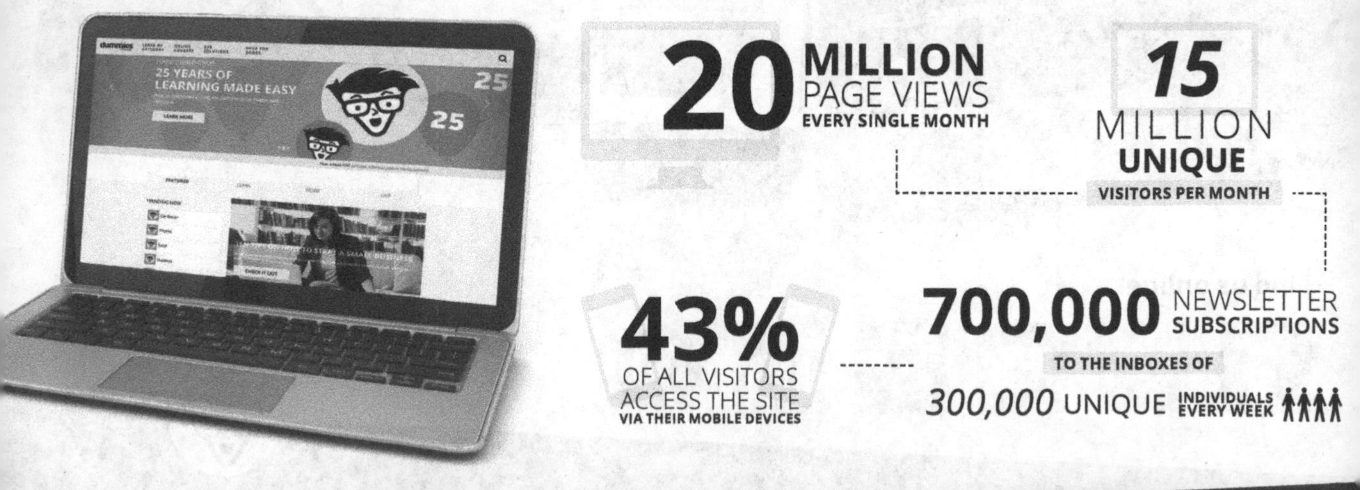

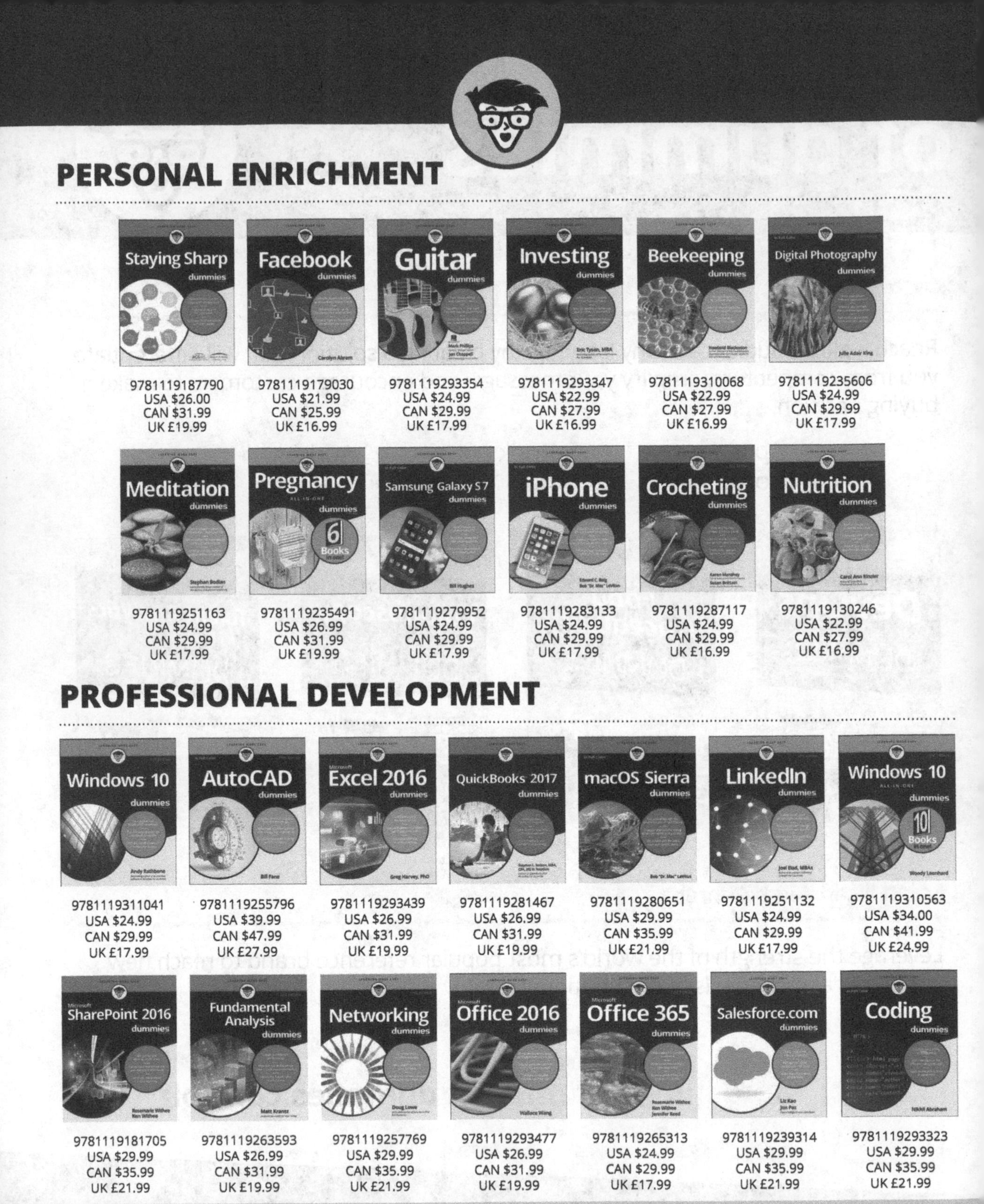
PERSONAL ENRICHMENT
Staying Sharp dummies
9781119187790
USA $26.00
CAN $31.99
UK £19.99
Facebook dummies
9781119179030
USA $21.99
CAN $25.99
UK £16.99
Guitar dummies
9781119293354
USA $24.99
CAN $29.99
UK £17.99
Investing dummies
9781119293347
USA $22.99
CAN $27.99
UK £16.99
Beekeeping dummies
9781119310068
USA $22.99
CAN $27.99
UK £16.99
Digital Photography dummies
9781119235606
USA $24.99
CAN $29.99
UK £17.99
Meditation dummies
9781119251163
USA $24.99
CAN $29.99
UK £17.99
Pregnancy dummies
9781119235491
USA $26.99
CAN $31.99
UK £19.99
Samsung Galaxy S7 dummies
9781119279952
USA $24.99
CAN $29.99
UK £17.99
iPhone dummies
9781119283133
USA $24.99
CAN $29.99
UK £17.99
Crocheting dummies
9781119287117
USA $24.99
CAN $29.99
UK £16.99
Nutrition dummies
9781119130246
USA $22.99
CAN $27.99
UK £16.99
PROFESSIONAL DEVELOPMENT
Windows 10 dummies
9781119311041
USA $24.99
CAN $29.99
UK £17.99
AutoCAD dummies
9781119255796
USA $39.99
CAN $47.99
UK £27.99
Excel 2016 dummies
9781119293439
USA $26.99
CAN $31.99
UK £19.99
QuickBooks 2017 dummies
9781119281467
USA $26.99
CAN $31.99
UK £19.99
macOS Sierra dummies
9781119280651
USA $29.99
CAN $35.99
UK £21.99
LinkedIn dummies
9781119251132
USA $24.99
CAN $29.99
UK £17.99
Windows 10 dummies
9781119310563
USA $34.00
CAN $41.99
UK £24.99
SharePoint 2016 dummies
9781119181705
USA $29.99
CAN $35.99
UK £21.99
Fundamental Analysis dummies
9781119263593
USA $26.99
CAN $31.99
UK £19.99
Networking dummies
9781119257769
USA $29.99
CAN $35.99
UK £21.99
Office 2016 dummies
9781119293477
USA $26.99
CAN $31.99
UK £19.99
Office 365 dummies
9781119265313
USA $24.99
CAN $29.99
UK £17.99
Salesforce.com dummies
9781119239314
USA $29.99
CAN $35.99
UK £21.99
Coding dummies
9781119293323
USA $29.99
CAN $35.99
UK £21.99

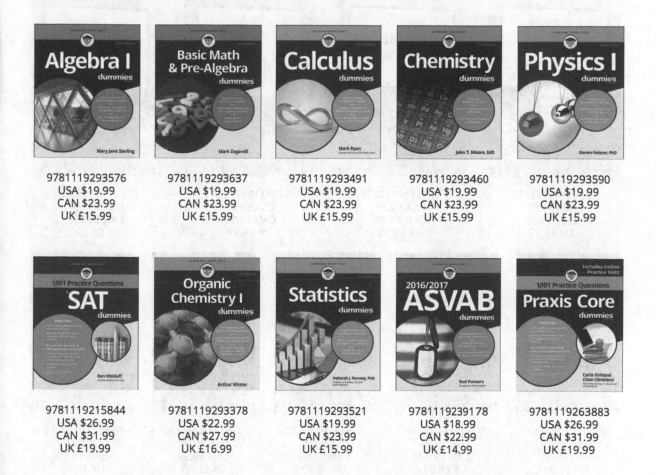